CRM Series in Mathematical Physics

Springer
New York
Berlin
Heidelberg
Hong Kong
London
Milan
Paris
Tokyo

CRM Series in Mathematical Physics

Conte, The Painlevé Property: One Century Later

Jackiw, Lectures on Fluid Dynamics: A Particle Theorist's View of Supersymmetric, Non-Abelian, Noncommutative Fluid Mechanics, and d-Branes

MacKenzie, Paranjape, and Zakrzewski, Solitons: Properties, Dynamics, Interactions, Applications

Saint-Aubin and Vinet, Algebraic Methods in Physics: A Symposium for the 60th Birthdays of Jiří Patera and Pavel Winternitz

Saint-Aubin and Vinet, Theoretical Physics at the End of the Twentieth Century: Lecture Notes of the CRM Summer School, Banff, Alberta

Sénéchal, Tremblay, and Bourbonnais, Theoretical Methods for Strongly Correlated Electrons

Semenoff and Vinet, Particles and Fields

van Diejen and Vinet, Calogero–Moser Sutherland Models

David Sénéchal
André-Marie Tremblay
Claude Bourbonnais
Editors

Theoretical Methods for Strongly Correlated Electrons

With 99 Illustrations

Springer

David Sénéchal
André-Marie Tremblay
Claude Bourbonnais
Département de Physique
Université de Sherbrooke
2500 boulevard de l'Université
Sherbrooke, Québec J1K 2R1
Canada
Contact: David.Senechal@USherbrooke.CA

PACS: 71.27

Library of Congress Cataloging-in-Publication Data
Theoretical methods for strongly correlated electrons / editors, David Sénéchal, André-Marie Tremblay, Claude Bourbonnais.
p. cm. — (The CRM series in mathematical physics)
Includes bibliographical references and index.
ISBN 0-387-00895-0 (alk. paper)
1. Electron configuration. 2. Condensed matter. 3. Mathematical physics. I. Sénéchal, David. II. Tremblay, André-Marie. III. Bourbonnais, Claude. IV. Series.
QC176.8.E4T448 2003
530.4′11—dc21 2003045457

ISBN 0-387-00895-0 Printed on acid-free paper.

Printed in the United States of America.

9 8 7 6 5 4 3 2 1 SPIN 10922040

www.springer-ny.com

Springer-Verlag New York Berlin Heidelberg
A member of BertelsmannSpringer Science+Business Media GmbH

Series Preface

The Centre de recherches mathématiques (CRM) was created in 1968 by the Université de Montréal to promote research in the mathematical sciences. It is now a national institute that hosts several groups and holds special theme years, summer schools, workshops, and a postdoctoral program. The focus of its scientific activities ranges from pure to applied mathematics and includes statistics, theoretical computer science, mathematical methods in biology and life sciences, and mathematical and theoretical physics. The CRM also promotes collaboration between mathematicians and industry. It is subsidized by the Natural Sciences and Engineering Research Council of Canada, the Fonds FCAR of the Province de Québec, and the Canadian Institute for Advanced Research and has private endowments. Current activities, fellowships, and annual reports can be found on the CRM Web page at www.CRM.UMontreal.CA.

The CRM Series in Mathematical Physics includes monographs, lecture notes, and proceedings based on research pursued and events held at the Centre de recherches mathématiques.

Yvan Saint-Aubin
Montreal, Quebec, Canada

Preface

The workshop *Méthodes théoriques pour les fermions fortement corrélés/ Theoretical methods for strongly correlated electrons* was held from Wednesday May 26 to Sunday May 30, 1999, at the Centre de recherches mathématiques, Université de Montréal, Québec, Canada.

Despite numerous conferences and workshops on strongly correlated electrons, one rarely finds workshops dedicated only to purely theoretical aspects of the problem. According to the speakers, this workshop was a first. It brought together experts on various approaches, focused around a unique problem, namely, the solution of models of the Hubbard type (problems where both localized and delocalized aspects are present) in low dimension. Relatively long pedagogical introductions (one and a half hours) during the morning, allowed everyone to appreciate the strengths and weaknesses of each approach as well as to understand open problems for each of the important methods. Short presentations in the afternoon provided a broad overview of current problems. One and a half hours a day were reserved for posters. Since these were exhibited during the whole workshop in a common area they also provided stimuli for informal discussions during coffee breaks and early in the morning. The workshop was open to theorists as well as to advanced students and postdocs.

These proceedings do not do full justice to this workshop, but they do strive to give the same mix of pedagogical review and outlook on current problems that was at the heart of this meeting. By avoiding exhaustive lists of short summaries to concentrate on only a few *in-depth* articles, we do hope that these proceedings will have lasting value for the student and the researcher alike. A list of talks and of participants follows this preface. They are provided both as a memento and as a reference for those who wish to contact an outstanding researcher on a topic not covered by these proceedings. The fact that a given talk or topic does *not* appear in these proceedings is uncorrelated with the quality of the presentation. Look for correlations only in electronic properties!

The contents of these proceedings can, roughly speaking, be divided into three parts that cover an impressive range of methods.

Part I, *Numerical Methods*, deals with two of the most widely used numerical methods in strongly correlated electrons. Both of these methods have found applications in areas spanning condensed-matter, high-energy, and sometimes nuclear or statistical physics. The first one, Density Matrix

Renormalization Group is, by now, one of the leading numerical methods for one-dimensional and quasi-one-dimensional problems. The paper by Karen Halberg covers the different areas where the method is applied and also gives the reader a general overview on the subject. Much activity has focused on extensions of the method to dynamical quantities, finite temperatures, disorder, phonons, etc. They are all discussed. The extensive list of references will be extremely valuable to the reader. The second method, Quantum Monte Carlo, is introduced in a pedagogical chapter by Shiwei Zhang. Most of the aspects of this methodology are covered, including primarily auxiliary-field (or determinantal) methods, but also configuration-space methods and Constrained Path Monte Carlo. There is a focus on the latter method that allows one to free oneself from the infamous *sign problem.* In this chapter, Shiwei has developed a formalism that unifies the different methods and allows for a systematic understanding of their strengths, weaknesses, and common features.

Problems in one dimension, or weakly coupled chains, have lended themselves to the development of a variety of analytical methods. These analytical methods have allowed detailed understanding of the fascinating physics that arises in the presence of interactions, such as spin-charge separation. And these methods continue to produce new results, some of which find extensions to higher dimension.

Part II, *Lagrangian, Functional Integral, Renormalization Group, Conformal and Bosonization Methods*, contains three reviews that cover the most widely used methods. The chapter by C. Bourbonnais, B. Guay, and R. Wortis reviews the renormalization group method and scaling concepts for interacting fermions. Peculiarities of the two-loop calculation are clarified for the first time. Dimensional crossover, multiple fixed points, and Kohn–Luttinger mechanisms in different channels are subjects of current interest that are also discussed. The chapter by D. Sénéchal contains a pedagogical review of bosonization methods and conformal invariance. These methods are applied, as an example, to the solution of the Tomonaga–Luttinger model, which embodies the physics of spin-charge separation. Non-Abelian bosonization as well as a variety of applications, such as edge states in the quantum Hall effect, are also discussed. The last chapter of Part II was written by T. Giamarchi and E. Orignac. Using one-dimensional spinless fermions as a pedagogical example, the authors explain various methods, in particular the replica method and the Gaussian Variational Method, to treat the elastic disordered theory that describes a large class of disordered fermionic systems. Extensions to higher dimension of problems such as Wigner crystal, Charge Density Waves, and Bose glass are also presented. These fall into the general class of disordered quantum solids. The authors also investigate in detail the interesting example of a disordered Mott insulator and argue that intermediate disorder can lead to a novel phase, the Mott glass, intermediate between a Mott and an Anderson insulator.

Part III, *Functional Derivatives, Mean-Field, Self-Consistent Methods, Slave-Bosons, and Extensions*, begins with a review by E. Bickers of Baym-Kadanoff, or Φ-derivable, approximations. Functional integrals results are also used to establish the connection between conventional mean-field theory and higher-order Baym-Kadanoff approximations. The Φ derivability criterion for thermodynamic consistency is discussed and contrasted with parquet, or crossing-symmetric, approximations. Instabilities of the electronic normal state and numerical techniques for the solution of self-consistent field approximations are reviewed, with particular emphasis on renormalization group methods for frequency and momentum space. This leads to the next chapter by J. Kroha and P. Wölfle. They review a new systematic many-body method capable of describing both Fermi liquid and non-Fermi liquid behavior of quantum impurity models at low temperatures on the same footing. The method covers the crossover to the high temperature local moment regime as well. In more technical terms, this chapter deals with the method of auxiliary particles introduced to effect the projection in Hilbert space while keeping most of the desirable features of renormalized perturbation theory. After a pedagogical introduction, approximations are derived from a generating Luttinger-Ward functional, Φ, in terms of renormalized perturbation theory in the hybridization V. The conserving T-matrix approximation (CTMA), discussed in the previous chapter, is used here again, but for the auxiliary particles. The results are compared with the non-crossing approximation (NCA) and with data obtained by the numerical renormalization group and the Bethe ansatz. Generalizations are discussed as well. The last chapter by S. Allen, A.-M. S. Tremblay, and Y. Vilk presents a formal derivation of a non-diagrammatic approach that was developed a few years ago. The derivation makes the analogies and differences with Φ-derivable approximations clearer. The two-particle self-consistent approach presented in this chapter has produced results that are more accurate, both quantitatively and qualitatively, than other methods when compared with Quantum Monte Carlo calculations, in particular for the so-called pseudogap problem. But its extensions to different problems are not as obvious as with Φ-derivable approximations. This is one of the many areas presented in this book where the reader will find challenges for the future.

In conclusion, we thank the Centre de recherches mathématiques and its skillful and courteous staff that made this event run smoothly and allowed the organizers to participate in the event instead of being caught up in logistics. We are indebted also to Luc Vinet and Yvan Saint-Aubin, who encouraged us to organize this workshop and applied for grants that made it possible. Yvan Saint-Aubin also gave valuable advice on the philosophy behind these proceedings. Finally, we acknowledge the financial support of the Natural Sciences and Engineering Research Council (NSERC) of Canada, the U.S. National Science Foundation (NSF), and the Fonds pour la formation de Chercheurs et l'Aide à la Recherche (FCAR) of the Québec government.

Claude Bourbonnais, Université de Sherbrooke
David Sénéchal, Université de Sherbrooke
Andrei Ruckenstein, Rutgers University
André-Marie S. Tremblay, Université de Sherbrooke

May 2003

Contents

Contributors

Steve Allen, Centre de recherche mathématiques, Université de Montréal, C.P. 6128, succ. Centre-ville, Montréal, QC H3C 3J7, Canada; allen@crm.umontreal.ca

N. Eugene Bickers, Department of Physics & Astronomy, University of Southern California, Los Angeles, CA 90089-0484, USA; bickers@usc.edu

Claude Bourbonnais, Département de physique, Université de Sherbrooke, 2500, boulevard de l'Université, Sherbrooke, QC J1K 2R1, Canada; claude.bourbonnais@physique.usherb.ca

Thierry Giamarchi, Laboratoire de Physique des Solides, Université Paris-Sud, Bâtiment 510, 91405 Orsay Cedex, France; giamarchi@lps.u-psud.fr

Benoit Guay, Département de physique, Université de Sherbrooke, 2500, boulevard de l'Université, Sherbrooke, QC J1K 2R1, Canada

Current address: 357, rue Michaud, Coaticook, QC J1A 1B2, Canada

Karen Hallberg, Centro Atómico Bariloche and Instituto Balseiro, 8400 San Carlos de Bariloche, Río Negro, Argentina; karen@cab.cnea.gov.ar

Johann Kroha, Institut für Theorie der Kondensierten Materie, Universität Karlsruhe, Wolfgang-Gaede-Strasse 1, Postfach 6980, 76128 Karlsruhe, Germany; kroha@tkm.physik.uni-karlsruhe.de

Edmond Orignac, Département de Physique École Normale Supérieure, 24, rue Lhomond, 75231 Paris Cedex 05, France; orignac@physique.ens.fr

David Sénéchal, Département de physique, Université de Sherbrooke, 2500, boulevard de l'Université, Sherbrooke, QC J1K 2R1, Canada; david.senechal@usherbrooke.ca

André-Marie Tremblay, Département de physique, Université de Sherbrooke, 2500, boulevard de l'Université, Sherbrooke, QC J1K 2R1, Canada;
tremblay@physique.usherb.ca

Yuri M. Vilk, 2100 Valencia Dr. #406, Northbrook, IL 60062, USA

Peter Wölfle, Institut für Theorie der Kondensierten Materie, Universität Karlsruhe, Wolfgang-Gaede-Strasse 1, Postfach 6980, 76128 Karlsruhe, Germany;
woelfle@tkm.physik.uni-karlsruhe.de

Rachel Wortis, Department of Physics, Trent University, 1600 West Bank Dr., Peterborough, ON K9J 7B8, Canada;
r-wortis@pobox.com

Shiwei Zhang, Department of Physics, College of William and Mary, Williamsburg, VA 23187, USA;
shiwei@physics.wm.edu

Part I

Numerical Methods

1

Density Matrix Renormalization

Karen Hallberg

ABSTRACT The Density Matrix Renormalization Group (DMRG) has become a powerful numerical method that can be applied to low-dimensional strongly correlated fermionic and bosonic systems. It allows for a very precise calculation of static, dynamical, and thermodynamical properties. Its field of applicability has now extended beyond Condensed Matter, and is successfully used in statistical mechanics and high-energy physics as well. In this chapter, we briefly review the main aspects of the method. We also comment on some of the most relevant applications so as to give an overview on the scope and possibilities of DMRG and mention the most important extensions of the method, such as the calculation of dynamical properties, the application to classical systems; inclusion of temperature, phonons and disorder, field theory; time-dependent properties, and the *ab initio* calculation of electronic states in molecules.

1 Introduction

The basics of the Density Matrix Renormalization Group were developed by S. White in 1992 [1] and since then DMRG has proved to be a very powerful method for low-dimensional interacting systems. Its remarkable accuracy can be seen, for example, in the spin-1 Heisenberg chain: for a system of hundreds of sites a precision of 10^{-10} for the ground-state energy can be achieved. Since then it has been applied to a great variety of systems and problems, including, among others, spin chains and ladders, fermionic and bosonic systems, disordered models, impurities and molecules, and two-dimensional (2D) electrons in high magnetic fields. It has also been improved substantially in several directions like 2D (and 3D) classical systems, stochastic models, the presence of phonons, quantum chemistry, field theory, the inclusion of temperature and the calculation of dynamical and time-dependent properties. Some calculations have also been performed in 2D quantum systems. All these topics are treated in detail and in a pedagogical way in [2], where the reader can find an extensive review on DMRG. In this chapter we will attempt to cover the different areas where it has been applied, going into detail in just a few cases, where we have chosen some representative contributions. We suggest that the interested reader look for further information in the referenced work. Our aim here is to give a general overview on the subject.

One of the most important limitations of numerical calculations in finite systems is the great number of states that have to be considered and their exponential growth with system size. Several methods have been introduced to reduce the size of the Hilbert space to be able to reach larger systems, such as Monte Carlo, renormalization group (RG), and DMRG. Each method considers a particular criterion of keeping the relevant information.

The DMRG was originally developed to overcome the problems that arise in interacting systems in 1D when standard RG procedures were applied. Consider a block B (a block is a collection of sites) where the Hamiltonian H_{B} and end-operators are defined. These traditional methods consist in putting together two or more blocks (e.g., B-B$'$, which we will call the superblock), connected using end-operators, in a basis that is a direct product of the basis of each block, forming $H_{\mathrm{BB}'}$. This Hamiltonian is then diagonalized, the superblock is replaced by a new effective block $\mathrm{B}_{\mathrm{new}}$ formed by a certain number m of lowest-lying eigenstates of $H_{\mathrm{BB}'}$, and the iteration is continued (see [3]). Although it has been used successfully in certain cases, this procedure, or similar versions of it, has been applied to several interacting systems with poor performance. For example, it has been applied to the 1D Hubbard model keeping $m \simeq 1{,}000$ states. For 16 sites, an error of 5–10% was obtained [4]. Other results [5] were also discouraging. A better performance was obtained [6] by adding a single site at a time rather than doubling the block size. However, there is one case where a similar version of this method applies very well: the Kondo (and Anderson) model. Wilson [7] mapped the one-impurity problem onto a one-dimensional lattice with exponentially descreasing hoppings. The difference with the method explained above is that in this case, one site (equivalent to an "onion shell") is added at each step and, due to the exponential decrease of the hopping, very accurate results can be obtained.

Returning to the problem of putting several blocks together, the main source of error comes from the election of eigenstates of $H_{\mathrm{BB}'}$ as representative states of a superblock. Since $H_{\mathrm{BB}'}$ has no connection to the rest of the lattice, its eigenstates may have unwanted features (like nodes) at the ends of the block and this can't be improved by increasing the number of states kept. Based on this consideration, Noack and White [8] tried including different boundary conditions and boundary strengths. This turned out to work well for single-particle and Anderson localization problems; however, it did not improve the results significantly for interacting systems. These considerations led to the idea of taking a larger superblock that includes the blocks BB$'$, diagonalizing the Hamiltonian in this large superblock, and then somehow projecting the most favorable states onto BB′. Then BB′ is replaced by B_{new}. In this way, awkward features in the boundary would vanish and a better representation of the states in the infinite system would be achieved. White [1,3] proposed the density matrix as the optimal way of projecting the best states onto part of the system and this will be discussed in the next section. Some considerations concerning the effect of boundary conditions in the nature of the states kept (and an analogy to the physics of black holes) are given in [9]. The justification of using the density matrix is given in detail in [2]. A very easy and pedagogical way of understanding the basic functioning of DMRG is applying it to the calculation of simple quantum problems like one particle in a tight binding chain [10, 11].

In the following section we will briefly describe the standard method, in Section 3 we will mention some of the most important applications, in Section 4 we review the most relevant extensions to the method, and finally in Section 5 we will concentrate on the way dynamical calculations can be performed within DMRG.

2 The Method

The DMRG allows for a systematic truncation of the Hilbert space by keeping the most probable states describing a wave function (e.g., the ground state) instead of the lowest energy states usually kept in previous real space renormalization techniques.

The basic idea consists of starting from a small system (e.g., with N sites) and then gradually increasing its size (to $N+2$, $N+4, \ldots$) until the desired length is reached. Let us call the collection of N sites the *universe* and divide it into two parts: the *system* and the *environment* (see Fig. 1.1). The Hamiltonian is constructed in the *universe* and its ground state $|\psi_0\rangle$ is obtained. This is considered as the state of the *universe* and called the *target state*. It has components on the *system* and the *environment*. We want to obtain the most relevant states of the *system*, i.e., the states of the *system* that have largest weight in $|\psi_0\rangle$. To obtain this, the *environment* is considered as a statistical bath and the density matrix [12] is used to obtain the desired information on the *system*. So instead of keeping eigenstates of the Hamiltonian in the block (*system*), we keep eigenstates of the density matrix. We will be more explicit below.

Let's define block [B] as a finite chain with l sites, having an associated Hilbert space with m states where operators are defined (in particular the Hamiltonian in this finite chain, H_{B}, and the operators at the ends of the block, useful for linking it to other chains or added sites). Except for the first iteration, the basis in this block isn't explicitly known due to previous basis rotations and reductions. The operators in this basis are matrices and

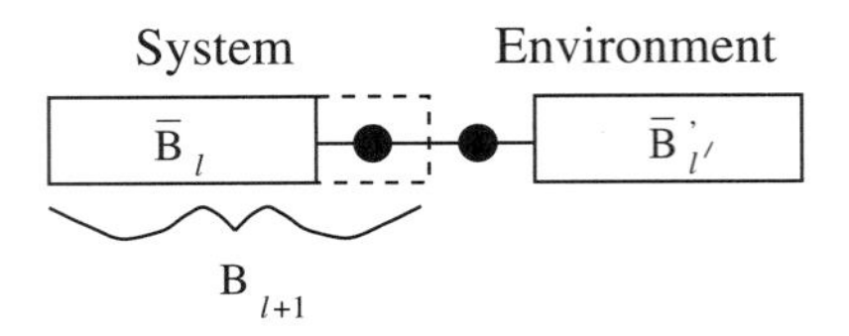

FIGURE 1.1. A scheme of the superblock configuration for the DMRG algorithm [3].

the basis states are characterized by quantum numbers (like S^z, charge, or number of particles, etc). We also define an added block or site as [a] having n states. A general iteration of the method is described below:

(i) Define the Hamiltonian $H_{\mathrm{BB'}}$ for the superblock (the *universe*) formed by putting together two blocks [B] and [B′] and two added sites [a] and [a′] in this way: [B a a′ B′] (the primes are only to indicate additional blocks, but the primed blocks have the same structure as the non-primed ones; this can vary, see the finite-size algorithm below). In general, blocks [B] and [B′] come from the previous iteration. The total Hilbert space of this superblock is the direct product of the individual spaces corresponding to each block and the added sites. In practice a quantum number of the superblock can be fixed (in a spin chain, for example, one can look at the total $S^z = 0$ subspace), so the total number of states in the superblock is much smaller than $(mn)^2$. In some cases, as the quantum number of the superblock consists of the sum of the quantum numbers of the individual blocks, each block must contain several subspaces (several values of S^z for example). Here periodic boundary conditions can be attached to the ends and a different block layout should be considered (e.g., [B a B′ a′]) to avoid connecting blocks [B] and [B′] which takes longer to converge. The boundary conditions are between [a′] and [B]. For closed chains the performance is poorer than for open boundary conditions [3, 13].

(ii) Diagonalize the Hamiltonian $H_{\mathrm{BB'}}$ to obtain the ground state $|\psi_0\rangle$ (target state) using Lanczos [14] or Davidson [15] algorithms. Other states could also be kept, such as the first excited ones: they are all called *target states*.

(iii) Construct the density matrix

$$\rho_{ii'} = \sum_j \psi_{0,ij}\psi_{0,i'j} \tag{2.1}$$

on block [B a], where $\psi_{0,ij} = \langle i \otimes j|\psi_0\rangle$, the states $|i\rangle$ belonging to the Hilbert space of the block [B a] and the states $|j\rangle$ to the block [B′ a′]. The density matrix considers the part [B a] as a system and [B′ a′], as a statistical bath. The eigenstates of ρ with the highest eigenvalues correspond to the most probable states (or equivalently the states with highest weight) of block [B a] in the ground state of the whole superblock. These states are kept up to a certain cutoff, keeping a total of m states per block.

The density matrix eigenvalues sum up to unity and the truncation error, defined as the sum of the density matrix eigenvalues corresponding to discarded eigenvectors, gives a qualitative indication of the accuracy of the calculation.

(iv) With these m states a rectangular matrix O is formed and it is used to change basis and reduce all operators defined in [B a]. This block [B a] is then renamed as block [$\mathrm{B_{new}}$] or simply [B] (for example, the Hamiltonian in block [B a], H_{Ba}, is transformed into H_{B} as $H_{\mathrm{B}} = O^{\dagger} H_{\mathrm{Ba}} O$).

(v) A new block [a] is added (one site in our case) and the new superblock [B a a′ B′] is formed as the direct product of the states of all the blocks.

(vi) This iteration continues until the desired length is achieved. At each step the length is $N = 2l + 2$ (if [a] consists of one site).

When more than one target state is used, i.e., when it is desirable that more than one state be well described, the density matrix is defined as

$$\rho_{ii'} = \sum_l p_l \sum_j \phi_{l,ij} \phi_{l,i'j}, \tag{2.2}$$

where p_l defines the probability of finding the system in the target state $|\phi_l\rangle$ (not necessarily eigenstates of the Hamiltonian).

The method described above is usually called the *infinite-system algorithm* since the system size increases at each iteration. There is a way to increase precision at each length N called the *finite-system algorithm.* It consists of fixing the lattice size and zipping a couple of times until convergence is reached. In this case and for the block configuration [B a a′ B′], $N = l+1+1+l'$, where l and l' are the number of sites in B and B′, respectively. In this step the density matrix is used to project onto the left $l+1$ sites. In order to keep N fixed, in the next block configuration, the right block B' should be defined in $l-1$ sites such that $N = (l+1)+1+1+(l-1)'$. The operators in this smaller block should be kept from previous iterations (in some cases from the iterations for the system size with $N-2$) [2].

The calculation of static properties like correlation functions is easily done by keeping the operators in question at each step and performing the corresponding basis change and reduction, in a similar manner as was done with the Hamiltonian in each block [3]. The energy and measurements are calculated in the superblock. In [16] an interpretation of the correlation functions of systems at criticality is given in terms of wave function entanglement, conjecturing a modification of DMRG for these cases that preserves the entanglement.

A faster convergence of Lanczos or Davidson algorithm is achieved by choosing a good trial vector [17, 18]. An interesting analysis on DMRG accuracy is done in [19]. Fixed points of the DMRG and their relation to matrix product wave functions were studied in [20] and an analytic formulation combining the block renormalization group with variational and Fokker–Planck methods in [21]. The connection of the method with quantum groups and conformal field theory is treated in [22]. There are also interesting connections between the density matrix spectra and integrable models [23] via corner transfer matrices. These articles give a deep insight into the essence of the DMRG method.

3 Applications

Since its development, the number of papers using DMRG has grown enormously and other improvements to the method have been devised. We would like to mention some applications where this method has proved to be useful. Other applications related to further developments of the DMRG will be mentioned in Section 4.

A very impressive result with unprecedented accuracy was obtained by White and Huse [24] when calculating the spin gap in a $S = 1$ Heisenberg chain obtaining $\Delta = 0.41050$ J. They also calculated very accurate spin correlation functions and excitation energies for one and several magnon states and performed a very detailed analysis of the excitations for different momenta. They obtained a spin correlation length of 6.03 lattice spacings. Simultaneously Sørensen and Affleck [25] also calculated the structure factor and spin gap for this system up to length 100 with very high accuracy, comparing their results with the nonlinear σ model. In a subsequent paper [26] they applied the DMRG to the anisotropic $S = 1$ chain, obtaining the values for the Haldane gap. They also performed a detailed study of the $S = \frac{1}{2}$ end excitations in an open chain. Thermodynamical properties in open $S = 1$ chains such as specific heat, electron paramagnetic resonance (EPR), and magnetic susceptibility calculated using DMRG gave an excellent fit to experimental data, confirming the existence of free spins $\frac{1}{2}$ at the boundaries [27]. A related problem, i.e., the effect of non-magnetic impurities in spin systems (dimerized, ladders, and 2D) was studied in [28, 29]. In addition, the study of magnon interactions and magnetization of $S = 1$ chains was done in [30], supersymmetric spin chains modeling plateau transitions in the integer quantum Hall effect in [31] and ESR studies in these systems was considered in [32]. For larger integer spins there have also been some studies. Nishiyama and coworkers [33] calculated the low-energy spectrum and correlation functions of the $S = 2$ antiferromagnetic Heisenberg open chain. They found $S = 1$ end excitations (in agreement with the Valence Bond Theory). Edge excitations for other values of S have been studied in [34]. Almost at the same time Schollwöck and Jolicœur [35] calculated the spin gap in the same system, up to 350 sites ($\Delta = 0.085$ J), correlation functions that showed topological order and a spin correlation length of 49 lattice spacings. More recent accurate studies of $S = 2$ chains are found in [36–38] and of $S = 1$ chains in staggered magnetic fields [39] including a detalied comparison to the non-linear sigma model in [40]. In [41] the dispersion of the single magnon band and other properties of the $S = 2$ antiferromagnetic Heisenberg chains were calculated.

Concerning $S = \frac{1}{2}$ systems, DMRG has been crucial for obtaining the logarithmic corrections to the $1/r$ dependence of the spin-spin correlation functions in the isotropic Heisenberg model [42]. For this, very accurate values for the energy and correlation functions were needed. For $N = 100$ sites an error of 10^{-5} was achieved keeping $m = 150$ states per block, comparing with the exact finite-size Bethe Ansatz results. For this model it was found that the data for the correlation function has a very accurate scaling behavior and advantage was taken of this to obtain the logarithmic corrections in the thermodynamic limit. Other calculations of the spin correlations have been performed for the isotropic [43, 44] and anisotropic cases [45]. Luttinger liquid behavior with magnetic fields have been studied in [46], field-induced gaps in [47], anisotropic systems in [48, 49] and the Heisenberg model with a weak link in [50]. An analysis of quantum critical points and critical behaviour in spin chains by combining DMRG with finite-size scaling was done in [51].

Similar calculations have been performed for the $S = \frac{3}{2}$ Heisenberg chain [52]. In this case a stronger logarithmic correction to the spin correlation function was found. For this model there was interest in obtaining the central charge c to elucidate whether this model corresponds to the same universality class as the $S = \frac{1}{2}$ case, where the central charge can be obtained from the finite-size scaling of the energy. Although there have been previous attempts [53], these calculations presented difficulties since they involved also a term $\sim 1/\ln^3 N$. With the DMRG the value $c = 1$ was clearly obtained.

In [54], DMRG was applied to an effective spin Hamiltonian obtained from an SU(4) spin-orbit critical state in 1D. Other applications were done to enlarged symmetry cases with SU(4) symmetry in order to study coherence in arrays of quantum dots [55], to obtain the phase diagram for 1D spin orbital models [56], and dynamical properties in a magnetic field [57].

Dimerization and frustration have been considered in [58–66] and alternating spin chains in [67].

The case of several coupled spin chains (ladder models) have been investigated in [68–72], spin ladders with cyclic four-spin exchanges in [73–76], and Kagome antiferromagnets in [77]. Zigzag spin chains have been considered in [78–80] and spin chains of coupled triangles in [81–83]. As the DMRG's performance is optimal in open systems, an interesting analysis of the boundary effect on correlation functions is done in [13]. Magnetization properties and plateaus for quantum spin ladder systems [84–86] have also been studied. An interesting review on the applications to some exact and analytical techniques for quantum magnetism in low dimension, including DMRG, is presented in [87].

There has been a great amount of applications to fermionic systems such as 1D Hubbard and t-J models [88–98], Luttinger liquids with boundaries [99], the Falicov–Kimball model [100], the quasiperiodic Aubry–André chain [101], and Fibonacci–Hubbard models [102]. It has also been applied to field theory [9, 103]. The method has been very successful for several band Hubbard models [104], Hubbard ladders [105–107] and t-J ladders [108]. Also several coupled chains at different dopings have been considered [109, 110] as well as flux phases in these systems [111]. Time reversal symmetry-broken fermionic ladders have been studied in [112] and power laws in spinless-fermion ladders in [113]. Long-range Coulomb interactions in the 1D electron gas and the formation of a Wigner crystal was studied in [114]. Several phases, including the Wigner crystal, incompressible and compressible liquid states, and stripe and pairing phases, have been found using DMRG for 2D electrons in high magnetic fields considering different Landau levels [115]. Persistent currents in mesoscopic systems have been considered in [116].

Quite large quasi-2D systems can be reached, for example in [117] where a 4×20 lattice was considered to study ferromagnetism in the infinite-U Hubbard model; the ground state of a 4-leg t-J ladder in [118]; the one- and two-hole ground state in 9×9 and 10×7 t-J lattices in [119]; a doped 3-leg t-J ladder in [120]; the study of striped phases in [121]; domain walls in 19×8 t-J systems in [122]; the 2D t-J model in [123]; and the magnetic polaron in a 9×9 t-J lattice in [124]. Also big CaV_4O_9 spin-$\frac{1}{2}$ lattices reaching 24×11 sites [17] have been studied. There have been some recent attempts to implement DMRG in two and higher dimensions [125–129] but the performance is still poorer than in 1D. A recent extension using a two-step DMRG algorithm for highly anisotropic spin systems has shown promising results [130].

Impurity problems have been studied for example in one- [131] and two-impurity [132] Kondo systems, in spin chains [133] and in Luttinger Liquids [134]. There have also been applications to Kondo and Anderson lattices [135–144], Kondo lattices with localized f^2 configurations [145], the two-channel Kondo lattice on a ladder [146], a t-J chain coupled to localized Kondo spins [147], and ferromagnetic Kondo models for manganites [148–150].

4 Other Extensions to DMRG

There have been several extensions to DMRG like the inclusion of symmetries such as spin and parity [151–153]. Total spin conservation and continuous symmetries have been treated in [143] and in interaction-round a face Hamiltonians [154], a formulation that can be applied to rotational-invariant sytems like $S = 1$ and 2 chains [37]. A momentum representation of this technique [110, 126, 155] that allows for a diagonalization in a fixed momentum subspace has been developed as well as applications in dimensions higher than one [17, 125, 126, 156] and Bethe lattices [157]. The inclusion of symmetries is essential to the method since it allows us to consider a smaller number of states, enhance precision, and obtain eigenstates with definite quantum numbers. Other recent applications have been in nuclear shell model calculations, where a two-level pairing model has been considered [158] and in the study of ultrasmall superconducting grains, in this case, using the particle (hole) states around the Fermi level as the system (environment) block [159].

A very interesting and successful application is a recent work in high-energy physics [160]. Here the DMRG is used in an asymptotically free model with bound states, a toy model for quantum chromodynamics, namely, the two-dimensional delta-function potential. For this case an algorithm similar to the momentum space DMRG [155] was used where the block and environment consist of low-, and high-energy states, respectively. The results obtained here are much more accurate than with the similarity renormalization group [161] and a generalization to field-theoretical models is proposed based on the discreet light-cone quantization in momentum space [162]. Below we briefly mention other important extensions, leaving the calculation of dynamical properties for the next section.

4.1 Classical Systems

The DMRG has been very successfully extended to study classical systems. For a detailed description we refer the reader to [163]. Since 1D quantum systems are related to 2D classical systems [164], it is natural to adapt DMRG to the classical 2D case. This method is based on the renormalization group transformation for the transfer matrix T (TMRG). It is a variational method that maximizes the partition function using a limited number of degrees of freedom, where the variational state is written as a product of local matrices [20]. For 2D classical systems, this algorithm is superior to the classical Monte Carlo method in accuracy, speed and in the possibility of treating much larger systems. A recent improvement of this method considering periodic boundary conditions is given in [165] and a detailed comparison between symmetric and asymmetric targetting is done in [166]. TMRG has also been successfully used to renormalize stochastic transfer matrices in a study of cellular automatons [167]. The calculation of thermodynamical properties of 3D classical statistical systems has been proposed [127] where the eigenstate of the transfer matrix with maximum eigenvalue is represented by the product of local tensors optimized using DMRG.

A further improvement to this method is based on the corner transfer matrix [168], the CTMRG [169–172], and can be generalized to any dimension [173].

It was first applied to the Ising model [163,174–176] and also to the Potts model [177], where very accurate density profiles and critical indices were calculated. Further applications have included non-Hermitian problems in equilibrium and non-equilibrium physics. In the first case, transfer matrices may be non-Hermitian and several situations have been considered: a model for the Quantum Hall effect [178], the q-symmetric Heisenberg chain related to the conformal series of critical models [179], and the anisotropic triangular nearest and next-nearest neighbor Ising models [83]. In the second case, the adaptation of the DMRG to non-equilibrium physics like the asymmetric exclusion problem [180] and reaction-diffusion problems [181, 182] has proved to be very successful. It has also been applied to stochastic lattice models as in [183] and to the 2D XY model [184].

4.2 Finite-Temperature DMRG

The adaptation of the DMRG method for classical systems paved the way for the study of 1D quantum systems at non-zero temperature, by using the Trotter–Suzuki method [185–189]. In this case the system is infinite and the finiteness is in the level of the Trotter approximation. Standard DMRG usually produces its best results for the ground-state energy and less accurate results for higher excitations. A different situation occurs here: the lower the temperature, the less accurate the results.

Very nice results have been obtained for the dimerized, $S = \frac{1}{2}$, XY model, where the specific heat was calculated involving an extremely small basis set [185] ($m = 16$), the agreement with the exact solution being much better in the case where the system has a substantial gap. It has also been used to calculate thermodynamical properties of the anisotropic $S = \frac{1}{2}$ Heisenberg model, with relative errors for the spin susceptibility of less than 10^{-3} down to temperatures of the order of $0.01J$ keeping $m = 80$ states [187]. A complete study of thermodynamical properties like magnetization, susceptibility, specific heat and temperature-dependent correlation functions for the $S = \frac{1}{2}$ and $\frac{3}{2}$ Heisenberg models was done in [190]. Other applications have been the calculation of the temperature dependence of the charge and spin gap in the Kondo insulator [191], the calculation of thermodynamical properties of ferrimagnetic chains [192] and spin ladders [86], the study of impurity properties in spin chains [193, 194], frustrated quantum spin chains [195], t-J [196] and spin ladders [197], and dimerized frustrated Heisenberg chains [198].

An alternative way of incorporating temperature into the DMRG procedure was developed by Moukouri and Caron [199]. They considered the standard DMRG taking into account several low-lying target states (see

(2.2)) to construct the density matrix, weighted with the Boltzmann factor (β is the inverse temperature):

$$\rho_{ii'} = \sum_l e^{-\beta E_l} \sum_j \phi_{l,ij}\phi_{l,i'j} \tag{4.1}$$

With this method they performed reliable calculations of the magnetic susceptibility of quantum spin chains with $S = \frac{1}{2}$ and $\frac{3}{2}$, showing excellent agreement with Bethe Ansatz exact results. They also calculated low temperature thermodynamical properties of the 1D Kondo Lattice Model [200] and of organic conductors [201]. Zhang et al. [202] applied the same method in the study of a magnetic impurity embedded in a quantum spin chain.

4.3 Phonons, Bosons, and Disorder

A significant limitation to the DMRG method is that it requires a finite basis, and calculations in problems with infinite degrees of freedom per site require a large truncation of the basis states [203]. However, Jeckelmann and White developed a way of including phonons in DMRG calculations by transforming each boson site into several artificial interacting two-state pseudo-sites and then applying DMRG to this interacting system [204] (called the "pseudo-site system"). The idea is based on the fact that DMRG is much better able to handle several few-states sites than few many-state sites [205]. The key idea is to substitute each boson site with 2^N states into N pseudo-sites with 2 states [206]. They applied this method to the Holstein model for several hundred sites (keeping more than a hundred states per phonon mode) obtaining negligible error. In addition, to date, this method is the most accurate one to determine the ground-state energy of the polaron problem (Holstein model with a single electron).

An alternative method (the "optimal phonon basis") [207] is a procedure for generating a controlled truncation of a large Hilbert space, which allows the use of a very small optimal basis without significant loss of accuracy. The system here consists of only one site and the environment has several sites, both having electronic and phononic degrees of freedom. The density matrix is used to trace out the degrees of freedom of the environment and extract the most relevant states of the site in question. In following steps, more bare phonons are included to the optimal basis obtained in this way. This method was successfully applied to study the interactions induced by quantum fluctuations in quantum strings, as an application to cuprate stripes [208] and the dissipative two-state system [209]. A variant of this scheme is the "four block method," as described in [210], where Bursill et al. very accurately obtain the Luttinger liquid-CDW insulator transition in the 1D Holstein model for spinless fermions.

The method has also been applied to pure bosonic systems such as the disordered bosonic Hubbard model [211], where gaps, correlation functions, and superfluid density are obtained. The phase diagram for the non-disordered Bose–Hubbard model, showing a reentrance of the superfluid phase into the insulating phase, was calculated in [212]. It has also been used to study a chain of oscillators with optical phonon spectrum [213] and optical phonons in the quarter-filled Hubbard model for organic conductors [214].

The DMRG has also been generalized to 1D random and disordered systems, and applied to the random antiferromagnetic and ferromagnetic Heisenberg chains [215], including quasiperiodic exchange modulation [216] and a detailed study of the Haldane [217] and Griffiths phase [218] in these systems. Strongly disordered spin ladders have been considered in [219]. It has also been used in disordered Fermi systems such as the spinless model [220, 221]. In particular, the transition from the Fermi glass to the Mott insulator and the strong enhancement of persistent currents in the transition was studied in correlated one-dimensional disordered rings [222]. Disorder-induced crossover effects at quantum critical points were studied in [223].

4.4 Molecules and Quantum Chemistry

There have been several applications to molecules and polymers, such as the Pariser–Parr–Pople (PPP) Hamiltonian for a cyclic polyene [224] (where long-range interactions are included), magnetic Keplerate molecules [225], molecular iron rings [226], and polyacenes considering long-range interactions [227]. It has also been applied to conjugated organic systems (polymers), adapting the DMRG to take into account the most important symmetries in order to obtain the desired excited states [151]. Also conjugated one-dimensional semiconductors [228] have been studied, in which the standard approach can be extended to complex 1D oligomers where the fundamental repeat is not just one or two atoms, but a complex molecular building block. Relatively new fields of application are the calculation of dynamical properties in the Rubinstein-Duke model for reptons [229] and excitons in dendrimer molecules [230].

Recent attempts to apply DMRG to the *ab initio* calculation of electronic states in molecules have been successful [231–234]. Here, DMRG is applied within the conventional quantum chemical framework of a finite basis set with non-orthogonal basis functions centered on each atom. After the standard Hartree-Fock (HF) calculation in which a Hamiltonian is produced within the orthogonal HF basis, DMRG is used to include correlations beyond HF, where each orbital is treated as a "site" in a 1D lattice. One important difference with standard DMRG is that, as the interactions are long-range, several operators must be kept, making the calculation somewhat cumbersome. However, very accurate results have been obtained in a check performed in a water molecule (keeping up to 25 orbitals and $m \simeq 200$

states per block), obtaining an offset of 0.00024 Hartrees with respect to the exact ground-state energy [235], a better performance than any other approximate method [231].

In order to avoid the non-locality introduced in the treatment explained above, White introduced the concept of *orthlets*, local, orthogonal, and compact wave functions that allow prior knowledge about singularities to be incorporated into the basis and an adequate resolution for the cores [232]. The most relevant functions in this basis are chosen via the density matrix. An application based on the combination with the momentum version of DMRG is used in [236] to calculate the ground state of several molecules.

5 Dynamical Correlation Functions

The DMRG was originally developed to calculate static ground-state properties and low-lying energies. However, it can also be useful to calculate dynamical response functions. These are of great interest in condensed matter physics in connection with experiments such as nuclear magnetic resonance (NMR), neutron scattering, optical absorption, photoemission, etc. We will describe three different methods in this section.

A recent development for calculating response functions in single impurity systems in the presence of a magnetic field was done in [237] by using the DMRG within Wilson's NRG to obtain the Green function.

An interesting extension of DMRG to tackle time-dependent quantum many-body systems out of equilibrium was considered by Cazalilla and Marston [238].

5.1 Lanczos and Correction Vector Techniques

An effective way of extending the basic ideas of this method to the calculation of dynamical quantities is described in [239]. It is important to notice here that due to the particular real-space construction, it is not possible to fix the momentum as a quantum number. However, we will show that by keeping the appropriate target states, a good momentum value can be obtained.

We want to calculate the following dynamical correlation function at $T = 0$:

$$C_A(t - t') = \langle \psi_0 | A^\dagger(t) A(t') | \psi_0 \rangle, \tag{5.1}$$

where $A^\dagger$ is the Hermitian conjugate of the operator A, $A(t)$ is the Heisenberg representation of A, and $|\psi_0\rangle$ is the ground state of the system. Its Fourier transform is

$$C_A(\omega) = \sum_n \left| \langle \psi_n | A | \psi_0 \rangle \right|^2 \delta\big(\omega - (E_n - E_0)\big), \tag{5.2}$$

where the summation is taken over all the eigenstates $|\psi_n\rangle$ of the Hamiltonian H with energy E_n, and E_0 is the ground-state energy.

Defining the Green function

$$G_A(z) = \langle\psi_0|A^\dagger(z-H)^{-1}A|\psi_0\rangle, \tag{5.3}$$

the correlation function $C_A(\omega)$ can be obtained as

$$C_A(\omega) = -\frac{1}{\pi}\lim_{\eta\to 0^+} \operatorname{Im} G_A(\omega + i\eta + E_0). \tag{5.4}$$

The function G_A can be written in the form of a continued fraction:

$$G_A(z) = \cfrac{\langle\psi_0|A^\dagger A|\psi_0\rangle}{z - a_0 - \cfrac{b_1^2}{z - a_1 - \cfrac{b_2^2}{z - \cdots}}} \tag{5.5}$$

The coefficients a_n and b_n can be obtained using the following recursion equations [240, 241]:

$$|f_{n+1}\rangle = H|f_n\rangle - a_n|f_n\rangle - b_n^2|f_{n-1}\rangle \tag{5.6}$$

where

$$\begin{aligned} |f_0\rangle &= A|\psi_0\rangle \\ a_n &= \langle f_n|H|f_n\rangle / \langle f_n|f_n\rangle, \\ b_n^2 &= \langle f_n|f_n\rangle / \langle f_{n-1}|f_{n-1}\rangle; \quad b_0 = 0. \end{aligned} \tag{5.7}$$

For finite systems the Green function $G_A(z)$ has a finite number of poles, so only a certain number of coefficients a_n and b_n have to be calculated. The DMRG technique presents a good framework to calculate such quantities. With it, the ground state, Hamiltonian, and the operator A required for the evaluation of $C_A(\omega)$ are obtained. An important requirement is that the reduced Hilbert space should also describe with great precision the relevant excited states $|\psi_n\rangle$. This is achieved by choosing the appropriate target states. For most systems it is enough to consider as target states the ground state $|\psi_0\rangle$ and the first few $|f_n\rangle$ with $n = 0, 1, dotsc$ and $|f_0\rangle = A|\psi_0\rangle$ as described above. In doing so, states in the reduced Hilbert space relevant to the excited states connected to the ground state via the operator of interest A are included. The fact that $|f_0\rangle$ is an excellent trial state, in particular, for the lowest triplet excitations of the two-dimensional antiferromagnet was shown in [242]. Of course, if the number m of states kept per block is fixed, the more target states considered, the less precisely each one of them is described. An optimal number of target states and m have to be found for each case. Due to this reduction, the algorithm can be applied up to certain lengths, depending on the states involved. For longer chains, the higher-energy excitations will become inaccurate. Proper sum rules have to be calculated to determine the errors in each case.

As an application of the method we calculate

$$S^{zz}(q,\omega) = \sum_n \left|\langle\psi_n|S^z_q|\psi_0\rangle\right|^2 \delta\big(\omega - (E_n - E_0)\big), \tag{5.8}$$

for the 1D isotropic Heisenberg model with spin $S = \frac{1}{2}$.

The spin dynamics of this model has been extensively studied. The lowest excited states in the thermodynamic limit are the des Cloiseaux-Pearson triplets [243], having total spin $S^T = 1$. The dispersion of this spin-wave branch is $\omega^l_q = \frac{J\pi}{2}|\sin(q)|$. Above this lower boundary there exists a two-parameter continuum of excited triplet states that have been calculated using the Bethe Ansatz approach [244] with an upper boundary given by $\omega^u_q = J\pi|\sin(q/2)|$. It has been shown [245], however, that there are excitations above this upper boundary due to higher-order scattering processes, with a weight that is at least one order of magnitude lower than the spin-wave continuum.

In Fig. 1.2 we show the spectrum for $q = \pi$ and $N = 24$ for different values of m, where exact results are available for comparison. The delta peaks of (5.8) are broadened by a Lorentzian for visualizing purposes. As expected, increasing m gives more precise results for the higher excitations. This spectra has been obtained using the infinite-system method and more precise results are expected using the finite-system method, as described later.

In Fig. 1.3 we show the spectrum for two systems lengths and $q = \pi$ and $q = \pi/2$ keeping $m = 200$ states and periodic boundary conditions. For this case it was enough to take three target states, i.e., $|\psi_0\rangle$, $|f_0\rangle = S^z_\pi|\psi_0\rangle$ and $|f_1\rangle$. Here we have used ~ 40 pairs of coefficients a_n and b_n, but we noticed that if we considered only the first (~ 10) coefficients a_n and b_n, the spectrum at low energies remains essentially unchanged. Minor differences arise at $\omega/J \simeq 2$. This is another indication that only the first $|f_n\rangle$ are relevant for the low-energy dynamical properties for finite systems.

In the inset of Fig. 1.3 the spectrum for $q = \pi/2$ and $N = 28$ is shown. For this case we considered five target states, i.e., $|\psi_0\rangle$, $|f_0\rangle = S^z_{\pi/2}|\psi_0\rangle$, $|f_n\rangle$ $n = 1, 3$ and $m = 200$. Here, and for all the cases considered, we have verified that the results are very weakly dependent on the weights p_l of the target states (see (2.2)) as long as the appropriate target states are chosen. For lengths where this value of q is not defined we took the nearest value.

Even though we are including states with a given momentum as target states, due to the particular real-space construction of the reduced Hilbert space, this translational symmetry is not fulfilled and the momentum is not fixed. To check how the reduction on the Hilbert space influences the momentum q of the target state $|f_0\rangle = S^z_q|\psi_0\rangle$, we calculated the expectation values $\langle\psi_0|S^z_{-q'}S^z_q|\psi_0\rangle$ for all q'. If the momenta of the states were well

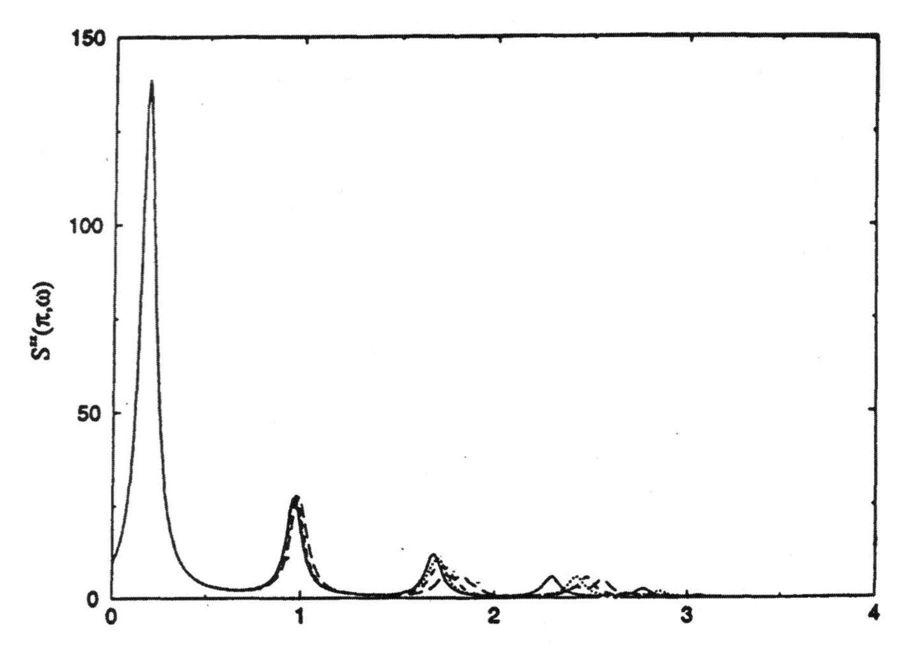

FIGURE 1.2. Spectral function for a Heisenberg chain with $N = 24$ and $q = \pi$. Full line: exact result [246]. The rest are calculated using DMRG with $m = 100$ (long-dashed line), $m = 150$ (dashed line) and $m = 200$ (dotted line).

defined, this value is proportional to $\delta_{q-q'}$ if $q \neq 0$. For $q = 0$, $\sum_r S_r^z = 0$.

The momentum distribution for $q = \pi$ is shown in Fig. 1.4 in a semilogarithmic scale where the y-axis has been shifted by .003 so as to have well-defined logarithms. We can see here that the momentum is better defined, even for much larger systems, but, as expected, more weight on other q' values arises for larger N.

As a check of the approximation we calculated the sum rule

$$\frac{1}{4\pi^2}\int_0^\infty d\omega \int_{q=0}^{2\pi} S^{zz}(q,\omega) \equiv \langle\psi_0|(S_{r=0}^z)^2|\psi_0\rangle = \tfrac{1}{4} \tag{5.9}$$

for $N = 28$, five target states, and $m = 200$. We obtain a relative error of 0.86%.

Recently, important improvements to this method have been published [247]: By considering the finite system method in open chains, Kühner and White obtained a higher precision in dynamical responses of spin chains. In order to define a momentum in an open chain and to avoid end effects, they introduce a filter function with weight centered in the middle of the chain and zero at the boundaries.

Recent applications of this method include the calculation of excitations in spin-orbital models (SU(4)) in a magnetic field [57], spin dynamics in

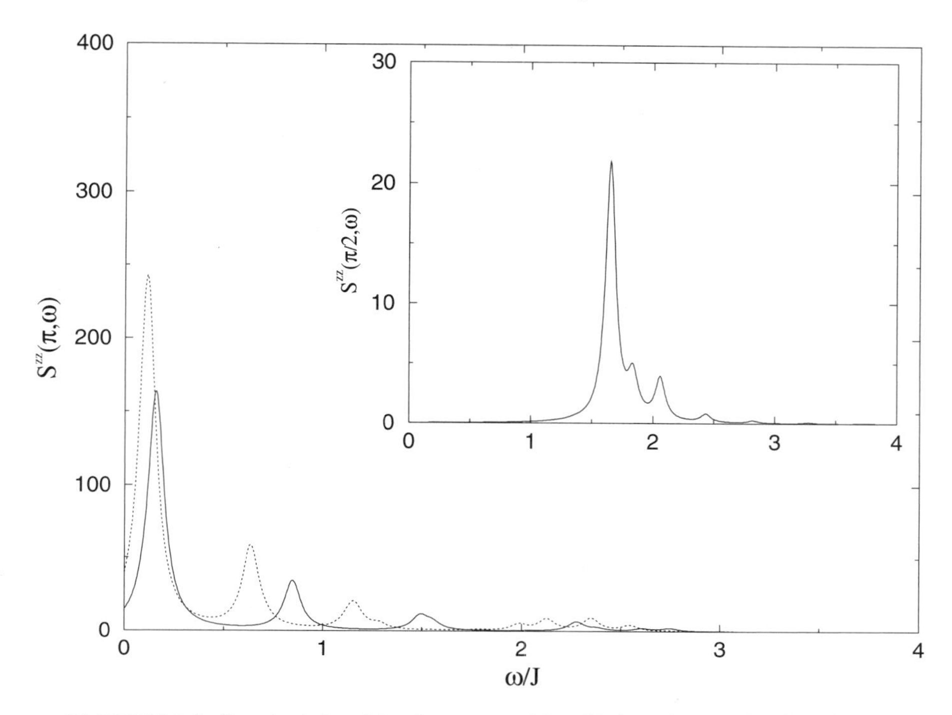

FIGURE 1.3. Spectral densities for $q = \pi$, $N = 28$ (continuous line) and $N = 40$ (dotted line). Inset: Spectral density for $q = \pi/2$ for $N = 28$ ($\eta = 0.05$).

models for cuprate spin ladders including cyclic spin exchange [75], optical conductivity of the ionic Hubbard model [248], excitations in the one-dimensional Bose–Hubbard model [249], and the optical response in 1D Mott insulators [250].

In this section we have presented a method of calculating dynamical responses with DMRG. Although the basis truncation is big, this method keeps only the most relevant states and, for example, even by considering a 0.1% of the total Hilbert space (for $N = 28$ only $\sim 40{,}000$ states are kept) a reasonable description of the low-energy excitations is obtained. We show that it is also possible to obtain states with well-defined momenta if the appropriate target states are used.

5.1.1 Correction Vector Technique

Introduced in [251] in the DMRG context and improved in [247], this method focuses on a particular energy or energy window, allowing a more precise description in that range and the possibility of calculating spectra for higher energies. Instead of using the tridiagonalization of the Hamiltonian, but maintaining a similar spirit regarding the important target states to be kept, the spectrum can be calculated for a given $z = w + i\eta$ by using a

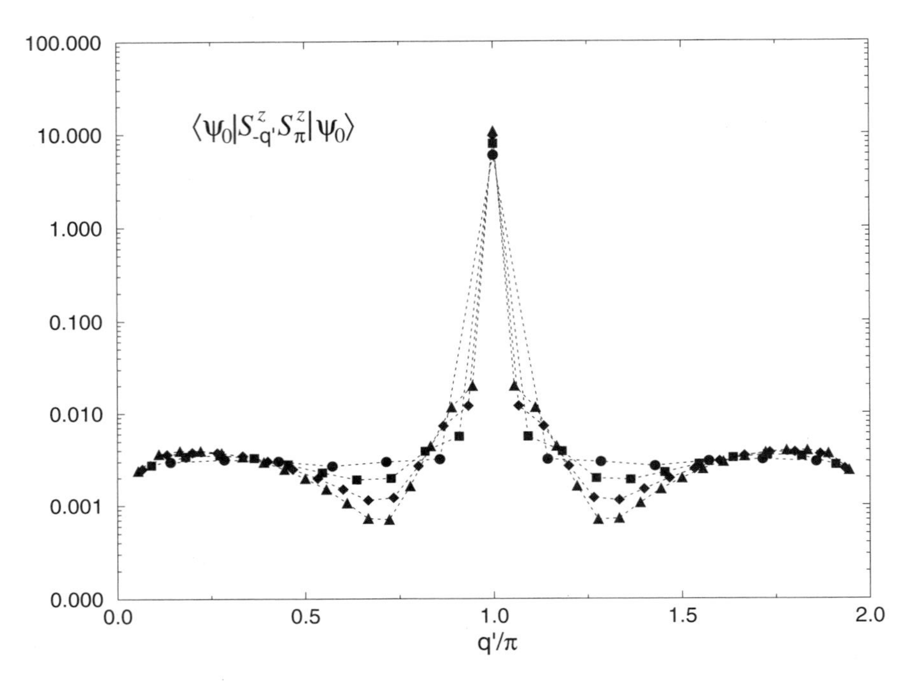

FIGURE 1.4. Momentum weights of a target state with $q = \pi$ for $N = 28$ (circles), $N = 44$ (squares), $N = 60$ (diamonds) and $N = 72$ (triangles). The dotted lines are a guide for the eye.

correction vector (related to the operator A that can depend on momentum q).

Following (5.3), the (complex) correction vector $|x(z)\rangle$ can be defined as

$$|x(z)\rangle = \frac{1}{z - H} A|\psi_0\rangle; \tag{5.10}$$

so the Green function can be calculated as

$$G(z) = \langle\psi_0|A^\dagger|x(z)\rangle. \tag{5.11}$$

Separating the correction vector in real and imaginary parts $|x(z)\rangle = |x^r(z)\rangle + i|x^i(z)\rangle$, we obtain

$$\left((H - w)^2 + \eta^2\right)|x^i(z)\rangle = -\eta A|\psi_0\rangle \tag{5.12}$$

and

$$|x^r(z)\rangle = \frac{1}{\eta}(w - H)|x^i(z)\rangle. \tag{5.13}$$

The former equation is solved using the conjugate gradient method. In order to keep the information on the excitations at this particular energy, the following states are targeted in the DMRG iterations: the ground state

$|\psi_0\rangle$, the first Lanczos vector $A|\psi_0\rangle$, and the correction vector $|x(z)\rangle$. Even though only a certain energy is focused on, DMRG gives the correct excitations for an energy range surrounding this particular point so that by running several repetitions for nearby frequencies, an approximate spectrum can be obtained for a wider region [247].

A variational formulation of the correction vector technique which leads to more accurate excited energies and spectral weights has been developed in [252]. It has been successfully applied to calculate the optical conductivity of Mott insulators [253].

5.2 Moment Expansion

This method [254] relies on a moment expansion of the dynamical correlations using sum rules that depend only on static correlation functions which can be calculated with DMRG. With these moments, the Green functions can be calculated using the maximum entropy method.

The first step is the calculation of sum rules. As an example, and following [254], the spin-spin correlation function $S^z(q,w)$ of the Heisenberg model is calculated where the operator A of (5.1) is $S^z(q) = N^{-1/2}\sum S^z(l)\exp(iql)$ and the sum rules are [255]

$$\begin{aligned} m_1(q) &= \int_0^\infty \frac{dw}{\pi}\frac{S^z(q,w)}{w} = \tfrac{1}{2}\chi(q,w=0) \\ m_2(q) &= \int_0^\infty \frac{dw}{\pi}w\frac{S^z(q,w)}{w} = \tfrac{1}{2}S^z(q,t=0) \\ m_3(q) &= \int_0^\infty \frac{dw}{\pi}w^2\frac{S^z(q,w)}{w} = -\tfrac{1}{2}\big\langle[[H,S^z(q)],S^z(-q)]\big\rangle \\ &= 2[1-\cos(q)]\sum_i\langle S_i^+S_{i+1}^- + S_i^-S_{i+1}^+\rangle \end{aligned} \tag{5.14}$$

where $\chi(q,w=0)$ is the static susceptibility. These sum rules can be easily generalized to higher moments:

$$\begin{aligned} m_l(q) &= \int_0^\infty \frac{dw}{\pi}w^{l-1}\frac{S^z(q,w)}{w} \\ &= -\tfrac{1}{2}\big\langle[[H,\ldots,[H,S^z(q)]\ldots],S^z(-q)]\big\rangle \end{aligned} \tag{5.15}$$

for l odd. A similar expression is obtained for l even, where the outer square bracket is replaced by an anticommutator and the total sign is changed. Here H appears in the commutator $l-2$ times.

Apart from the first moment which is given by the static susceptibility, all the other moments can be expressed as equal time correlations (using a symbolic manipulator). The static susceptibility χ is calculated by applying a small field $h_q\sum_i n_i\cos(qi)$ and calculating the density response $\langle n_q\rangle = 1/N\sum_i\langle n_q\rangle\cos(qi)$ with DMRG. Then $\chi = \langle n_q\rangle/h_q$ for $h_q \to 0$.

These moments are calculated for several chain lengths and extrapolated to the infinite system. Once the moments are calculated, the final spectra is constructed via the maximum entropy method (ME), which has become a standard way to extract maximum information from incomplete data (for details see [254] and references therein). Reasonable spectra are obtained for the XY and isotropic models, although information about the exact position of the gaps has to be included. Otherwise, the spectra are only qualitatively correct. This method requires the calculation of a large amount of moments in order to get good results: The more information given to the ME equations, the better the result.

5.3 Finite Temperature Dynamics

In order to include temperature in the calculation of dynamical quantities, the Transfer Matrix RG described above (TMRG [185, 187, 189]) was extended to obtain imaginary time correlation functions [256–258]. After Fourier transformation in the imaginary time axis, analytic continuation from imaginary to real frequencies is done using maximum entropy (ME). The combination of the TMRG and ME is free from statistical errors and the negative sign problem of Monte Carlo methods. Since we are dealing with the transfer matrix, the thermodynamic limit can be discussed directly without extrapolations. However, in the present scheme, only local quantities can be calculated.

A systematic investigation of local spectral functions is done in [258] for the anisotropic Heisenberg antiferromagnetic chain. The authors obtain good qualitative results, especially for high temperatures, but a quantitative description of peaks and gaps are beyond the method, due to the severe intrinsic limitation of the analytic continuation. This method was also applied with great success to the 1D Kondo insulator [257]. The temperature dependence of the local density of states and local dynamic spin and charge correlation functions were calculated.

6 Conclusions

We have presented here a very brief description of the Density Matrix Renormalization Group technique and its applications and extensions. The aim of this article is to give the inexperienced reader an idea of the possibilities and scope of this powerful, though relatively simple, method. The experienced reader can find here an extensive (however incomplete) list of references covering most applications to date using DMRG in a great variety of fields, such as condensed matter, statistical mechanics, and high-energy physics.

Acknowledgments: The author acknowledges the hospitality of the Centre de recherches mathématiques, Université de Montréal, and of the Physics Department of the University of Buenos Aires, Argentina, where this work has been performed. We thank S. White for a critical reading of the manuscript and all those authors that have updated references and sent instructive comments. K.H. is a fellow of CONICET, Argentina. Grants: PICT 03-00121-02153 and PICT 03-00000-00651.

7 References

[1] S. White, Phys. Rev. Lett. **69**, 2863 (1992).

[2] I. Peschel, X. Wang, M. Kaulke and K. Hallberg (eds.), *Density Matrix Renormalization—A New Numerical Method in Physics*, Lecture Notes in Phys., vol. 528 (Springer, Berlin, 1999).

[3] S. White, Phys. Rev. B **48**, 10345 (1993).

[4] J. Bray and S. Chui, Phys. Rev. B **19**, 4876 (1979).

[5] C. Pan and X. Chen, Phys. Rev. B **36**, 8600 (1987); M. Kovarik, Phys. Rev. B **41**, 6889 (1990).

[6] T. Xiang and G. Gehring, Phys. Rev. B **48**, 303 (1993).

[7] K. Wilson, Rev. Mod. Phys. **47**, 773 (1975).

[8] S. White and R. Noack, Phys. Rev. Lett. **68**, 3487 (1992); R. Noack and S. White, Phys. Rev. B **47**, 9243 (1993).

[9] J. Gaite, Mod. Phys. Lett. A **16**, 1109 (2001).

[10] R. Noack and S. White in [2, Chapter 2 (Part I)].

[11] M. Martín-Delgado, G. Sierra, and R. Noack, J. Phys. A: Math. Gen. **32**, 6079 (1999).

[12] R. Feynman, *Statistical Mechanics: A Set of Lectures* (Benjamin, Reading, MA, 1972).

[13] S.R. White, I. Affleck and D. Scalapino, Phys. Rev. B **65**, 165122 (2002).

[14] See E. Dagotto, Rev. Mod. Phys. **66**, 763 (1994).

[15] E.R. Davidson, J. Comput. Phys. **17**, 87 (1975); E.R. Davidson, Comput. Phys. **7**, No. 5, 519 (1993).

[16] T. Osborne and M. Nielsen, Quantum Information Processing, Volume 1, Issue 1-2, 45 (2002) (available on-line: `http://www.kluwer-online.com/issn/1570-0755/`).

[17] S. White, Phys. Rev. Lett. **77**, 3633 (1996).

[18] T. Nishino and K. Okunishi, J. Phys. Soc. Jap. **64**, 4084 (1995); U. Schollwöck, Phys. Rev. B **58**, 8194 (1998) and Phys. Rev. B **59**, 3917 (1999) (erratum).

[19] Ö. Legeza and G. Fáth, Phys. Rev. B **53**, 14349 (1996); M.-B. Lepetit and G. Pastor, Phys. Rev. B **58**, 12691 (1998).

[20] S. Östlund and S. Rommer, Phys. Rev. Lett. **75**, 3537 (1995); Phys. Rev. B **55**, 2164 (1997); M. Andersson, M. Boman and S. Östlund, Phys. Rev. B **59**, 10493 (1999); H. Takasaki, T. Hikihara and T. Nishino, J. Phys. Soc. Jap. **68**, 1537 (1999); K. Okunishi, Y. Hieida and Y. Akutsu, Phys. Rev. E **59**, R6227 (1999).

[21] M.A. Martín-Delgado and G. Sierra, Int. J. Mod. Phys. A **11**, 3145 (1996).

[22] G. Sierra and M.A. Martín-Delgado, in *The Exact Renormalization Group*, eds. A. Krasnitz, R. Potting, Y.A. Kubyshin and P.S. de Sa (World Scientific, Singapore, 1999) (cond-mat/9811170).

[23] I. Peschel, M. Kaulke and Ö. Legeza, Annalen Phys. **8**, 153 (1999), (cond-mat/9810174).

[24] S.R. White and D. Huse, Phys. Rev. B **48**, 3844 (1993).

[25] E.S. Sørensen and I. Affleck, Phys. Rev. B **49**, 13235 (1994).

[26] E.S. Sørensen and I. Affleck, Phys. Rev. B **49**, 15771 (1994); E. Polizzi, F. Mila and E. Sørensen, Phys. Rev. B **58**, 2407 (1998).

[27] C. Batista, K. Hallberg, and A. Aligia, Phys. Rev. B **58**, 9248 (1998); Phys. Rev. B **60**, 12553 (1999); Physica B **259**, 1017 (1999); E. Jannod, C. Payen, K. Schoumacker, C. Batista, K. Hallberg, and A. Aligia, Phys. Rev B **62**, 2998 (2000).

[28] M. Laukamp et al., Phys. Rev. B **5**, 10755 (1998).

[29] T-K. Ng, J. Lou, and Z. Su, Phys. Rev. B **61**, 11487 (2000).

[30] J. Lou, S. Qin, T-K. Ng, Z. Su, and I. Affleck, Phys. Rev. B **62**, 3786 (2000).

[31] S-W Tsai and J.B. Marston, Annalen Phys. **8**, Special Issue, 261 (1999).

[32] M. Sieling, U. Löw, B. Wolf, S. Schmidt, S. Zvyagin, and B.Lüthi, Phys. Rev. B **61**, 88 (2000).

[33] Y. Nishiyama, K. Totsuka, N. Hatano, and M. Suzuki, J. Phys. Soc. Jap **64**, 414 (1995).

[34] S. Qin, T. Ng, and Z.-B. Su, Phys. Rev. B **52**, 12844 (1995).

[35] U. Schollwöck and T. Jolicœur, Europhys. Lett. **30**, 493 (1995).

[36] X. Wang, S. Qin, and Lu Yu, Phys. Rev. B **60**, 14529 (1999).

[37] W. Tatsuaki, Phys. Rev. E **61**, 3199 (2000); T. Wada and T. Nishino, cond-mat/0103508 (Proceedings of the Conference on Computational Physics 2000 (CCP2000), Gold Coast, Queensland, Australia, 3-8 December 2000).

[38] M. Capone and S. Caprara, Phys. Rev. B **64**, 184418 (2001).

[39] J. Lou, X. Dai, S. Qin, Z. Su, and L. Yu, Phys. Rev. B **60**, 52 (1999).

[40] E. Ercolessi, G. Morandi, P. Pieri, and M. Roncaglia, Europhys. Lett. **49**, 434 (2000).

[41] S. Qin, X. Wang, and L. Yu, Phys. Rev. B **56**, R14251 (1997).

[42] K. Hallberg, P. Horsch, and G. Martínez, Phys. Rev. B **52**, R719 (1995).

[43] H.E. Boos, V.E. Korepin, Y. Nishiyama, and M. Shiroishi, J. Phys. A **35**, 4443 (2002); V.E. Korepin, S. Lukyanov, Y. Nishiyama, and M. Shiroishi, cond-mat/0210140.

[44] M. Shiroishi, M. Takahashi, and Y. Nishiyama, J. Phys. Soc. Jap. **70**, 3535 (2001).

[45] T. Hikihara and A. Furusaki, Phys. Rev. B **58**, R583 (1998).

[46] G. Fath, cond-mat/0208580.

[47] J. Lou, S. Qin, C. Chen, Z. Su, and L. Yu, Phys. Rev. B **65**, 064420 (2002).

[48] F. Capraro and C. Gros, cond-mat/0207279.

[49] Y. Hieida, K. Okunishi, and Y. Akutsu, Phys. Rev. B **64**, 224422 (2001).

[50] T. Byrnes, R. Bursill, H.-P. Eckle, C. Hamer, and A., Sandvik, cond-mat/0205140.

[51] S.-W. Tsai and J.B. Marston, Phys. Rev. B **62**, 5546 (2000).

[52] K. Hallberg, X. Wang, P. Horsch, and A. Moreo, Phys. Rev. Lett. **76**, 4955 (1996).

[53] A. Moreo, Phys. Rev. B **35**, 8562 (1987); T. Ziman and H. Schulz, Phys. Rev. Lett **59**, 140 (1987).

[54] Y. Yamashita, N. Shibata, and K. Ueda, Phys. Rev. B **58**, 9114 (1998); J. Phys. Soc. Jap. Vol.69, 242-247 (2000); Phys. Rev. B **61**, 4012 (2000).

[55] A. Onufriev and B. Marston, Phys. Rev. B **59**, 12573 (1999).

[56] C. Itoi, S. Qin, and I. Affleck, Phys. Rev. B **61**, 6747 (2000).

[57] W. Yu and S. Haas, Phys. Rev. B **63**, 024423 (2001).

[58] R.J. Bursill, T. Xiang, and G.A. Gehring, J. Phys. A **28** 2109 (1994).

[59] R.J. Bursill, G.A. Gehring, D.J.J. Farnell, J.B. Parkinson, T. Xiang, and C. Zeng, J. Phys. C **7** 8605 (1995).

[60] U. Schollwöck, Th. Jolicœur, and T. Garel, Phys. Rev. B **53**, 3304 (1996).

[61] R. Chitra, S. Pati, H.R. Krishnamurthy, D. Sen, and S. Ramasesha, Phys. Rev. B **52**, 6581 (1995); S. Pati, R. Chitra, D. Sen, H.R. Krishnamurthy, and S. Ramasesha, Europhys. Lett. **33**, 707 (1996); J. Malek, S. Drechsler, G. Paasch, and K. Hallberg, Phys. Rev. B **56**, R8467 (1997); E. Sørensen et al. in [2, Chapter 1.2 (Part II) and references therein]; D. Augier, E. Sørensen, J. Riera, and D. Poilblanc, Phys. Rev. B **60**, 1075 (1999).

[62] Y. Kato and A. Tanaka, J. Phys. Soc. Jap. **63**, 1277 (1994).

[63] S.R. White and I. Affleck, Phys. Rev. B **54**, 9862 (1996).

[64] G. Bouzerar, A. Kampf, and G. Japaridze, Phys. Rev. B **58**, 3117 (1998); G. Bouzerar, A. Kampf, and F. Schönfeld, cond-mat/9701176, unpublished; M.-B. Lepetit, and G. Pastor, Phys. Rev. B **56**, 4447 (1997).

[65] M. Kaburagi, H. Kawamura, and T. Hikihara, J. Phys. Soc. Jap. **68**. 3185 (1999).

[66] T. Hikihara, J. Phys. Soc. Jap. **71**, 319 (2002); T. Hikihara, M. Kaburagi, H. Kawamura, and T. Tonegawa, Phys. Rev. B **63**, 174430 (2001).

[67] S. Pati, S. Ramasesha, and D. Sen, Phys. Rev. B **55**, 8894 (1996); J. Phys. Condens. Matter **9**, 8707 (1997); T. Tonegawa et al., J. Magn. Magn. Mater. **177-181**, 647 (1998) (cond-mat/9712298).

[68] M. Azzouz, L. Chen, and S. Moukouri, Phys. Rev. B **50** 6223 (1994); S.R. White, R.M. Noack, and D.J. Scalapino, Phys. Rev. Lett. **73** 886 (1994); K. Hida, J. Phys. Soc. Jap. **64** 4896 (1995); T. Narushima, T. Nakamura, and S. Takada, J. Phys. Soc. Jap. **64** 4322 (1995); U. Schollwöck and D. Ko, Phys. Rev. B **53** 240 (1996); G. Sierra, M.A. Martín-Delgado, S. White, and J. Dukelsky, Phys. Rev. B **59**, 7973 (1999); S. White, Phys. Rev. B **53**, 52 (1996).

[69] A. Kawaguchi, A. Koga, K. Okunishi, and N. Kawakami, cond-mat/0205635.

[70] A. Trumper and C. Gazza, Phys. Rev. B **64**, 134408 (2001).

[71] M. Roger, Phys. Rev. B **63**, 144433 (2001).

[72] T. Hikihara and A. Furusaki, Phys. Rev. B **63**, 134438 (2001).

[73] A. Laeuchli, G. Schmida, and M. Troyer, cond-mat/0206153.

[74] T. Hikihara, T. Momoi, and X. Hu, cond-mat/0206102.

[75] T. Nunner, P. Brune, T. Kopp, M. Windt, and M. Grüninger, cond-mat/0203472.

[76] Y. Honda and T. Horiguchi, cond-mat/0106426.

[77] S.K. Pati and R.R.P. Singh, Phys. Rev. B **60**, 7695 (1999); S.R. White and R. Singh, Phys. Rev. Lett. **85**, 3330 (2000).

[78] N. Maeshima, M. Hagiwara, Y. Narumi, K. Kindo, T.C. Kobayashi, and K. Okunishi, cond-mat/0208373; N. Maeshima and K. Okunishi, Phys. Rev. B **62**, 934–939 (2000).

[79] C. Itoi and S. Qin, Phys. Rev. B **63**, 224423 (2001).

[80] J. Lou, J. Dai, S. Qin, Z. Su, and L. Yu, Phys. Rev. B **62**, 8600 (2000).

[81] C. Raghu, I. Rudra, S. Ramasesha, and D. Sen, Phys. Rev. B **62**, 9484 (2000).

[82] A. Honecker, M. Kaulke, and K.D. Schotte, Eur. Phys. J. B **15**, 423 (2000).

[83] A. Gendiar and A. Surda, Phys. Rev. B **62**, 3960 (2000).

[84] K. Tandon et al., Phys. Rev. B **59**, 396 (1999); R. Citro, E. Orignac, N. Andrei, C. Itoi, and S. Qin, J. Phys. Condens. Matter **12**, 3041 (2000).

[85] J. Lou, C. Chen, and S. Qin, Phys. Rev. B **64**, 144403 (2001).

[86] X. Wang and L. Yu, Phys. Rev. Lett. **84**, 5399 (2000).

[87] S.K. Pati, S. Ramasesha, and D. Sen, in *Magnetism: Molecules to Materials.* IV, eds. J.S. Miller and M. Drillon (Wiley-VCH Weinheim, 2002), Chapter 4 (cond-mat/0106621).

[88] L. Chen and S. Moukouri, Phys. Rev. B **53** 1866 (1996); S.J. Qin, S.D. Liang, Z.B. Su, and L. Yu, Phys. Rev. B **52** R5475 (1995); R. Noack in [2, Chapter 1.3 (Part II)]; S. Daul and R. Noack, Phys. Rev. B **61**, 12361 (2000); M. Vojta, R. Hetzel, and R. Noack, Phys. Rev. B **60**, R8417 (1999).

[89] K. Penc, K. Hallberg, F. Mila, and H. Shiba, Phys. Rev. Lett. **77**, 1390 (1996); Phys. Rev. B **55**, 15475 (1997).

[90] N. Bulut, cond-mat/0207186, to appear in Adv. Phys. **51**, no. 6 (2002).

[91] A. Malvezzi, T. Paiva, and R. dos Santos, cond-mat/0205266.

[92] E. Jeckelmann, cond-mat/0204244.

[93] Y. Zhang, C. Wu, and H.Q. Lin, Phys. Rev. B **65**, 115101 (2002).

[94] C. Aebischer, D. Baeriswyl, and R.M. Noack, Phys. Rev. Lett. **86**, 468 (2001).

[95] R. Arita and H. Aoki, Phys. Rev. B **61**, 12261 (2000); R. Arita, Y. Shimoi, K. Kuroki, and H. Aoki, Phys. Rev. B **57**, 10609 (1998).

[96] S. Daul, cond-mat/9911361.

[97] S. Qin, J. Lou, T. Xiang, G.-S. Tian, and Z. Su, cond-mat/0004162.

[98] A. Aligia, K. Hallberg, C. Batista, and G. Ortiz, Phys. Rev. B **61**, 7883 (2000).

[99] V. Meden, W. Metzner, U. Schollwöck, O. Schneider, T. Stauber, and K. Schoenhammer, , Eur. Phys. J. B **16**, 631 (2000).

[100] P. Farkasovsky, Phys. Rev. B **65**, 081102 (2002).

[101] C. Schuster, R. Roemer, and M. Schreiber, Phys. Rev. B **65**, 115114 (2002).

[102] K. Hida, Phys. Rev. Lett. **86**, 1331 (2001).

[103] W. Lay and J. Rudnick, Phys. Rev. Lett. **88**, 057203 (2002); Y. Nishiyama, J. Phys. A **34**, 11215 (2001); S.G. Chung, Phys. Rev. E **62**, 3262 (2000).

[104] B. Srinivasan and M.-B. Lepetit, Phys. Rev. B **66**, 024421 (2002); H. Sakamoto, T. Momoi, and K. Kubo, Phys. Rev. B **65**, 224403 (2002).

[105] K. Hamacher, C. Gros, and W. Wenzel, Phys. Rev. Lett. **88**, 217203 (2002).

[106] M. Vojta, A. Huebsch, and R.M. Noack, Phys. Rev. B **63**, 045105 (2001); Z. Weihong, J. Oitmaa, C.J. Hamer, and R.J. Bursill, J. Phys. C **13**, 433 (2001).

[107] Y. Park, S. Liang, and T.K. Lee, Phys. Rev. B **59**, 2587 (1999).

[108] S. Rommer, S.R. White, and D.J. Scalapino, Phys. Rev. B **61**, 13424 (2000).

[109] R.M. Noack, S.R. White, and D.J. Scalapino, Phys. Rev. Lett. **73**, 882 (1994); S.R. White, R.M. Noack, and D.J. Scalapino, J. Low Temp. Phys. **99**, 593 (1995); R.M. Noack, S.R. White, and D.J. Scalapino, Europhys. Lett. **30**, 163 (1995); C.A. Hayward, D. Poilblanc, R.M. Noack, D.J. Scalapino, and W. Hanke, Phys. Rev. Lett. **75**, 926 (1995); S. White and D. Scalapino, Phys. Rev. Lett. **81**, 3227 (1998); E. Jeckelmann, D. Scalapino, and S. White, Phys. Rev. B **58**, 9492 (1998).

[110] S. Nishimoto, E. Jeckelmann, and D. Scalapino, cond-mat/0208189.

[111] J.B. Marston, J.O. Fjaerestad, and A. Sudbo, Phys. Rev. Lett. **89**, 056404 (2002).

[112] U. Schollwöck, S. Chakravarty, J.O. Fjaerestad, J.B. Marston, and M. Troyer, cond-mat/0209444.

[113] L.G. Caron and C. Bourbonnais, Phys. Rev. B **66**, 045101 (2002).

[114] G. Fano, F. Ortolani, A. Parola, and L. Ziosi, Phys. Rev. B **60**, 15654 (1999).

[115] N. Shibata and D. Yoshioka, Phys. Rev. Lett. **86**, 5755 (2001); Physica E **12**, 43 (2002) D. Yoshioka and N. Shibata, J. Phys. Soc. Jap. **70**, 3690 (2001).

[116] V. Meden and U. Schollwöck, cond-mat/0209588.

[117] S. Liang and H. Pang, Europhys. Lett. **32**, 173 (1995).

[118] S. White and D. Scalapino, Phys. Rev. B **55**, 14701 (1997).

[119] S. White and D. Scalapino, Phys. Rev. B **55**, 6504 (1997).

[120] S. White and D. Scalapino, Phys. Rev. B **57**, 3031 (1998).

[121] E. Arrigoni, A.P. Harju, W. Hanke, B. Brendel, and S.A. Kivelson, Phys. Rev. B **65**, 134503 (2002); J. Bonca, J.E. Gubernatis, M. Guerrero, E. Jeckelmann, and S.R. White, Phys. Rev. B **61**, 3251 (2000); A.L. Chernyshev, S. White, and A.H. Castro Neto, Phys. Rev. B **65**, 214527 (2002); S. White and D. Scalapino, Phys. Rev. B **61**, 6320 (2000).

[122] S. White and D. Scalapino, Phys. Rev. Lett. **80**, 1272 (1998) (see also cond-mat/9907375).

[123] I.P. McCulloch, A.R. Bishop, and M. Gulacsi, Phil. Mag. B **81**, 1603 (2001).

[124] S.R. White and I. Affleck, Phys. Rev. B **64**, 024411 (2001).

[125] T. Xiang, J. Lou, and Z. Su, Phys. Rev. B **64**, 104414 (2001); P. Henelius, Phys. Rev. B **60**, 9561 (1999).

[126] S. Nishimoto, E. Jeckelmann, F. Gebhard, and R. Noack, Phys. Rev. B **65**, 165114 (2002).

[127] N. Maeshima, Y. Hieida, Y. Akutsu, T. Nishino, and K. Okunishi, Phys. Rev. E **64** (2001) 016705.

[128] M.A. Martin-Delgado, J. Rodriguez-Laguna, and G. Sierra, Nucl. Phys. B **601**, 569 (2001).

[129] P. Henelius, Phys. Rev. B **60**, 9561 (1999).

[130] S. Moukouri and L.G. Caron, cond-mat/0210668.

[131] T.A. Costi, P. Schmitteckert, J. Kroha, and P. Wölfle, Phys. Rev. Lett. **73**, 1275 (1994); S. Eggert and I. Affleck, Phys. Rev. Lett. **75**, 934 (1995); E.S. Sørensen and I. Affleck, Phys. Rev. B **51**. 16115 (1995); X.Q. Wang and S. Mallwitz, Phys. Rev. B **53**, R492 (1996); W. Wang, S.J. Qin, Z.Y. Lu, L. Yu, and Z.B. Su, Phys. Rev. B **53**, 40 (1996); C.C. Yu and M. Guerrero, Phys. Rev. B **54**, 15917 (1996); A. Furusaki and T. Hikihara, Phys. Rev. B **58**, 5529 (1998).

[132] K. Hallberg and R. Egger, Phys. Rev. B **55**, 8646 (1997).

[133] C. Schuster and U. Eckern, Annalen Phys. **11**, 901 (2002) (cond-mat/0201390); W. Zhang, J. Igarashi and P. Fulde, Phys. Rev. B **56**, 654 (1997).

[134] V. Meden, W. Metzner, U. Schollwöck, and K. Schoenhammer, Phys. Rev. B **65**, 045318 (2002).

[135] C.C. Yu and S.R. White, Phys. Rev. Lett. **71**, 3866 (1993); C.C. Yu and S.R. White, Phys. B **199**, 454 (1994).

[136] S. Moukouri and L.G. Caron, Phys. Rev. B **52**, 15723 (1995); N. Shibata, T. Nishino, K. Ueda, and C. Ishii, Phys. Rev. B **53**, R8828 (1996); M. Guerrero and R.M. Noack, Phys. Rev. B **53**, 3707 (1996).

[137] H. Otsuka and T. Nishino, Phys. Rev. B **52**, 15066 (1995); S. Moukouri, L.G. Caron, C. Bourbonnais, and L. Hubert, Phys. Rev. B **51**, 15920 (1995).

[138] J.C. Xavier, E. Novais, and E. Miranda, Phys. Rev. B **65**, 214406 (2002).

[139] T. Yamamoto, R. Manago, and Y. Mori, cond-mat/0204588.

[140] I.P. McCulloch, A. Juozapavicius, A. Rosengren, and M. Gulacsi, Phys. Rev. B **65**, 052410 (2002).

[141] M. Guerrero and R.M. Noack, Phys. Rev. B **63**, 144423 (2001).

[142] S. Watanabe, J. Phys. Soc. Jap. **69**, 2947 (2000).

[143] I.P. McCulloch, M. Gulacsi, S. Caprara, and A. Juozapavicius, J. Low Temp. Phys. **117**, 323 (1999).

[144] J. Xavier, R. Pereira, E. Miranda, and I. Affleck, cond-mat/0209623.

[145] S. Watanabe, Y. Kuramoto, T. Nishino, and N. Shibata, J. Phys. Soc. Jap **68**, 159 (1999).

[146] J. Moreno, S. Qin, P. Coleman, and L. Yu, Phys. Rev. B **64**, 085116 (2001).

[147] S. Moukouri, L. Chen, and L.G. Caron, Phys. Rev. B **53**, R488 (1996).

[148] J. Riera, K. Hallberg, and E. Dagotto, Phys. Rev. Lett. **79**, 713 (1997); E. Dagotto et al., Phys. Rev. B **58**, 6414 (1998).

[149] D. García, K. Hallberg, C. Batista, M. Avignon, and B. Alascio, Phys. Rev. Lett. **85**, 3720 (2000); D.J. García, K. Hallberg, C.D. Batista, S. Capponi, D. Poilblanc, M. Avignon and B. Alascio, Phys. Rev. B **65**, 134444 (2002).

[150] B. Ammon and M. Imada, J. Phys. Soc. Jap. **70**, 547 (2001).

[151] S. Ramasesha et al., Phys. Rev. B **54**, 7598 (1996); Synth. Metals **85**, 1019 (1997).

[152] E.S. Sørensen and I. Affleck, Phys. Rev. B **51**, 16115 (1995).

[153] I. McCulloch and M. Gulacsi, Europhys. Lett. **57**, 852 (2002).

[154] G. Sierra and T. Nishino, Nucl. Phys. B **495**, 505 (1997) (cond-mat/9610221).

[155] T. Xiang, Phys. Rev. B **53**, R10445 (1996).

[156] M.S.L. du Croo de Jongh and J.M.J. van Leeuwen, Phys. Rev. B **57**, 8494 (1998).

[157] M.-B. Lepetit, M. Cousy, and G. Pastor, Eur. Phys. J. B **13**, 421 (2000); H. Otsuka, Phys. Rev. B **53**, 14004 (1996).

[158] J. Dukelsky and G. Dussel, Phys. Rev. B **59**, R3005 (1999) and references therein.

[159] J. Dukelsky and G. Sierra, Phys. Rev. Lett. **83**, 172 (1999).

[160] M.A. Martín-Delgado and G. Sierra, Phys. Rev. Lett. **83**, 1514 (1999).

[161] S. Glazek and K. Wilson, Phys. Rev. D **48**, 5863 (1993); Phys. Rev. D **49**, 4214 (1994).

[162] T. Eller, H.-C. Pauli, and S. Brodsky, Phys. Rev. D **35**, 1493 (1987).

[163] T. Nishino, J. Phys. Soc. Jap. **64**, 3598 (1995); see also T. Nishino in [2, Chapter 5 (Part I)]; T. Nishino and K. Okunishi in *Strongly Correlated Magnetic and Superconducting Systems*, eds. G. Sierra and M.A. Martín-Delgado, Lecture Notes in Phys., vol. 478 (Springer, Berlin, 1997).

[164] H. Trotter, Proc. Am. Math. Soc. **10**, 545 (1959); M. Suzuki, Prog. Theor. Phys. **56**, 1454 (1976); R. Feynman and A. Hibbs *Quantum Mechanics and Path Integrals* (McGraw-Hill, New York, 1965).

[165] A. Gendiar and A. Surda, Phys. Rev. B **63**, 014401 (2001).

[166] T. Nishino and N. Shibata, J. Phys. Soc. Jap. **68**, 3501 (1999); T. Enss and U. Schollwöck, J. Phys. A **34**, 7769 (2001).

[167] A. Kemper, A. Schadschneider, and J. Zittartz, J. Phys. A **34**, L279 (2001); T. Enss and U. Schollwöck, J Phys A **34**, 7769 (2001).

[168] R. Baxter, J. Math. Phys. **9**, 650 (1968); J. Statist. Phys. **19**, 461 (1978).

[169] T. Nishino and K. Okunishi, J. Phys. Soc. Jap. **65**,891 (1996); J. Phys. Soc. Jap. **66**, 3040 (1997); T. Nishino, K. Okunishi and M. Kikuchi, Phys. Lett. A **213**, 69 (1996).

[170] H. Takasaki, T. Nishino, and Y. Hieida, J. Phys. Soc. Jap. **70**, 1429 (2001).

[171] T. Nishino, Y. Hieida, K. Okunishi, N. Maeshima, Y. Akutsu, and A. Gendiar, Prog. Theor. Phys. **105**, 409 (2001).

[172] C. Ritter and G. von Gehlen, in *Quantization, Gauge Theory and Strings*, eds. A. Semikhatov et al., Vol. I, pp. 563–578 (Scientific World, Singapore, 2001) (cond-mat/0009255).

[173] T. Nishino and K. Okunishi, J. Phys. Soc. Jap. **67**, 3066 (1998); T. Nishino, K. Okunishi, Y. Hieida, N. Maeshima, and Y. Akutsu, Nucl. Phys. B **575**, 504 (2000).

[174] E. Carlon and A. Drzewiński, Phys. Rev. Lett. **79**, 1591 (1997); Phys. Rev. E **57**, 2626 (1998); E. Carlon, A. Drzewiński, and J. Rogiers, Phys. Rev. B **58**, 5070 (1998); A. Drzewiński, A. Ciach, and A. Maciolek, Eur. Phys. J. B **5**, 825 (1998); Phys. Rev. E **60**, 2887 (1999).

[175] A. Drzewinski, A. Maciolek, and R. Evans, Phys. Rev. Lett. **85**, 3079 (2000); A. Drzewinski, Phys. Rev. E **62**, 4378 (2000).

[176] M.-C. Chung, M. Kaulke, I. Peschel, M. Pleimling, and W. Selke, Eur. Phys. J. B **18**, 655 (2000).

[177] E. Carlon and F. Iglói, Phys. Rev. B **57**, 7877 (1998); Phys. Rev. B **59**, 3783 (1999); E. Carlon, C. Chatelain and B. Berche, Phys. Rev. B **60**, 12974 (1999).

[178] J. Kondev and J. Marston, Nucl. Phys. B **497**, 639 (1997); T. Senthil, B. Marston, and M. Fisher, Phys. Rev. B **60**, 4245 (1999); J. Marston and S. Tsai, Phys. Rev. Lett. **82**, 4906 (1999); S. Tsai and J. Marston, Annalen Phys. **8**, Special Issue, 261 (1999).

[179] M. Kaulke and I. Peschel, Eur. Phys. J. B **5**, 727 (1998).

[180] Y. Hieida, J. Phys. Soc. Jap. **67**, 369 (1998).

[181] I. Peschel and M. Kaulke in [2, Chapter 3.1 (Part II)].

[182] E. Carlon, M. Henkel, and U. Schollwöck, Eur. Phys. J. B **12**, 99 (1999); E. Carlon, M. Henkel, and U. Schollwöck, Phys. Rev. E **63**, 036101 (2001).

[183] J. Hooyberghs, E. Carlon, and C. Vanderzande, Phys. Rev. E **62**, 036124 (2001).

[184] S.G. Chung, Phys. Rev. B **60**, 11761 (1999).

[185] R. Bursill, T. Xiang, and G. Gehring, J. Phys. C **8**, L583 (1996).

[186] H. Trotter, Proc. Am. Math. Soc. **10**, 545 (1959); M. Suzuki, Prog. Theor. Phys. **56**, 1454 (1976).

[187] X. Wang and T. Xiang, Phys. Rev. B **56**, 5061 (1997).

[188] N. Shibata and K. Ueda, J. Phys. Soc. Jap. **70**, 3690 (2001).

[189] N. Shibata, J. Phys. Soc. Jap. **66**, 2221 (1997).

[190] T. Xiang, Phys. Rev. B **58**, 9142 (1998).

[191] N. Shibata, B. Ammon, T. Troyer, M. Sigrist, and K. Ueda, J. Phys. Soc. Jap. **67**, 1086 (1998).

[192] K. Maisinger, U. Schollwöck, S. Brehmer, H.-J. Mikeska, and S. Yamamoto, Phys. Rev. B **58**, R5908 (1998).

[193] S. Rommer and S. Eggert, Phys. Rev. B **59**, 6301 (1999).

[194] I. Maruyama, N. Shibata, and K. Ueda, Phys. Rev. B **65**, 174421 (2002).

[195] K. Maisinger and U. Schollwöck, Phys. Rev. Lett. **81**, 445 (1999).

[196] B. Ammon, M. Troyer, T. Rice, and N. Shibata, Phys. Rev. Lett. **82**, 3855 (1999); N. Shibata and H. Tsunetsugu, J. Phys. Soc. Jap. **68**, 3138 (1999).

[197] F. Naef and X. Wang, Phys. Rev. Lett. **84**, 1320 (2000).

[198] A. Klümper, R. Raupach and F. Schönfeld, Phys. Rev. B **59**, 3612 (1999).

[199] S. Moukouri and L. Caron, Phys. Rev. Lett. **77**, 4640 (1996).

[200] S. Moukouri and L. Caron, see [2, Chapter 4.5 (Part II)].

[201] S. Moukouri, cond-mat/0011169.

[202] W. Zhang, J. Igarashi, and P. Fulde, J. Phys. Soc. Jap. **66**, 1912 (1997).

[203] L. Caron and S. Moukouri, Phys. Rev. Lett. **76**, 4050 (1996); Phys. Rev. B **56**, R8471 (1997).

[204] E. Jeckelmann and S. White, Phys. Rev. B **57**, 6376 (1998).

[205] R. Noack, S. White, and D. Scalapino in *Computer Simulations in Condensed Matter Physics*. VII, eds. D. Landau, K.-K. Mon and H.-B. Schüttler (Springer-Verlag, Heidelberg and Berlin, 1994).

[206] E. Jeckelmann, C. Zhang, and S. White in [2, Chapter 5.1 (Part II)].

[207] C. Zhang, E. Jeckelmann, and S. White, Phys. Rev. Lett. **80**, 2661 (1998); E. Jeckelmann, C. Zhang, and S.R. White, Phys. Rev. B **60**, 7950 (1999).

[208] Y. Nishiyama, cond-mat/0102123.

[209] Y. Nishiyama, Eur. Phys. J. B **12**, 547 (1999).

[210] R. Bursill, Y. McKenzie, and C. Hammer, Phys. Rev. Lett. **80**, 5607 (1998); Phys. Rev. Lett. **83**, 408 (1999); R. Bursill, Phys. Rev. B **60**, 1643 (1999).

[211] R. Pai, R. Pandit, H. Krishnamurthy, and S. Ramasesha, Phys. Rev. Lett. **76**, 2937 (1996) (see also the comment by N.V. Prokof′ev and B.V. Svistunov, Phys. Rev. Lett. **80**, 4355 (1998)); S. Rapsch, U. Schollwöck, and W. Zwerger, Europhys. Lett. **46**, 559 (1999).

[212] T. Kühner and H. Monien, Phys. Rev. B **58**, R14741 (1998).

[213] I. Peschel and M.-C. Chung, J. Phys. A: Math. Gen. **32**, 8419 (1999).

[214] P. Maurel and M.-B. Lepetit, Phys. Rev. B **62**, 10744 (2000); P. Maurel, M.-B. Lepetit, and D. Poilblanc, Eur. Phys. J. B **21**, 481 (2001).

[215] K. Hida, J. Phys. Soc. Jap. **65**, 895 (1996) and 3412 (1996) (erratum); J. Phys. Soc. Jap. **66**, 330 (1997); J. Phys. Soc. Jap. **66**, 3237 (1997); Prog. Theor. Phys. Suppl. **145**, 320 (2002).

[216] K. Hida, J. Phys. Soc. Jap. **68**, 3177 (1999).

[217] K. Hida, Phys. Rev. Lett. **83**, 3297 (1999).

[218] F. Igloi, R. Juhasz, and P. Lajko, Phys. Rev. Lett. **86**, 1343 (2001).

[219] R. Mélin, Y-C. Lin, P. Lajkó, H. Rieger, and F. Iglói, Phys. Rev. B **65**, 104415 (2002).

[220] P. Schmitteckert, T. Schulze, C. Schuster, P. Schwab, and U. Eckern, Phys. Rev. Lett. **80**, 560 (1998); P. Schmitteckert and U. Eckern, Phys. Rev. B **53**, 15397 (1996).

[221] D. Weinmann, P. Schmitteckert, R. Jalabert, and J. Pichard, Eur. Phys. J. B **19**, 139–156 (2001).

[222] P. Schmitteckert, R. Jalabert, D. Weinmann, and J.L. Pichard, Phys. Rev. Lett. **81**, 2308 (1998).

[223] E. Carlon, P. Lajko, and F. Iglói, Phys. Rev. Lett. **87**, 277201 (2001).

[224] G. Fano, F. Ortolani, and L. Ziosi, J. Chem. Phys. **108**, 9246 (1998) (cond-mat/9803071); R. Bursill and W. Barford, Phys. Rev. Lett. **82**, 1514 (1999).

[225] M. Exler and J. Schnack, cond-mat/0205068.

[226] B. Normand, X. Wang, X. Zotos, and D. Loss, Phys. Rev. B **63**, 184409 (2001)

[227] C. Raghu, Y. Anusooya, S. Pati, and S. Ramasesha, J. Phys. A **34**, 11215 (2001).

[228] W. Barford and R. Bursill, Chem. Phys. Lett. **268**, 535 (1997); W. Barford, R. Bursill, and M. Lavrentiev, J. Phys. Condens. Matter **10**, 6429 (1998); W. Barford in [2, Chapter 2.3 (Part II) and references therein]; M. Lavrentiev, W. Barford, S. Martin, H. Daly, and R. Bursill, Phys. Rev. B **59**, 9987 (1999).

[229] E. Carlon, A. Drzewinski, and J. van Leeuwen, J. Chem. Phys. **117**, 2425 (2002); Phys. Rev. E **64**, 010801 (R) (2001); M. Paessens and G. Schütz, Phys. Rev. E **66**, 021806 (2002).

[230] M.A. Martín-Delgado, J. Rodriguez-Laguna, and G. Sierra, Phys. Rev. B **65**, 155116 (2002).

[231] S. White and R. Martin, J. Chem. Phys. **110**, 4127 (1999); see also S. White in [2, Chapter 2.1].

[232] S. White in [2, Chapter 2.1].

[233] Ö. Legeza, J. Röder, and B.A. Hess, cond-mat/0208187.

[234] S. Daul, I. Ciofini, C. Daul, and S.R. White, Int. J. Quant. Chem. **79**, 331 (2000) (cond-mat/9912348).

[235] C. Bauschlicher and P. Taylor, J. Chem. Phys. **85**, 2779 (1986).

[236] Ö. Legeza, J. Röder, and B.A. Hess, cond-mat/0204602.

[237] W. Hofstetter, Phys. Rev. Lett. **85**, 1508 (2000) .

[238] M.A. Cazalilla and J.B. Marston, Phys. Rev. Lett. **88**, 256403 (2002).

[239] K. Hallberg, Phys. Rev. B **52**, 9827 (1995).

[240] E.R. Gagliano and C.A. Balseiro, Phys. Rev. Lett. **59**, 2999 (1987).

[241] G. Grosso and G. Partori Parravicini, in *Memory Function Approaches to Stochastic Problems in Condensed Matter*, Adv. Chem. Phys., vol. 62, 133 (Wiley, New York, 1985).

[242] P. Horsch and W. von der Linden, Z. Phys. B **72** 181 (1981).

[243] J. des Cloiseaux and J.J. Pearson, Phys. Rev. **128**, 2131 (1962).

[244] T. Yamada, Prog. Theor. Phys. Jap. **41**, 880 (1969); L.D. Fadeev and L.A. Takhtajan, Phys. Lett. A **85**, 375 (1981).

[245] G. Müller, H. Thomas, H. Beck, and J. Bonner, Phys. Rev. B **24**, 1429 (1981) and references therein.

[246] S. Haas, J. Riera, and E. Dagotto, Phys. Rev. B **48**, 3281 (1993).

[247] T. Kühner and S. White, Phys. Rev. B **60**, 335 (1999).

[248] P. Brune, G. Japaridze, and A.P. Kampf, cond-mat/0106007 (unpublished).

[249] T.D. Kühner, S.R. White, and H. Monien, Phys. Rev. B **61**, 12474 (2000).

[250] S.S. Kancharla and C.J. Bolech, Phys. Rev. B **64**, 085119 (2001).

[251] Y. Anusooya, S. Pati, and S. Ramasesha, J. Chem. Phys. **106**, 1 (1997); S. Ramasesha, K. Tandon, Y. Anusooya, and S. Pati, Proc. SPIE Int. Soc. Opt. Eng. **3145**, 282 (1997); S. Ramasesha, Z. Shuai, and J. Brédas, Chem. Phys. Lett. **245**, 224 (1995).

[252] E. Jeckelmann, Phys. Rev. B **66**, 045114 (2002).

[253] E. Jeckelmann and F. Gebhard, Phys. Rev. Lett. **85**, 3910 (2000); F. Essler, F. Gebhard, and E. Jeckelmann, Phys. Rev. B **64**, 125119 (2001); E. Jeckelmann, cond-mat/0208480.

[254] H.B. Pang, H. Akhlaghpour, and M. Jarrell, Phys. Rev. B **53** 5086 (1996).

[255] P. Hohenberg and W. Brinkman, Phys. Rev. B **10**, 128 (1974).

[256] X. Wang, K. Hallberg, and F. Naef in [2, Chapter 7 (Part I)].

[257] T. Mutou, N. Shibata, and K. Ueda, Phys. Rev. Lett **81**, 4939 (1998); Phys. Rev. Lett. **82**, 3727 (1999) (erratum).

[258] F. Naef, X. Wang, X. Zotos, and W. von der Linden, Phys. Rev. B **60**, 359 (1999).

2

Quantum Monte Carlo Methods for Strongly Correlated Electron Systems

Shiwei Zhang

ABSTRACT We review some of the recent development in quantum Monte Carlo (QMC) methods for models of strongly correlated electron systems. QMC is a promising general theoretical tool to study many-body systems, and has been widely applied in areas spanning condensed-matter, high-energy, and nuclear physics. Recent progress has included two new methods, the ground-state and finite-temperature constrained path Monte Carlo methods. These methods significantly improve the capability of numerical approaches to lattice models of correlated electron systems. They allow calculations without any decay of the sign, making possible calculations for large system sizes and low temperatures. The methods are approximate. Benchmark calculations show that accurate results on energy and correlation functions can be obtained. This chapter gives a pedagogical introduction to quantum Monte Carlo, with a focus on the constrained path Monte Carlo methods.

1 Introduction

In order to understand properties of correlated electron systems and their theoretical and technological implications, accurate calculations are necessary to treat microscopic correlations correctly. This crucial need is manifested in the daily demand to solve model systems, which are often intractable analytically, in order to develop and benchmark theory and compare with experiments.

To effectively calculate electron correlation effects is an extremely challenging task. It requires the solution of the Schrödinger equation or evaluation of the density matrix for a many-body system at equilibrium. Explicit numerical approaches are an obvious possibility; these have found wide applications and proved very valuable, e.g., exact diagonalization in the study of lattice models for high-temperature superconductivity and for magnetism, and configuration interaction in quantum chemistry. However, since the dimensionality involved grows exponentially with system size, they all suffer from *exponential* complexity, i.e., exponential scaling of the required computer time with system size or accuracy. As an example, with the most powerful computers we now can treat a lattice of about 50 sites in exact diagonalization of the two-dimensional spin-$\frac{1}{2}$ Heisenberg model, up from ~ 20 sites twenty years ago. During this period, the peak speed of computers has been doubling roughly every two years.

Monte Carlo methods offer a promising alternative as a numerical approach to study many-body systems. Its required computer time scales algebraically (as opposed to exponentially in exact diagonalization) with system size. Rather than explicitly integrating over phase space, Monte Carlo methods sample it. The central limit theorem dictates that the statistical error in a Monte Carlo calculation decays algebraically with computer time. With the rapid advent of scalable parallel computers, these methods offer the possibility to study large system sizes systematically and extract information about the thermodynamic limit.

For fermion systems, however, Monte Carlo methods suffer from the so-called "sign" problem [1–3]. In these systems, the Pauli exclusion principle requires that the states be anti-symmetric under interchange of two particles. As a consequence, negative signs appear, which cause cancellations among contributions of the Monte Carlo samples of phase space. In fact, as the temperature is lowered or the system size is increased, such cancellation becomes more and more complete. The net signal thus decays *exponentially*. The algebraic scaling is then lost, and the method breaks down. Clearly the impact of this problem on the study of correlated electron systems is extremely severe.

To date most applications of QMC methods to strongly correlated electron models have either lived with the sign problem or relied on some form of approximation to overcome the exponential scaling. The former has difficulties reaching large system sizes or low temperatures. The latter loses the exactness of QMC and the results are biased by the approximation. Despite these limitations, QMC methods have proved a very useful theoretical tool in the study of strongly correlated electron systems. In many cases they have provided very accurate and reliable numerical results which are sometimes the only ones available for the system in question. Especially recently, progress has been rapid in the development of approximate QMC methods for lattice models, which has led to a growing number of applications. For example, with the constrained path Monte Carlo method (CPMC) discussed below, a system of 220 electrons on a 16×16 lattice in the Hubbard model can be studied with moderate computer time [4]. The dimension of the Hilbert space for this system exceeds 10^{150}.[1]

In this chapter, we review some of the recent progress in the study of models for strongly correlated electron systems with quantum Monte Carlo

[1] The size for a 16×16 system with 109 electrons with up spins and 109 with down spins is about 10^{149}.

methods. The chapter is not meant to be a comprehensive review. Our focus is on the ground-state and finite-temperature constrained path Monte Carlo methods [3, 5–7]. We will, however, include in the Appendix some discussions of variational Monte Carlo and Green's function Monte Carlo (GFMC) [8] methods, which have also seen extensive applications in the study of correlated electron models. We do so to illustrate that, while GFMC is very different from the auxiliary-field-based methods we focus on, the underlying ideas and the nature of the sign problems have much in common. Our goal is to highlight these ideas and show the capabilities that QMC methods in general bring as a theoretical tool, as well as the current algorithmic difficulties.

2 Preliminaries

2.1 Starting Point of Quantum Monte Carlo (QMC)

Most ground-state QMC methods are based on

$$|\Psi_0\rangle \propto \lim_{\tau\to\infty} e^{-\tau\widehat{H}}|\Psi_T\rangle; \tag{2.1}$$

that is, the ground state $|\Psi_0\rangle$ of a many-body Hamiltonian $\widehat{H}$ can be projected from any known trial state $|\Psi_T\rangle$ that satisfies $\langle\Psi_T \mid \Psi_0\rangle \neq 0$. In a numerical method, the limit can be obtained iteratively by

$$|\Psi^{(n+1)}\rangle = e^{-\Delta\tau\widehat{H}}|\Psi^{(n)}\rangle, \tag{2.2}$$

where $|\Psi^{(0)}\rangle = |\Psi_T\rangle$. Ground-state expectation $\langle\widehat{O}\rangle$ of a physical observable $\widehat{O}$ is given by

$$\langle\widehat{O}\rangle = \lim_{n\to\infty} \frac{\langle\psi^{(n)}|\widehat{O}|\psi^{(n)}\rangle}{\langle\psi^{(n)} \mid \psi^{(n)}\rangle}. \tag{2.3}$$

For example, the ground-state energy can be obtained by letting $\widehat{O} = \widehat{H}$. A so-called mixed estimator exists, however, which is exact for the energy (or any other $\widehat{O}$ that commutes with $\widehat{H}$) and can lead to considerable simplifications in practice:

$$E_0 = \lim_{n\to\infty} \frac{\langle\psi_T|\widehat{H}|\psi^{(n)}\rangle}{\langle\psi_T \mid \psi^{(n)}\rangle}. \tag{2.4}$$

Finite-temperature QMC methods use the density matrix. The expectation value of $\widehat{O}$ is

$$\langle\widehat{O}\rangle = \frac{\mathrm{Tr}(\widehat{O}e^{-\beta\widehat{H}})}{\mathrm{Tr}(e^{-\beta\widehat{H}})}, \tag{2.5}$$

where $\beta = 1/kT$ is the inverse temperature. In other words, $\langle \widehat{O} \rangle$ is simply a weighted average with respect to the density matrix $e^{-\beta \widehat{H}}$. In a numerical method, the partition function in the denominator of (2.5) is written as

$$Z \equiv \mathrm{Tr}(e^{-\beta \widehat{H}}) = \mathrm{Tr}[\underbrace{e^{-\Delta\tau \widehat{H}} \cdots e^{-\Delta\tau \widehat{H}} e^{-\Delta\tau \widehat{H}}}_{L}], \tag{2.6}$$

where $\Delta\tau = \beta/L$ is the "time" step, and L is the number of time slices.

Quantum Monte Carlo methods carry out the iteration in (2.2) or the trace in (2.6)—both requiring integration in many-dimensional spaces—by Monte Carlo sampling. The difference in the different classes of methods amounts primarily to the space that is used to represent the wave function or density matrix and to carry out the integration. The ground-state and finite-temperature constrained path Monte Carlo methods work in second quantized representation and in an auxiliary-field space, while Green's function Monte Carlo works in first-quantized representation and in configuration space.

2.2 Basics of Monte Carlo Techniques

We list a few key elements from standard Monte Carlo techniques. In addition to serving as a brief reminder to the reader, they will help to introduce several results and notations that will be useful in our discussion of the QMC methods. Excellent books exist [9] for extensive discussions of Monte Carlo methods.

Monte Carlo methods are often used to compute many-dimensional integrals of the form

$$G = \frac{\int_{\Omega_0} f(\vec{x}) g(\vec{x})\, d\vec{x}}{\int_{\Omega_0} f(\vec{x})\, d\vec{x}}, \tag{2.7}$$

where $\vec{x}$ is a vector in a many-dimensional space and Ω_0 is a domain in this space. We will assume that $f(\vec{x}) \geq 0$ on Ω_0 and that it is normalizable; i.e., the denominator is finite. A familiar example of the integral in (2.7) comes from classical statistical physics, where $f(\vec{x})$ is the Boltzmann distribution.

To compute G by Monte Carlo, we *sample* $\vec{x}$ from a probability density function (PDF) proportional to $f(\vec{x})$, i.e., the PDF $\bar{f}(\vec{x}) \equiv f(\vec{x}) / \int_{\Omega_0} f(\vec{x})\, d\vec{x}$. This means we generate a sequence $\{\vec{x}_1, \vec{x}_2, \dots, \vec{x}_i, \dots\}$ so that the probability that any $\vec{x}_i$ is in the sub-domain $(\vec{x}, \vec{x} + d\vec{x})$ is

$$\mathrm{Prob}\{\vec{x}_i \in (\vec{x}, \vec{x} + d\vec{x})\} = \bar{f}(\vec{x}) d\vec{x} \tag{2.8}$$

Below when we refer to sampling a function $f(\vec{x})$, it should be understood as sampling the corresponding PDF $\bar{f}(\vec{x})$. There are different techniques to sample a many-dimensional function $f(\vec{x})$. The most general and perhaps most often used is the Metropolis algorithm, which creates a Markov chain random walk in $\vec{x}$-space whose equilibrium distribution is the desired function.

Given $\mathcal{M}$ independent samples from $f(\vec{x})$, the integral in (2.7) is estimated by

$$G_{\mathcal{M}} = \frac{1}{\mathcal{M}} \sum_{i=1}^{\mathcal{M}} g(\vec{x}_i). \tag{2.9}$$

The error in the estimate decays algebraically with the number of samples: $|G_{\mathcal{M}} - G| \propto 1/\sqrt{\mathcal{M}}$.

Using the results above, we can compute

$$G' = \frac{\int_{\Omega_0} f(\vec{x}) g(\vec{x}) h(\vec{x})\, d\vec{x}}{\int_{\Omega_0} f(\vec{x}) h(\vec{x})\, d\vec{x}}, \tag{2.10}$$

if the function $h(\vec{x})$ is such that both the numerator and denominator exist. Formally

$$G'_{\mathcal{M}} = \frac{\sum_{i=1}^{\mathcal{M}} g(\vec{x}_i) h(\vec{x}_i)}{\sum_{i=1}^{\mathcal{M}} h(\vec{x}_i)}, \tag{2.11}$$

although, as we will see, difficulties arise when $h(\vec{x})$ can change sign and is rapidly oscillating.

Integral equations are another main area of applications of Monte Carlo methods. For example [9], the integral equation

$$\Psi'(\vec{x}) = \int_{\Omega_0} K(\vec{x}, \vec{y}) w(\vec{y}) \Psi(\vec{y})\, d\vec{y}, \tag{2.12}$$

can be viewed in terms of a *random walk* if it has the following properties: $\Psi(\vec{y})$ and $\Psi'(\vec{x})$ can be viewed as PDF's (in the sense of f in (2.7)), $w(\vec{y}) \geq 0$, and $K(\vec{x}, \vec{y})$ is a PDF for $\vec{x}$ conditional on $\vec{y}$. Then, given an ensemble $\{\vec{y}_i\}$ sampling $\Psi(\vec{y})$, the following two steps will allow us to generate an ensemble that samples $\Psi'(\vec{x})$. First an absorption/branching process is applied to each $\vec{y}_i$ according to $w(\vec{y}_i)$. For example, we can make $\mathrm{int}(w(\vec{y}_i) + \xi)$ copies of $\vec{y}_i$, where ξ is a uniform random number on $(0, 1)$. Second we randomly walk each new $\vec{y}_j$ to an $\vec{x}_j$ by sampling the PDF $K(\vec{x}, \vec{y}_j)$. The resulting $\{\vec{x}_j\}$ are Monte Carlo samples of $\Psi'(\vec{x})$. We emphasize that the purpose of our discussion of random walks here is to illustrate the basic concept, which we will use later. The simple procedure described above is thus not meant as an accurate account of the technical details necessary for an efficient implementation.

2.3 *Slater Determinant Space*

In the auxiliary-field-based QMC [10–13] method, the Monte Carlo algorithm works in a space of Slater determinants. The building blocks of Slater determinants are single-particle basis states. The single-particle basis states can be plane waves, or lattice sites in the Hubbard model, or energy eigenstates in a mean-field potential. Often the space of single-particle basis states is truncated. Single-particle wave functions (orbitals) are formed with the basis states. Slater determinants are then built from the single-particle orbitals.

We first define some notations that we will use throughout the discussion of standard auxiliary-field quantum Monte Carlo (AFQMC) and then later the constrained path Monte Carlo (CPMC) methods.

N: number of single-electron basis states. For example, N can be the number of lattice sites ($L \times L$) in the two-dimensional Hubbard model.

$|\chi_i\rangle$: the ith single-particle basis state ($i = 1, 2, \ldots, N$). For example, $|\chi_{\mathbf{G}}\rangle$ can be a plane wave basis state with $\chi_{\mathbf{G}}(\mathbf{r}) \propto e^{i\mathbf{G}\cdot\mathbf{r}}$, where $\mathbf{r}$ is a real-space coordinate.

$c_i^\dagger$ and c_i: creation and annihilation operators for an electron in $|\chi_i\rangle$. $n_i \equiv c_i^\dagger c_i$ is the corresponding number operator.

M: number of electrons (if we omit spin index, e.g., if the system is fully polarized). In the more general case, M_σ is the number of electrons with spin σ ($\sigma =\uparrow$ or $\downarrow$). Of course, the choice of N above must ensure that $M_\sigma \leq N$.

φ_m: single-particle orbital (we include an index m for discussions below to distinguish different single-particle orbitals). A single-particle orbital φ_m, given in terms of the single-particle basis states $\{|\chi_i\rangle\}$ as $\sum_i \varphi_{i,m}|\chi_i\rangle = \sum_i c_i^\dagger \varphi_{i,m}|0\rangle$, can be conveniently expressed as an N-dimensional vector:

$$\begin{pmatrix} \varphi_{1,m} \\ \varphi_{2,m} \\ \vdots \\ \varphi_{N,m} \end{pmatrix}$$

Given M different single-particle orbitals, we form a many-body wave function from their anti-symmetrized product:

$$|\phi\rangle \equiv \widehat{\varphi}_1^\dagger \widehat{\varphi}_2^\dagger \cdots \widehat{\varphi}_M^\dagger |0\rangle \tag{2.13}$$

where the operator

$$\widehat{\varphi}_m^\dagger \equiv \sum_i c_i^\dagger \varphi_{i,m} \tag{2.14}$$

creates an electron in the mth single-particle orbital $\{\varphi_{1,m}, \varphi_{2,m}, \ldots, \varphi_{N,m}\}$. The many-body state $|\phi\rangle$ in (2.13) can be conveniently expressed as an

$N \times M$ matrix:

$$\Phi \equiv \begin{pmatrix} \varphi_{1,1} & \varphi_{1,2} & \cdots & \varphi_{1,M} \\ \varphi_{2,1} & \varphi_{2,2} & \cdots & \varphi_{2,M} \\ \vdots & \vdots & & \vdots \\ \varphi_{N,1} & \varphi_{N,2} & \cdots & \varphi_{N,M} \end{pmatrix}$$

Each column of this matrix represents a single-particle orbital that is completely specified by its N-dimensional vector.

If the real-space co-ordinates of the electrons are $R = \{\mathbf{r}_1, \mathbf{r}_2, \ldots, \mathbf{r}_M\}$, the many-body state in (2.13) gives

$$\langle R \mid \phi \rangle = \phi(R) = \det \begin{pmatrix} \varphi_1(\mathbf{r}_1) & \varphi_2(\mathbf{r}_1) & \cdots & \varphi_M(\mathbf{r}_1) \\ \varphi_1(\mathbf{r}_2) & \varphi_2(\mathbf{r}_2) & \cdots & \varphi_M(\mathbf{r}_2) \\ \vdots & \vdots & & \vdots \\ \varphi_1(\mathbf{r}_M) & \varphi_2(\mathbf{r}_M) & \cdots & \varphi_M(\mathbf{r}_M) \end{pmatrix},$$

where $\varphi_m(\mathbf{r}) = \sum_i \varphi_{i,m}\chi_i(\mathbf{r})$.

The many-body state $|\phi\rangle$ is known as a Slater determinant. One example of a Slater determinant is the Hartree–Fock (HF) solution $|\phi_{\mathrm{HF}}\rangle = \prod_\sigma |\phi^\sigma_{\mathrm{HF}}\rangle$, where each $|\phi^\sigma_{\mathrm{HF}}\rangle$ is defined by a matrix $\Phi^\sigma_{\mathrm{HF}}$ whose columns are the N_σ lowest HF eigenstates. We can now add the following to our list of notations above:

$|\phi\rangle$: a many-body wave function which can be written as a Slater determinant.

Φ: an $N \times M$ matrix which represents the Slater determinant $|\phi\rangle$. Φ_{ij} will denote the matrix element of the matrix Φ in the ith row and jth column. For example, $\Phi_{ij} = \varphi_{i,j}$ above in Φ. Below when a Slater determinant $|\phi\rangle$ is referred to, it will often be helpful to think in terms of the matrix representation Φ operationally.

$|\Psi\rangle$ (upper case): a many-body wave function which is not necessarily a single Slater determinant, e.g., $|\Psi^{(n)}\rangle$ in (2.2).

Several properties of the Slater determinant are worth mentioning. For any two real nonorthogonal Slater determinants, $|\phi\rangle$ and $|\phi'\rangle$, it can be shown that their overlap integral is

$$\langle \phi \mid \phi' \rangle = \det(\Phi^{\mathrm{T}}\Phi'). \tag{2.15}$$

The operation on any Slater determinant by any operator $\widehat{B}$ of the form

$$\widehat{B} = \exp\left(\sum_{ij} c_i^\dagger U_{ij} c_j\right) \tag{2.16}$$

simply leads to another Slater determinant [14], i.e.,

$$\widehat{B}|\phi\rangle = \hat{\phi}_1'^{\dagger}\hat{\phi}_2'^{\dagger}\cdots\hat{\phi}_M'^{\dagger}|0\rangle \equiv |\phi'\rangle \tag{2.17}$$

with $\hat{\phi}'^{\dagger}_m = \sum_j c_j^\dagger \Phi'_{jm}$ and $\Phi' \equiv e^U \Phi$, where U is a square matrix whose elements are given by U_{ij} and $B \equiv e^U$ is therefore an $N \times N$ square matrix as well. In other words, the operation of $\widehat{B}$ on $|\phi\rangle$ simply involves multiplying an $N \times N$ matrix to an $N \times M$ matrix.

The many-body trace of an operator which is an exponential of one-body operators (i.e., of the form $\widehat{B}$ in (2.16)) or a product of exponentials of one-body operators can be conveniently evaluated. The grand-canonical trace, which involves summing over a complete basis for a fixed number of particles M as well as over different numbers of particles (from 0 to N), has a particularly simple form [10, 12]:

$$\mathrm{Tr}(\widehat{B}) = \det[I + B], \tag{2.18}$$

where I is the $N \times N$ unit matrix and B once again is the corresponding matrix of the operator $\widehat{B}$. The canonical trace, which is for a fixed number of particles M, is given by

$$\mathrm{Tr}_M(\widehat{B}) = \frac{1}{M!} \frac{d^M}{d\epsilon^M} \det[I + \epsilon B]\big|_{\epsilon=0}. \tag{2.19}$$

Note that this is simply the sum of all rank-M diagonal minors of the matrix B. As we would expect, the sum over all possible values of M of (2.19), with the appropriate factor for the chemical potential, recovers (2.18).

For an operator $\widehat{O}$, we can define its expectation with respect to a pair of Slater determinants

$$\overline{\langle \widehat{O} \rangle} \equiv \frac{\langle \phi | \widehat{O} | \phi' \rangle}{\langle \phi \mid \phi' \rangle} \tag{2.20}$$

or with respect to a propagator B under the finite-temperature grand-canonical formalism of (2.18)

$$\overline{\langle \widehat{O} \rangle} \equiv \frac{\mathrm{Tr}(\widehat{O}\widehat{B})}{\mathrm{Tr}(\widehat{B})}. \tag{2.21}$$

The "bar" distinguishes these from the true interacting many-body expectations in (2.3) and (2.5). The latter are of course what we wish to compute with QMC.

The simplest example of (2.20) and (2.21) is the single-particle Green's function $G_{ij} \equiv \overline{\langle c_i c_j^\dagger \rangle}$. In the ground-state formalism,

$$G_{ij} \equiv \frac{\langle \phi | c_i c_j^\dagger | \phi' \rangle}{\langle \phi \mid \phi' \rangle} = \delta_{ij} - [\Phi'(\Phi^{\mathrm{T}} \Phi')^{-1} \Phi^{\mathrm{T}}]_{ij}. \tag{2.22}$$

In the finite-temperature grand-canonical formalism

$$G_{ij} \equiv \frac{\mathrm{Tr}(c_i c_j^\dagger \widehat{B})}{\mathrm{Tr}(\widehat{B})} = (I + B)^{-1}_{ij}. \tag{2.23}$$

Given the Green's function G, the general expectation defined in (2.20) and (2.21) can be computed for most operators of interest. This is an important property that will be used in ground-state and finite-temperature QMC calculations. For example, we can calculate the expectation of a general two-body operator, $\widehat{O} = \sum_{ijkl} O_{ijkl} c_i^\dagger c_j^\dagger c_k c_l$, under definitions (2.20) and (2.21):

$$\overline{\langle \widehat{O} \rangle} = \sum_{ijkl} O_{ijkl} (G'_{jk} G'_{il} - G'_{ik} G'_{jl}), \tag{2.24}$$

where the matrix G' is defined as $G' \equiv I - G$.

2.4 Hubbard–Stratonovich Transformation

In order to carry out (2.2) and (2.6) in the Slater-determinant space we have introduced above, we write the many-body propagator $e^{-\Delta\tau \widehat{H}}$ in single-particle form. For simplicity we will treat the system as spin-polarized and suppress the spin index in most of the discussions below. It is, however, straightforward to generalize the discussions to include spins. Assuming that the many-body Hamiltonian involves two-body interactions only, we can write it as

$$\widehat{H} = \widehat{K} + \widehat{V} = \sum_{i,j} K_{ij} (c_i^\dagger c_j + c_j^\dagger c_i) + \sum_{ijkl} V_{ijkl} c_i^\dagger c_j^\dagger c_k c_l. \tag{2.25}$$

For example, $\widehat{K}$ and $\widehat{V}$ can be the kinetic and potential energy operators, respectively. With a small $\Delta\tau > 0$, the Trotter approximation can be used:

$$e^{-\Delta\tau \widehat{H}} \approx e^{-\Delta\tau \widehat{K}} e^{-\Delta\tau \widehat{V}}, \tag{2.26}$$

which introduces a Trotter error. For actual implementation of the algorithms we discuss here, higher order Trotter break-ups are often used. The Trotter error can be further reduced with an extrapolation procedure after separate calculations have been done with different values of $\Delta\tau$.

The $\widehat{K}$ part of the propagator in (2.26) is the exponential of a one-body operator. The $\widehat{V}$ part is not. It is, however, possible to rewrite $e^{-\Delta\tau \widehat{V}}$ in this desired form through a so-called Hubbard–Stratonovich transformation [15]. As we will show below, $\widehat{V}$ can be written as a sum of terms of the form $\lambda \hat{v}^2/2$, where λ is a constant and $\hat{v}$ is a one-body operator similar to $\widehat{K}$. The Hubbard–Stratonovich transformation then allows us to write

$$e^{-\Delta\tau/2\, \lambda\, \hat{v}^2} = \int_{-\infty}^{\infty} dx \, \frac{e^{-x^2/2}}{\sqrt{2\pi}} e^{x\sqrt{-\Delta\tau\lambda}\, \hat{v}}, \tag{2.27}$$

where x is an auxiliary-field variable. The constant in the exponent on the right-hand side can be real or imaginary dependent on the sign of λ. The key is that the quadratic form (in $\hat{v}$) on the left is replaced by a linear one on the right.

We now show one way to write $\widehat{V}$ as a sum of $\lambda\hat{v}^2/2$. With the most general form of $\widehat{V}$ in (2.25) we can cast V_{ijkl} in the form of a Hermitian matrix by introducing two indices $\alpha = (i,l)$ and $\beta = (j,k)$ and letting $\mathcal{V}_{\alpha\beta} = \mathcal{V}_{(i,l),(j,k)} = V_{ijkl}$. The Hermitian matrix $\mathcal{V}$ can then be diagonalized and written as $\mathcal{V} = R\Lambda R^{\mathrm{T}}$, where R is a matrix whose columns are the eigenvectors of $\mathcal{V}$ and Λ is a diagonal matrix containing the corresponding eigenvalues λ_α. That is

$$\mathcal{V}_{\alpha\beta} = \sum_\gamma R_{\alpha\gamma}\lambda_\gamma R^*_{\beta\gamma}. \tag{2.28}$$

The two-body operator $\widehat{V}$ can therefore be written as

$$\begin{aligned}\widehat{V} &= \sum_{ijkl} V_{ijkl}\, c_i^\dagger c_l c_j^\dagger c_k - \sum_{ijkl} V_{ijkl}\, c_i^\dagger c_k\, \delta_{jl} \\ &= \sum_\gamma \lambda_\gamma \Big(\sum_{il} R_{(i,l)\gamma} c_i^\dagger c_l\Big)\Big(\sum_{jk} R^*_{(j,k)\gamma} c_j^\dagger c_k\Big) - \sum_{ik}\Big(\sum_j V_{ijkj}\Big) c_i^\dagger c_k.\end{aligned}$$

Noting that $\widehat{V}$ is Hermitian, we can put the above in a more symmetric form

$$\widehat{V} = \frac{1}{2}\sum_\gamma \lambda_\gamma \{\hat{\rho}_\gamma, \hat{\rho}_\gamma^\dagger\} + \hat{\rho}_0, \tag{2.29}$$

where the one-body operators are defined as $\hat{\rho}_\gamma \equiv \sum_{il} R_{(i,l)\gamma} c_i^\dagger c_l$ and $\hat{\rho}_0 \equiv -\sum_{ik}[\sum_j (V_{ijkj} + V_{jijk})/2] c_i^\dagger c_k$. Since

$$\{\hat{\rho}_\gamma, \hat{\rho}_\gamma^\dagger\} = \tfrac{1}{2}[(\hat{\rho}_\gamma + \hat{\rho}_\gamma^\dagger)^2 - (\hat{\rho}_\gamma - \hat{\rho}_\gamma^\dagger)^2], \tag{2.30}$$

we have succeeded in writing $\widehat{V}$ in the desired form.

The way to decompose $\widehat{V}$ above leads to approximately $2N^2$ auxiliary fields. Often the interaction simplifies $\widehat{V}$ and the number of auxiliary fields can be much reduced. In fact, certain type of interactions have particularly simple forms of Hubbard–Stratonovich transformations. For example, for the repulsive on-site interaction $Un_{i\uparrow}n_{i\downarrow}$ ($\uparrow$ and $\downarrow$ denote electron spin) in the Hubbard model , an *exact*, discrete Hubbard–Stratonovich transformation [16] exists:

$$e^{-\Delta\tau U n_{i\uparrow}n_{i\downarrow}} = e^{-\Delta\tau U(n_{i\uparrow}+n_{i\downarrow})/2} \sum_{x_i=\pm1} \frac{1}{2}\, e^{\gamma x_i(n_{i\uparrow}-n_{i\downarrow})}, \tag{2.31}$$

where the constant γ is determined by $\cosh(\gamma) = \exp(\Delta\tau U/2)$. Similarly, for an attractive interaction $Vn_{i\uparrow}n_{i\downarrow}$ with $V < 0$:

$$e^{-\Delta\tau V n_{i\uparrow}n_{i\downarrow}} = e^{-\Delta\tau V(n_{i\uparrow}+n_{i\downarrow}-1)/2} \sum_{x_i=\pm1} \frac{1}{2}\, e^{\gamma x_i(n_{i\uparrow}+n_{i\downarrow}-1)}, \tag{2.32}$$

where $\cosh(\gamma) = \exp(\Delta\tau|V|/2)$

If we denote the collection of auxiliary fields by $\vec{x}$ and combine one-body terms from $\widehat{K}$ and $\widehat{V}$, we obtain the following compact representation of the outcome of the HS transformation:

$$e^{-\Delta\tau\widehat{H}} = \int d\vec{x}\, p(\vec{x})\widehat{B}(\vec{x}), \tag{2.33}$$

where $p(\vec{x})$ is a probability density function (e.g., a multi-dimensional Gaussian). The propagator $\widehat{B}(\vec{x})$ in (2.33) has a *special form*, namely, it is a product of operators of the type in (2.16), with U_{ij} depending on the auxiliary field. The matrix representation of $\widehat{B}(\vec{x})$ will be denoted by $B(\vec{x})$.

In essence, the HS transformation replaces the two-body interaction by one-body interactions with a set of random external auxiliary fields. In other words, it converts an interacting system into many *noninteracting* systems living in fluctuating external auxiliary-fields. The sum over all configurations of auxiliary fields recovers the interaction.

Different forms of the HS transformation exist [17,18]. It is reasonable to expect that they can affect the performance of the QMC method. Indeed experience shows that they can not only impact the statistical accuracy, but also lead to different quality of approximations under the constrained path methods that we discuss below. Therefore, although we discuss the algorithms under the generic form of (2.33), we emphasize that it is worthwhile to explore different forms of the HS transformation in an actual implementation.

3 Standard Auxiliary-Field Quantum Monte Carlo

In this section we briefly describe the standard ground-state [11,13] and finite-temperature [10,12] auxiliary-field quantum Monte Carlo (AFQMC) methods. We will rely on the machinery established in the previous section. Our goal is to illustrate the essential ideas, in a way which will facilitate our discussion of the sign problem and help introduce the framework for the constrained path Monte Carlo methods. We will not go into details such as how to sample the auxiliary fields efficiently or how to stabilize the matrices. This is described in the literature. In addition, although the approaches are not identical, we will see these issues manifested in the next section and gain sufficient understanding of them when we discuss the corresponding constrained path Monte Carlo methods.

3.1 Ground-State Method

The ground-state expectation $\langle \widehat{O} \rangle$ can now by computed with (2.2) and (2.33). The denominator is

$$
\begin{aligned}
\langle \psi^{(0)} | e^{-n\Delta\tau \widehat{H}} \, e^{-n\Delta\tau \widehat{H}} | \psi^{(0)} \rangle \\
&= \int \langle \psi^{(0)} | \left[\prod_{l=1}^{2n} d\vec{x}^{(l)} p(\vec{x}^{(l)}) \widehat{B}(\vec{x}^{(l)}) \right] | \psi^{(0)} \rangle \\
&= \int \left[\prod_{l} d\vec{x}^{(l)} p(\vec{x}^{(l)}) \right] \det \left([\Psi^{(0)}]^{\mathrm{T}} \prod_{l} B(\vec{x}^{(l)}) \Psi^{(0)} \right).
\end{aligned}
\tag{3.1}
$$

In the standard ground-state AFQMC method [10], a (sufficiently large) value of n is first chosen and fixed throughout the calculation. If we use X to denote the collection of the auxiliary-fields $X = \{\vec{x}^{(1)}, \vec{x}^{(2)}, \ldots, \vec{x}^{(2n)}\}$ and $D(X)$ to represent the integrand in (3.1), we can write the expectation value of (2.3) as

$$
\langle \widehat{O} \rangle = \frac{\int \overline{\langle \widehat{O} \rangle} D(X) \, dX}{\int D(X) \, dX} = \frac{\int \overline{\langle \widehat{O} \rangle} |D(X)| s(X) \, dX}{\int |D(X)| s(X) \, dX}, \tag{3.2}
$$

where

$$
s(X) \equiv D(X)/|D(X)| \tag{3.3}
$$

measures the "sign" of $D(X)$. The noninteracting expectation for a given X is that defined in (2.20):

$$
\overline{\langle \widehat{O} \rangle} \equiv \frac{\langle \phi_L | \widehat{O} | \phi_R \rangle}{\langle \phi_L \mid \phi_R \rangle} \tag{3.4}
$$

with

$$
\begin{aligned}
\langle \phi_L | &= \langle \psi^{(0)} | \widehat{B}(\vec{x}^{(2n)}) \widehat{B}(\vec{x}^{(2n-1)}) \cdots \widehat{B}(\vec{x}^{(n+1)}) \\
| \phi_R \rangle &= \widehat{B}(\vec{x}^{(n)}) \widehat{B}(\vec{x}^{(n-1)}) \cdots \widehat{B}(\vec{x}^{(1)}) | \psi^{(0)} \rangle,
\end{aligned}
$$

which are both Slater determinants.

$D(X)$ as well as $\langle \phi_L |$ and $| \phi_R \rangle$ are completely determined by the path X in auxiliary-field space. The expectation in (3.2) is therefore in the form of (2.10), with $f(X) = |D(X)|$ and $g(X) = \overline{\langle \widehat{O} \rangle}$. The important point is that, for each X, $|D(X)|$ is a number and $g(X)$ can be evaluated using (2.22) or (2.24). Often the Metropolis Monte Carlo algorithm [9] is used to sample auxiliary-fields X from $|D(X)|$. Any $\langle \widehat{O} \rangle$ can then be computed following the procedure described by (2.9).

The Metropolis Monte Carlo algorithm allows one to, starting from an (arbitrary) initial configuration of X, create a random walk whose equilibrium distribution is $f(X)$. A kernel $K(X', X)$ must be chosen prior to the calculation which is a probability density function of X' (conditional on

X). The only other condition on K is ergodicity; i.e., it allows the random walk eventually to reach any position from any other position in X-space. For example, it can be a uniform distribution in a hyper-cube of linear dimension Δ centered at X

$$K(X', X) = \begin{cases} 1/\Delta^{2nd}, & \text{if } X' \text{ inside hyper-cube} \\ 0, & \text{otherwise,} \end{cases}$$

where Δ is a parameter and d is the dimensionality of the auxiliary-field $\vec{x}$.

The random walk resembles that in (2.12). A move from X to X' is proposed by sampling an X' according to $K(X', X)$. The key of the Metropolis algorithm is an acceptance probability

$$A(X', X) = \min\left[\frac{K(X, X')f(X')}{K(X', X)f(X)}, 1\right] \tag{3.5}$$

whose role is similar to that of w in (2.12). The one important difference is that, if the outcome of $A(X', X)$ is to reject X', X is kept and counted as one new Monte Carlo sample—and hence one new step in the random walk, even though the walker did not actually move.

The choice of K, which has a great deal of freedom, controls the efficiency of the random walk, i.e., how quickly it converges, how correlated successive steps of the random walk are, etc. A choice that is perhaps the most common in practice is to "sweep" through the field one component at a time. A change is proposed for only one component of the many-dimensional vector X, from a kernel $K(\vec{x}_i', \vec{x}_i)$ in similar form to K , while the rest of X is kept the same; the acceptance/rejection procedure is then applied. With the newly generated field, a change is proposed on the next ($(l+1)$th) component and the step above is repeated. The "local" nature of this choice of the kernel can often lead to simplifications in the computation of A.

3.2 Finite-Temperature Method

The standard finite-temperature auxiliary-field QMC method works in the grand-canonical ensemble. This means that the Hamiltonian contains an additional term $-\mu \sum_i n_i$ which involves the chemical potential μ. The term leads to an additional diagonal one-body operator in the exponent in (2.16). We will assume that the resulting term has been absorbed in $\widehat{B}$ in (2.33).

Substituting (2.33) into (2.6) leads to

$$Z = \int \left[\prod_l d\vec{x}_l p(\vec{x}_l)\right] \mathrm{Tr}[\widehat{B}(\vec{x}_L) \cdots \widehat{B}(\vec{x}_2)\widehat{B}(\vec{x}_1)]. \tag{3.6}$$

If we apply (2.18) to the above and again use X to denote a complete path in auxiliary-field space, $X \equiv \{\vec{x}_1, \vec{x}_2, \ldots, \vec{x}_L\}$, we obtain

$$Z = \int \det[I + B(\vec{x}_L) \cdots B(\vec{x}_2)B(\vec{x}_1)]\, p(X)\, dX. \tag{3.7}$$

Again representing the integrand by $D(X)$, we see that the finite-temperature expectation in (2.5) reduces to (3.2), exactly the same expression as in ground-state AFQMC. The noninteracting expectation $\overline{\langle \widehat{O} \rangle}$ is now

$$\overline{\langle \widehat{O} \rangle} \equiv \frac{\mathrm{Tr}[\widehat{O}\widehat{B}(\vec{x}_L) \cdots \widehat{B}(\vec{x}_2)\widehat{B}(\vec{x}_1)]}{\mathrm{Tr}[\widehat{B}(\vec{x}_L) \cdots \widehat{B}(\vec{x}_2)\widehat{B}(\vec{x}_1)]}. \tag{3.8}$$

Note that, as in the ground-state algorithm, this can be computed for each X. Equation (2.23) remains valid. The matrix B must be replaced by the ordered product of the corresponding matrices, $B(\vec{x}_L) \cdots B(\vec{x}_2)B(\vec{x}_1)$. "Wrapping" this product (without changing the ordering) gives the equal time Green's function at different imaginary times.

Thus we have unified the ground-state and finite-temperature AFQMC methods with (3.2). Although the actual forms for $D(X)$ and $\overline{\langle \widehat{O} \rangle}$ are different, the finite-temperature algorithm is the same as that of the ground state described earlier. Applying the same procedure, we can sample auxiliary-fields X from $|D(X)|$, and then compute the desired many-body expectation values from (3.2).

4 Constrained Path Monte Carlo Methods—Ground-State and Finite-Temperature

In this section, we review the ground-state and finite-temperature constrained path Monte Carlo (CPMC) methods. These methods are free of any decay of the average sign, while retaining many of the advantages of the standard AFQMC methods. The methods are approximate, relying on what we will generally refer to as the constrained path approximation. Below, other than presenting a way to make a more formal connection between the ground-state and finite-temperature methods, we will largely follow [5, 7] in discussing these methods.

4.1 Why and How Does the Sign Problem Occur?

In the standard AFQMC methods, the sign problem occurs because $D(X)$ is not always positive.[2] The Monte Carlo samples of X therefore have to

[2]In fact $\widehat{B}(\vec{x})$ is complex for a general HS transformation when the interaction is repulsive. $D(X)$ is therefore complex as well. We will limit our discussions here of the

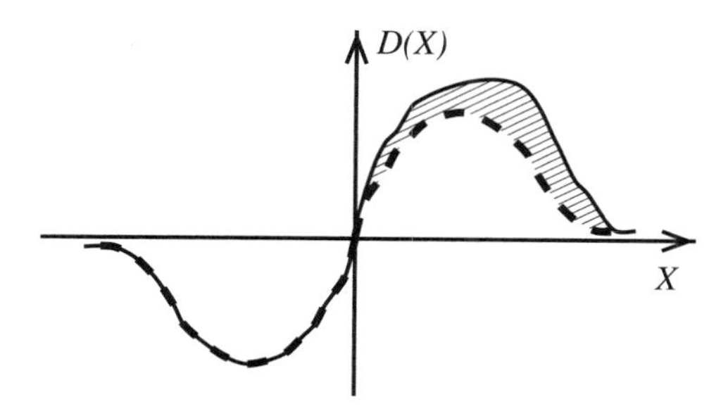

FIGURE 2.1. Schematic illustration of the sign problem. The X-axis represents an abstraction of the many-dimensional auxiliary-field paths X; each point denotes a collection of X's. The sign problem occurs because the contributing part (shaded area) is exponentially small compared to what is sampled, namely, $|D(X)|$.

be drawn from the probability distribution function defined by $|D(X)|$. As Fig. 2.1 illustrates, however, $D(X)$ approaches an anti-symmetric function exponentially as N or the the length of the path is increased. That is, its average sign, i.e., the denominator in (3.2), vanishes at large system size or low temperature. Thus the variance of the calculation in (2.11) grows exponentially, which negates the advantage of the "square-root" behavior of a Monte Carlo calculation (cf. (2.9)). Fig. 2.2 shows exponential decay of the average sign $\langle s(X)\rangle$ of the Monte Carlo samples as a function of inverse temperature. This problem was largely uncontrolled in both the ground-state and finite-temperature algorithms, preventing general simulations at low temperatures or large system sizes.

Here we will develop a more detailed picture for the behavior of $D(X)$ and hence for how the sign problem occurs. Under this picture the constrained path approximation, which will eliminate the decay of the denominator in (3.2) as a function of n or lattice size, emerges naturally.

For simplicity, we will now use one common symbol L to denote the length of the path X. That is, in the ground-state AFQMC algorithm in the previous section, we let $L = 2n$. $D(X) = D(\vec{x}_1, \vec{x}_2, \dots, \vec{x}_L)$ will denote the corresponding integrand in either the ground-state or finite-temperature AFQMC algorithm. We recall that the goal is to sample X according to $D(X)$ (although we had to resort to sampling $|D(X)|$ in the previous section).

To gain insight we conduct the following thought experiment [7]. We imagine sampling the complete path X by L successive steps, from $\vec{x}_1$ to $\vec{x}_L$. We consider the contribution in (3.2) by an individual *partial path*

sign problem and the constrained path approximation to the case where $D(X)$ is real, although generalizations are possible.

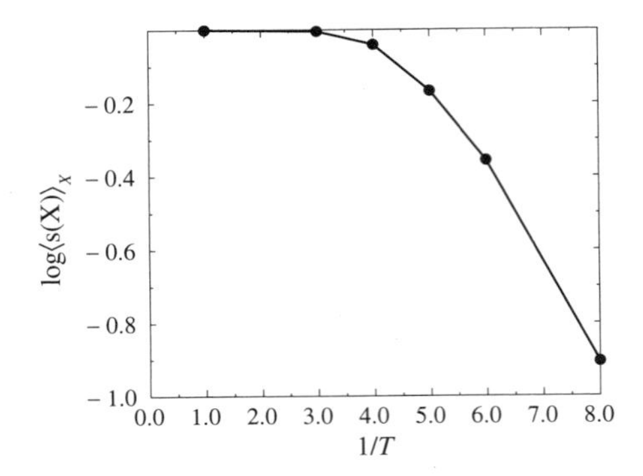

FIGURE 2.2. Decay of the average sign as a function of inverse temperature. Note the logarithmic scale on the vertical axis. The calculation is for a 4×4 Hubbard model using the finite-temperature formalism. The chemical potential is such that the average electron density is 0.875 (corresponding to $7 \uparrow 7 \downarrow$ electrons). The on-site repulsion $U = 4$. (Data courtesy of Richard Scalettar [19].)

$\{\vec{x}_1, \vec{x}_2, \ldots, \vec{x}_l\}$:

$$\mathcal{P}_l(\{\vec{x}_1, \vec{x}_2, \ldots, \vec{x}_l\}) \equiv \int D(\vec{x}_1, \vec{x}_2, \ldots, \vec{x}_l, \vec{x}_{l+1}, \ldots, \vec{x}_L)\, d\vec{x}_{l+1} \cdots d\vec{x}_L. \quad (4.1)$$

Note that this is the same as replacing $e^{\Delta\tau\widehat{H}}$ by (2.33) only in the first l time slices. For simplicity we will use $\widehat{\mathcal{B}}$ to denote the many-body propagator $e^{\Delta\tau\widehat{H}}$ in our discussion below, i.e, $\widehat{\mathcal{B}} \equiv e^{\Delta\tau\widehat{H}}$. Under the ground-state formalism,

$$\mathcal{P}_l(\{\vec{x}_1, \vec{x}_2, \ldots, \vec{x}_l\}) = \langle\psi^{(0)}[\underbrace{\widehat{\mathcal{B}}\widehat{\mathcal{B}}\cdots\widehat{\mathcal{B}}}_{L-l}\ \widehat{B}(\vec{x}_l)\cdots\widehat{B}(\vec{x}_2)\widehat{B}(\vec{x}_1)]|\psi^{(0)}\rangle p(\vec{x}_l)\cdots p(\vec{x}_1), \quad (4.2)$$

while under the finite-temperature formalism

$$\mathcal{P}_l(\{\vec{x}_1, \vec{x}_2, \cdots, \vec{x}_l\}) = \mathrm{Tr}[\underbrace{\widehat{\mathcal{B}}\widehat{\mathcal{B}}\cdots\widehat{\mathcal{B}}}_{L-l}\,\widehat{B}(\vec{x}_l)\cdots\widehat{B}(\vec{x}_2)\widehat{B}(\vec{x}_1)]p(\vec{x}_l)\cdots p(\vec{x}_1). \quad (4.3)$$

We consider the case when $\mathcal{P}_l = 0$. This means that, after the remaining $L - l$ steps are finished, the collective contribution from *all* complete paths that result from this particular partial path will be precisely zero. In other words, all complete paths that have $\{\vec{x}_1, \vec{x}_2, \ldots, \vec{x}_l\}$ as their first l elements collectively make no contribution in the denominator in (3.2). This is because the path integral over all possible configurations for the rest

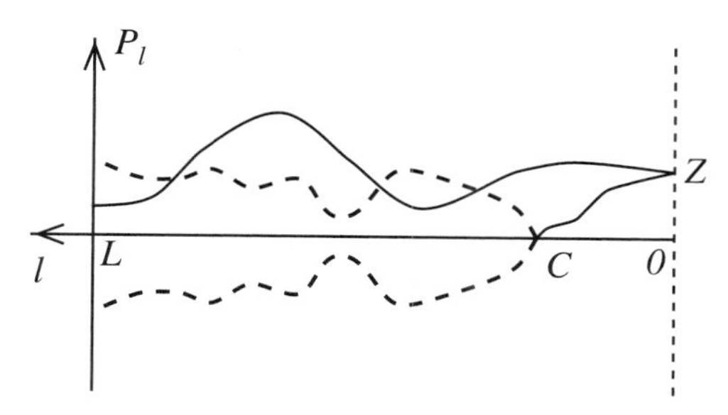

FIGURE 2.3. Schematic illustration of the boundary condition to control the sign problem in AFQMC. $\mathcal{P}_l$ (Eq. 4.1) is shown as a function of the length of the partial path, l, for several paths. The remainder of the path integral (of length $L-l$) is assumed to have been evaluated *analytically*. (See (4.1).) All paths start from the right ($l = 0$) at $\mathcal{P}_0$, which is either $\langle\psi_T|e^{L\Delta\tau\widehat{H}}|\psi_T\rangle > 0$ (ground-state) or Z (finite-temperature). When $\mathcal{P}_l$ becomes 0, ensuing paths (dashed lines) collectively contribute zero. Only complete paths with $\mathcal{P}_l > 0$ for all l (solid line) contribute in the denominator in (3.2).

of the path, $\{\vec{x}_{l+1}, \vec{x}_{l+2}, \ldots, \vec{x}_L\}$, is simply given by (4.1). In other words, the path integral simply reproduces the $\widehat{\mathcal{B}}$'s in (4.2) or (4.3), leading to $\mathcal{P}_l(\{\vec{x}_1, \vec{x}_2, \ldots, \vec{x}_l\})$ which is zero by assumption.

Thus, in our thought experiment any partial path that reaches the axis in Fig. 2.3 immediately turns into noise, regardless of what it does at future l's. A complete path which is in contact with the axis at any point belongs to the "anti-symmetric" part of $D(X)$ in Fig. 2.1, whose contributions cancel.

As L increases, it becomes more and more likely for a path to reach the axis at some point l. Therefore, the "noise" paths become an increasingly larger portion of all paths unless paths are completely prohibited from reaching the axis, such as in the half-filled Hubbard model with repulsive interaction, where particle-hole symmetry makes $D(X)$ positive. A complete path X which crosses the axis at least once, i.e., a noise path, can have a $D(X)$ which is either positive or negative. In the Monte Carlo sampling in AFQMC, we sample X according to $|D(X)|$, which makes no distinction of such paths from the contributing ones that stay completely above the axis. Asymptotically in L, the Monte Carlo samples consist of an equal mixture of positive and negative contributions in $D(X)$. The Monte Carlo signal is lost and the Monte Carlo estimate of (3.2) becomes increasingly noisy. This is the origin of the sign problem.

4.2 The Constrained-Path Approximation

$\mathcal{P}_0$ is positive, because $\langle\psi_T|e^{L\Delta\tau\widehat{H}}|\psi_T\rangle > 0$ (ground-state formalism) and $Z > 0$ (finite-temperature). $\mathcal{P}_l$ changes continuously with l at the limit $\Delta\tau \to 0$. Therefore a complete path contributes if and only if it stays entirely above the axis in Fig. 2.3. Thus, in our thought experiment, imposition of the boundary condition [5,7,20]

$$\begin{aligned} &\mathcal{P}_1(\{\vec{x}_1\}) > 0 \\ &\mathcal{P}_2(\{\vec{x}_1, \vec{x}_2\}) > 0 \\ &\vdots \\ &\mathcal{P}_L(\{\vec{x}_1, \vec{x}_2, \ldots, \vec{x}_L\}) > 0 \end{aligned} \tag{4.4}$$

will ensure all contributing complete paths to be selected while eliminating all noise paths. The axis acts as an infinitely absorbing boundary. A partial path is terminated and discarded as soon as it reaches the boundary. By discarding a path, we eliminate all of its future paths, including the ones that would eventually make positive contributions. The boundary condition makes the distribution of complete paths vanish at the axis, which accomplishes complete cancellation of the negative and the corresponding positive contributions in the anti-symmetric part of $D(X)$.

In practice $\widehat{\mathcal{B}}$ is of course not known. We replace it by a known trial propagator $\widehat{B}_T$. The boundary condition in (4.4) is then clearly approximate. Results from our calculation will depend on the trial propagator used in the constraint. We expect the quality of the approximation to improve with a better $\widehat{B}_T$. This is the basic idea of the constrained path approximation.

For ground-state calculations, let us imagine taking L to infinity. If $\widehat{B}_T$ is a mean-field propagator, the wave function

$$\langle \psi_T | = \lim_{L \to \infty} \langle \psi^{(0)} | \widehat{B}_T^{L-l} \tag{4.5}$$

can be thought of as the corresponding ground-state solution. The right-hand side of (4.2) can then be evaluated:

$$\mathcal{P}_l^T \equiv \langle \psi_T | \phi^{(l)} \rangle p(\vec{x}_l) \cdots p(\vec{x}_1), \tag{4.6}$$

where the Slater determinant that results from a particular partial path is defined as

$$|\phi^{(l)}\rangle \equiv \widehat{B}(\vec{x}_l) \cdots \widehat{B}(\vec{x}_2) \widehat{B}(\vec{x}_1) |\psi^{(0)}\rangle. \tag{4.7}$$

The approximate condition in (4.4) can then be written as

$$\langle \psi_T \mid \phi^{(l)} \rangle > 0. \tag{4.8}$$

Notice that the constraint now becomes independent of the length of the path l. It is expressed in terms of the overlap of a determinant with a trial wave function $|\psi_T\rangle$. For a choice of $|\psi_T\rangle$ that is a single Slater determinant or a linear combination of Slater determinants, (2.15) allows straightforward computation of the overlap in (4.8).

In principle, we could keep L finite and directly implement the constraint in the framework of the standard AFQMC, using the propagator $\widehat{B}_T$ instead of $|\psi_T\rangle$. But, as we will see below, the formalism above lends itself more naturally to the random walk realization of the ground-state method.

The constraint is manifested in one other way in the computation of expectations of operators $\widehat{O}$ that do not commute with $\widehat{H}$. Because we need to insert $\widehat{O}$ in the middle in the numerator of (3.4), we will have to "back-propagate" [3, 5] from the left to obtain a $\langle\phi_L|$. The constraint, however, is implemented in the opposite direction, as we sample the path from right to left. A set of determinants along the path of a random walk which does not violate the constraint at any step when going from right to left may violate it any even number of times when going from left to right. This sense of direction means that the expectation $\langle\widehat{O}\rangle$ we compute is not strictly $\langle\psi_0^c|\widehat{O}|\psi_0^c\rangle/\langle\psi_0^c \mid \psi_0^c\rangle$, with $|\psi_0^c\rangle$ the approximate ground-state wave function under the constraint. This difference is not crucial, however, since it is expected to be of the same order as the error due to the constrained path approximation itself.

For finite-temperature calculations, the right-hand side of (4.3) can be evaluated if $\widehat{B}_T$ is in the form of a single-particle propagator

$$\mathcal{P}_l^T \equiv \det[I + \underbrace{B_T B_T \cdots B_T}_{L-1} B(\vec{x}_1)] p(\vec{x}_l) \cdots p(\vec{x}_1), \tag{4.9}$$

where, following our convention, B_T is the matrix representation of $\widehat{B}_T$. Combining this with (4.4) leads to the following matrix representation of the approximate boundary condition:

$$\begin{aligned}
&\det[I + \underbrace{B_T B_T \cdots B_T}_{L-1} B(\vec{x}_1)] > 0 \\
&\det[I + \underbrace{B_T B_T \cdots B_T}_{L-2} B(\vec{x}_2) B(\vec{x}_1)] > 0 \\
&\vdots \\
&\det[I + B(\vec{x}_L) \cdots B(\vec{x}_2) B(\vec{x}_1)] > 0
\end{aligned} \tag{4.10}$$

Formally, the constrained path approximation in (4.8) and (4.10) has similarity to the fixed-node [21–23] (ground-state) or restricted-path [24] (finite-temperature) approximation in configuration space.[3] Significant differences exist, however, because of the difference between the real space and a Slater determinant space, which is nonorthogonal.

The goal of the new CPMC methods is to carry out the thought experiment stochastically. We wish to generate Monte Carlo samples of X which *both* satisfy the conditions in (4.8) or (4.10) *and* are distributed according to $D(X)$. The most natural way to accomplish this is perhaps to incorporate the boundary conditions into the standard AFQMC algorithm,

[3] See the Appendix for a brief description of the fixed-node approximation for lattice Fermion systems.

for example, as an additional acceptance condition. However, such an approach is likely to be inefficient: The boundary condition is *nonlocal* and breaks translational invariance in imaginary time, which requires simultaneous updating of the entire path. Without a scheme to propose paths that incorporates information on future contributions, it is difficult to find complete paths which satisfy all the constraints, especially as L increases.

We therefore seek to sample paths via a random walk whose time variable corresponds to the imaginary time l. We will see that we must introduce importance sampling schemes to guide the random walks by their projected contribution in $D(X)$. Below we discuss details of such schemes.

4.3 Ground-State Constrained Path Monte Carlo (CPMC) Method

We start from (2.2). Using (2.33), we write it as

$$|\psi^{(l+1)}\rangle = \int d\vec{x}\, p(\vec{x})\widehat{B}(\vec{x})|\psi^{(l)}\rangle. \tag{4.11}$$

In the random walk realization of this iteration, we represent the wave function at each stage by a finite ensemble of Slater determinants, i.e.,

$$|\psi^{(l)}\rangle \propto \sum_k w_k^{(l)}|\phi_k^{(l)}\rangle, \tag{4.12}$$

where k labels the Slater determinants and an overall normalization factor of the wave function has been omitted. A weight factor $w_k^{(l)}$ is introduced for each walker, even though in (4.11) the kernel p is normalized. This is because single-particle orbitals in a Slater determinants cease to be orthogonal to each other as a result of propagation by $\widehat{B}$. When they are re-orthogonalized (see Section 4.5), an overall factor appears, which we will view as the w term in the integral equation (2.12).

The structure of the random walk now resembles that of (2.12). For each random walker we sample an auxiliary-field configuration $\vec{x}$ from the probability density function $p(\vec{x})$ and propagate the walker to a new one via $\Phi_k^{(l+1)} = B(\vec{x})\Phi_k^{(l)}$. If necessary, a re-orthogonalization procedure (e.g., modified Gram–Schmidt) is applied to $\Phi_k^{(l)}$ prior to the propagation: $\Phi_k^{(l)} = [\Phi_k^{(l)}]'R$, where R is an $M \times M$ upper-triangular matrix. $[\Phi_k^{(l)}]'$ is then used in the propagation instead; the weight of the new walker is $w_k^{(l+1)} = w_k^{(l)}\det(R)$.

The simple random walk procedure is correct and can be very useful for thinking about many conceptual issues. It is, however, not efficient enough as a practical algorithm in most cases of interest, because the sampling of $\vec{x}$ is completely random with no regard to the potential contribution to $D(X)$. The idea of importance sampling is to iterate a modified equation with a modified wave function, without changing the underlying eigenvalue problem of (4.11). Specifically, for each Slater determinant $|\phi\rangle$, we define an importance function

$$O_T(\phi) \equiv \langle\psi_T \mid \phi\rangle, \tag{4.13}$$

which estimates its overlap with the ground-state wave function. We can then rewrite (4.11) as

$$|\tilde{\psi}^{(l+1)}\rangle = \int d\vec{x}_{l+1}\, \tilde{p}(\vec{x}_{l+1})\widehat{B}(\vec{x}_{l+1})|\tilde{\psi}^{(l)}\rangle, \tag{4.14}$$

where the modified "probability density function" is

$$\tilde{p}(\vec{x}_{l+1}) = \frac{O_T(\phi^{(l+1)})}{O_T(\phi^{(l)})}p(\vec{x}_{l+1}). \tag{4.15}$$

It is easy to see that the new kernel $\tilde{p}$ in (4.15) can be written in terms of the trial partial path contributions in (4.6):

$$\tilde{p}(\vec{x}_{l+1}) = \frac{\mathcal{P}_{l+1}^T}{\mathcal{P}_l^T}. \tag{4.16}$$

As we will see, this is formally analogous to the importance-sampling kernel in the finite-temperature algorithm we discuss next.

With the new kernel $\tilde{p}$, the probability distribution for $\vec{x}_{l+1}$ vanishes smoothly as $\mathcal{P}_{l+1}^T$ approaches zero, and the constraint is naturally imposed. The auxiliary field $\vec{x}_{l+1}$ is sampled according to our best estimate of the partial path contributions (from the trial propagator/wave function). As expected, $\tilde{p}(\vec{x})$ is a function of both the current and future positions in Slater-determinant space. Further, it modifies $p(\vec{x})$ such that the probability is increased when $\vec{x}$ leads to a determinant with larger overlap and is decreased otherwise. It is trivially verified that equations (4.11) and (4.14) are identical.

To see the effect of importance sampling better, we observe that if $|\psi_T\rangle = |\psi_0\rangle$, the normalization $\int \tilde{p}(\vec{x})\, d\vec{x}$ becomes a constant. Therefore the weights of walkers remain a constant and the random walk has no fluctuation.

In the random walk, the ensemble of walkers $\{|\phi_k^{(l)}\rangle\}$ now represents the modified wave function: $|\tilde{\psi}^{(l)}\rangle \propto \sum_k w_k^{(l)}|\phi_k^{(l)}\rangle$. The true wave function is then given formally by

$$|\psi^{(l)}\rangle \propto \sum_k w_k^{(l)}|\phi_k^{(l)}\rangle/O_T(\phi_k^{(l)}), \tag{4.17}$$

although in actual measurements it is $|\tilde{\psi}^{(l)}\rangle$ that is needed and division by O_T does not appear. The random walk process is similar to that discussed for (4.11), but with $p(\vec{x})$ replaced by $\tilde{p}(\vec{x})$. The latter is in general not a normalized probability density function, and we denote the normalization constant for walker k by $N(\phi_k^{(l)})$ and rewrite the iterative relation as

$$|\phi_k^{(l+1)}\rangle \leftarrow N(\phi_k^{(l)}) \int d\vec{x}\, \frac{\tilde{p}(\vec{x})}{N(\phi_k^{(l)})} B(\vec{x})|\phi_k^{(l)}\rangle. \tag{4.18}$$

This iteration now forms the basis of the ground-state constrained path Monte Carlo algorithm. For each walker $|\phi_k^{(l)}\rangle$, one step of the random walk consists of

1. sampling an $\vec{x}$ from the probability density function $\tilde{p}(\vec{x})/N(\phi_k^{(l)})$,
2. constructing the corresponding $B(\vec{x})$ and propagating the walker $\Phi_k^{(l)}$ by it to generate a new walker, and
3. assigning a weight $w_k^{(l+1)} = w_k^{(l)} N(\phi_k^{(l)})$ to the new walker.

Note that, in contrast with the primitive algorithm in (4.11), the weight factor of a walker does not need to be modified here when the re-orthogonalization procedure is applied. This is because the upper-triangular matrix R only contributes to the overlap O_T, which is already represented by the walker weight. After each modified Gram–Schmidt procedure, R can simply be discarded.

4.4 Finite-Temperature Method

We again construct an algorithm which builds directly into the sampling process both the constraints and some knowledge of the projected contribution. In terms of the trial projected partial contributions $\mathcal{P}_l^T$ defined in (4.9), the fermion determinant $D(X)$ can be written as

$$D(X) = \frac{\mathcal{P}_L^T}{\mathcal{P}_{L-1}^T} \frac{\mathcal{P}_{L-1}^T}{\mathcal{P}_{L-2}^T} \cdots \frac{\mathcal{P}_2^T}{\mathcal{P}_1^T} \frac{\mathcal{P}_1^T}{\mathcal{P}_0^T} \mathcal{P}_0^T. \tag{4.19}$$

We again construct the path X by a random walk of L steps, corresponding to stochastic representations of the L ratios in (4.19). In Fig. 2.4 the sampling procedure for one walker is illustrated schematically. At the $(l+1)$th step, we sample $\vec{x}_{l+1}$ from the conditional probability density function defined by $\tilde{p}(\vec{x}_{l+1}) = \mathcal{P}_{l+1}^T/\mathcal{P}_l^T$, which is precisely the same as (4.16) in the ground-state method. As in the ground-state method or in Green's function Monte Carlo, we simultaneously keep an ensemble of walkers, which undergo branching in their random walks (of L steps).

In the random walk, each walker is a product of B_T's and $B(\vec{x})$'s. They are initialized to $\mathcal{P}_0^T$, with overall weight 1. At the end of the lth step, each walker has the form

$$\underbrace{B_T B_T \cdots B_T}_{L-l} B(\vec{x}_l) \cdots B(\vec{x}_2) B(\vec{x}_1) \tag{4.20}$$

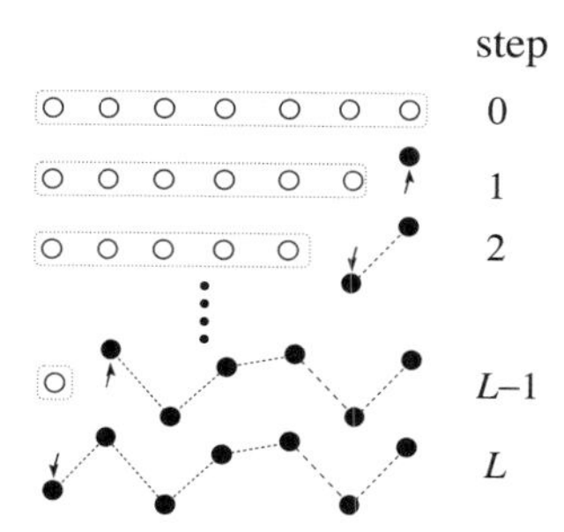

FIGURE 2.4. Illustration of the sampling procedure in the algorithm. Circles represent auxiliary-fields $\vec{x}_l$. A row shows the field configuration at the corresponding step number shown on the right. Within each row, the imaginary-time index l increases as we move to the left, i.e., the first circle is $\vec{x}_1$ and the last $\vec{x}_L$. Empty circles indicate fields which are not "activated" yet, i.e., $\widehat{B}_T$ is still in place of $\widehat{\mathcal{B}}$. Solid circles indicate fields that have been sampled, with the arrow indicating the one being sampled in the current step.

where $\{\vec{x}_l, \ldots, \vec{x}_2, \vec{x}_1\}$ is the partial path already sampled for this walker. Each step for a walker is formally identical to that in the ground-state method:

1. pick an $\vec{x}_{l+1}$ from the probability density function $\tilde{p}(\vec{x}_{l+1})$,
2. advance the walker by replacing the B_T next to $B(\vec{x}_l)$ by $B(\vec{x}_{l+1})$, and
3. multiply the overall weight of the walker by the normalization factor of $\tilde{p}(\vec{x}_{l+1})$.

The basic idea of the algorithm is, similar to the ground-state method, to select $\vec{x}_l$ according to the best estimate of its projected contribution in Z. Note that the probability distribution for $\vec{x}_l$ again vanishes smoothly as $\mathcal{P}_l^T$ approaches zero. This naturally imposes the boundary condition in (4.10). The Monte Carlo samples are generated from a probability distribution function of the contributing paths only (solid lines in Fig. 2.3), which is more efficient than sampling paths from $|D(X)|$ and imposing the boundary condition as an additional acceptance/rejection step.

4.5 Additional Technical Issues

4.5.1 Implementation of the Constraint at Finite $\Delta\tau$

In actual simulations $\Delta\tau$ is finite and paths are defined only at a discrete set of imaginary times. The boundary condition on the underlying continuous paths is the same, namely that the probability distribution must vanish at the axis in Fig. 2.3.

In Fig. 2.5, we illustrate how the boundary condition is imposed under the discrete representation. The "contact" point with the axis is likely to be

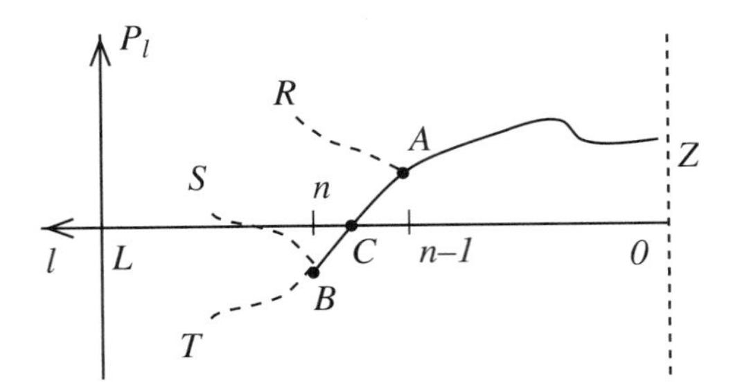

FIGURE 2.5. Imposition of the boundary condition at finite $\Delta\tau$. Paths are discrete. The point of contact, C (see Fig. 2.3), must be approximated, either by B (low-order algorithm) or by interpolation between B and A (higher order).

between time slices and not well defined, i.e., $\mathcal{P}_l$ may be zero at a noninteger value of l. To the lowest order, we can terminate a partial path when its $\mathcal{P}_l$ first turns negative. That is, we still impose (4.4) in our thought experiment to generate paths. In Fig. 2.5, this means terminating the path at $l = n$ (point B) and thereby discarding all its future paths (represented by the dashed lines "BS..." and 'BT...').

We actually use a higher order approximation, by terminating at either $l = n - 1$ or $l = n$, i.e., either point B or point A. The probability for terminating at A is chosen such that it approaches 1 smoothly as $\mathcal{P}_{n-1} \to 0$, for example, $p_{\text{A}} = 1/[1 + \mathcal{P}_{n-1}/|\mathcal{P}_n|]$. If A is chosen, all future paths from A are discarded (represented by "AR..." and "AB..."); otherwise we terminate at B as above.

It is important to note that, in both approaches, the finite-$\Delta\tau$ error in imposing the boundary condition vanishes as $\Delta\tau \to 0$. The latter algorithm, which resembles the use of mirror images to construct Green's functions that vanish at a boundary, is a higher-order approach and is sometimes referred to as a "mirror correction."

4.5.2 Population Control and Bias Correction

In the CPMC methods walkers carry weights. Often some form of population control is applied. The procedure we use to control the population is similar to that used in many GFMC calculations. First, a branching (or birth/death) scheme is applied, in which walkers with large weights are replicated and ones with small weights are eliminated by probabilities defined by the weights. There exist various ways to do this [25], with the guideline being that the process should not change the distribution statistically. In general, how this step is done only affects the efficiency of the algorithm, but does not introduce any bias.

Branching allows the total number of walkers to fluctuate and possibly become too large or too small. Thus as a second step, the population size is adjusted, if necessary, by rescaling the weights with an overall factor.

Re-adjusting the population size, i.e., changing the *overall* normalization of the population, does introduce a bias [25].

In the ground-state algorithm, we correct for this bias by carrying the m most recent overall rescaling factors and including them in the estimators when computing expectation values. In the calculation we keep a stack which stores the m latest factors, $f^{(j)}$ ($j = 1, m$). Suppose that at the current step the total number of walkers exceeds the pre-set upper bound. We modify the weight of each walker by a constant factor $f < 1$ which reduces the population size to near the expected number. We then replace the oldest $f^{(j)}$ in the stack by f. Whenever we compute expectation values, we multiply the weight of each walker by $1/\prod_j f^{(j)}$. In our calculations on the Hubbard model, m is typically between 5 and 10. As we include more such factors, i.e., increasing m, the bias is reduced, but the statistical error increases. On the other hand, as we reduce m, the statistical error becomes smaller, but the bias increases. The choice of m is thus a compromise between these two.

In finite-temperature calculations, we generally keep all overall rescaling factors throughout the L steps of the random walk. The global product of these rescaling factors are used in the computation of expectation values. Clearly, the overall factor only matters when we wish to combine the results from separate runs (each of L steps). It is advantageous to make the number of walkers as large as possible in finite-temperature calculations so as to reduce the appearance of these overall rescaling factors.

Another approach to eliminate—or at least to reduce—bias is to do several calculations with different (average) population sizes and extrapolate to the infinite population limit. Although this is in general less efficient, it is straightforward to implement.

Clearly, the schemes we have described to correct for bias have some arbitrariness and further improvements are possible. But this should not be a major factor in the calculation. In cases where the bias is *substantially* larger than can be handled by correction schemes in this spirit, it would most likely be more productive to attempt to improve the importance sampling and the algorithm, rather than details of the bias correction scheme.

4.5.3 Stabilizing Matrix Products

In either the ground-state or finite-temperature method, repeated multiplications of the matrix $B(\vec{x})$ can lead to a numerical instability. Eventually round-off errors will dominate. The product represents an unfaithful propagation of the single-particle propagators along the path. The same instability appears in the standard AFQMC methods. The problem is controlled [12] by a numerical stabilization technique that requires the periodic re-orthonormalization of the product. Our approach in the CPMC method is similar. The differences are a result of the different sampling procedures, namely the random walk formalism here versus Metropolis sampling in the standard methods where the entire path X is stored all at once.

In the ground-state method, we use a modified Gram–Schmidt procedure to stabilize the Slater determinants $|\phi_k^{(l)}\rangle$. The procedure is simpler than the corresponding ground-state AFQMC. For each walker $|\phi\rangle$, we factor the matrix Φ as $\Phi = QR$, where Q is a matrix whose columns are a set of orthonormal vectors representing the re-orthogonalized single-particle orbitals. R is a triangular matrix. Note that Q contains all the information about the walker $|\phi\rangle$. R, on the other hand, only contributes to the overlap of $|\phi\rangle$. As we mentioned at the end of Section 4.3, with importance sampling only Q needs to be kept to represent the walker.

The instability in the ground-state formalism is directly related to the collapse to a bosonic ground state. That is, if we let the propagation continue without any stabilization procedure, all single-particle orbitals each walker would become the same. In Green's function Monte Carlo this actually happens and is the first cause for the sign problem. Here the instability can be eliminated by the re-orthonormalization procedure because a walker is explicitly a Slater determinant, not a single point in configuration space. The procedure to stabilize the Slater determinants is like analytically canceling out the bosonic component. In GFMC it is possible to do such a cancellation stochastically, but the cancellation of walkers of positive and negative signs requires a large density of walkers and does not scale well with system size [26]. With Slater determinants, antisymmetry is automatically imposed in each walker. This would suggest that the sign problem is reduced in this formalism compared to GFMC. One trivial observation that is consistent with this is the case of a noninteracting system, where CPMC does not suffer from the sign problem and has zero-variance, while GFMC does have a sign problem and still requires the fixed-node approximation.

In the finite-temperature method, the objects of interest in the calculation are of the form in (4.20). We store the product of B_T's at regular intervals from 1 to L. For each walker, the $B_(\vec{x})$ part is stored as a product of three $N \times N$ square matrices: UDV, where U is full, D is diagonal, and V is upper-triangular. As a random walker takes a step, a $B(\vec{x}_l)$ is multiplied to the left of UDV. Periodically, the product UD is rewritten as $U'D'V''$ for each walker. $V''V$ is then combined to obtain a new upper-triangular matrix V'. The process can be carried out at regular imaginary-time intervals as required by the severity of the problem.

5 Illustrative Results

In a variety of benchmark calculations [3, 5, 7, 27], on the Hubbard model in one and two dimensions, the constrained path Monte Carlo methods have produced very accurate estimates of the energy as well as expectation values of other observables. Practically, running on current workstations the *systematic* error of CPMC, even for small (6×6) systems, is often smaller than the *statistical* error of the standard AFQMC. The statistical errors in the latter increase rapidly with system size and inverse temperature. (For example, at 12×12, to achieve the same statistical error as CPMC, ground-state AFQMC would have to run at least four orders of magnitude longer [5,28]).

The CPMC methods have seen various applications [4,27,29] to study the Hubbard model, the periodic Anderson model, and the three-band Hubbard model. It has also been applied [30] to systems including zeolites and organic superconductors, and in nuclear physics. Here, we show a few benchmark results on the two-dimensional Hubbard model. The model provides a good test case, with both its challenging nature and the availability of certain benchmark data. Our goal here is to illustrate the general behavior of the algorithms. Many more results from benchmark studies can be found in the references.

In Table 2.1 we show results from [5]. It compares ground-state energies from zero-temperature CPMC with available data from other numerical approaches, including stochastic diagonalization (SD) [31], AFQMC, and density-matrix renormalization group (DMRG) [32] methods. The trial Slater determinant $|\psi_T\rangle$ in CPMC is either a noninteracting (free) or an unrestricted Hartree–Fock (uHF) wave function. The SD method uses Monte Carlo methods to construct a basis for approximating the ground state wave function of the system. Since an explicit basis is used, no sign problem occurs; however, an exponential growth in computing time occurs reflecting the increased effort in selecting members of the basis as system size increases. In contrast, the AFQMC method is in principle exact, as we have seen, but suffers from exponential growth in computing time as system size increases because of the sign problem. Finally, the DMRG method is a variational method that has proved very effective for one-dimensional and quasi-one-dimensional systems.

In Table 2.2, we show results for a 4×4, $U = 4$ system. The average density is chosen such that the sign problem is the most severe. (See Fig. 2.2.) This limits the range of temperatures where accurate calculations can be carried out with the standard AFQMC algorithm. The system hence presents a challenging test case for the CPMC method. At high T, the CPMC algorithm gives results in excellent agreement with AFQMC results [19], which are exact. At low T, it reaches convergence and leads to results in good agreement with those from $T = 0\,\mathrm{K}$ exact diagonalization [34].

TABLE 2.1. Hubbard model ground-state energies from ground-state CPMC simulations compared with available results from other approaches. The first two columns indicate system size and electron filling ($L \times L$ and $M_\uparrow M_\downarrow$, respectively). The interaction strength U is 4. The stochastic diagonalization (SD) results are from [31]; the AFQMC results are from [28]; the density-matrix renormalization group (DMRG) results on two-chains are from [33]. The statistical errors are in the last one or two digits, as indicated.

system		$\|\psi_T\rangle$	$E_{\rm CPMC}$	$E_{\rm SD}$	$E_{\rm AFQMC}$
4×4	$5 \uparrow 5 \downarrow$	free	$-19.582(5)$	-19.58	$-19.58(1)$
6×6	$13 \uparrow 13 \downarrow$	free	$-42.34(2)$	-40.77	$-42.32(7)$
6×6	$14 \uparrow 14 \downarrow$	uHF	$-40.17(2)$		$-40.44(22)$
8×8	$25 \uparrow 25 \downarrow$	free	$-72.48(2)$	-67.00	$-72.80(6)$
8×8	$27 \uparrow 27 \downarrow$	uHF	$-67.46(4)$		$-67.55(19)$
10×10	$41 \uparrow 41 \downarrow$	free	$-109.55(3)$		$-109.7(6)$
12×12	$61 \uparrow 61 \downarrow$	free	$-153.43(5)$		$-151.4(1.4)$
system		$\|\psi_T\rangle$	$E_{\rm CPMC}$	$E_{\rm DMRG}$	
2×8	$7 \uparrow 7 \downarrow$	free	$-13.067(4)$	$-13.0664(2)$	
2×16	$14 \uparrow 14 \downarrow$	free	$-26.87(2)$	$-26.867(3)$	

TABLE 2.2. Comparison of finite-temperature and ground-state CPMC (indicated by FT and GS, respectively) with standard AFQMC and exact diagonalization (ED). $G(\mathbf{l})$ is the average Green's function $\langle c^\dagger_{i+\mathbf{l}\sigma} c_{i\sigma}\rangle$, and $P_d(\mathbf{l})$ the d-wave pairing correlation,[a] at separation $\mathbf{XSl} = (l_x, l_y)$. Numbers in parentheses indicate statistical errors in the last digit.

β		E/N	$G(1,0)$	$G(2,2)$	$P_d(2,1)$
3	CPMC-FT	$-0.9437(8)$	$0.1631(1)$	$-0.0415(1)$	$0.0625(2)$
	AFQMC	$-0.9434(3)$	$0.1631(1)$	$-0.0418(1)$	$0.0630(3)$
6	CPMC-FT	$-0.9648(6)$	$0.1663(3)$	$-0.0470(4)$	$0.077(2)$
	AFQMC	$-0.965(3)$	$0.1662(2)$	$-0.0465(2)$	$0.083(3)$
20	CPMC-FT	$-0.977(2)$	$0.166(1)$	$-0.050(1)$	$0.078(2)$
∞	CPMC-GS	$-0.9831(6)$	$0.167(1)$	$-0.051(1)$	$0.078(2)$
	exact	-0.9838	0.167	-0.051	??

[a] The definition we used for the d-wave pairing correlation is: $P_d(\mathbf{l}) \equiv \langle \Delta^\dagger_d(\mathbf{l}) \Delta_d(\mathbf{0})\rangle$, with $\Delta_d(\mathbf{l}) = \sum_{\boldsymbol{\delta}} f(\boldsymbol{\delta})\,(c_{\mathbf{l}\uparrow} c_{\mathbf{l}+\boldsymbol{\delta}\downarrow} - c_{\mathbf{l}\downarrow} c_{\mathbf{l}+\boldsymbol{\delta}\uparrow})$, where $\boldsymbol{\delta}$ is $(\pm 1, 0)$ and $(0, \pm 1)$, and $f(\boldsymbol{\delta})$ is 1 when $\boldsymbol{\delta} = (\pm 1, 0)$, and -1 otherwise.

6 Summary

In summary, we have reviewed quantum Monte Carlo methods, particularly auxiliary-field-based methods, for strongly correlated electron systems. We have developed a formalism here that unifies the different methods and allows for a systematic understanding of their strengths and weaknesses and common features. These methods have been widely applied to study lattice models for electron correlations. In addition, they have seen many applications in high-energy and nuclear physics. The standard auxiliary-field quantum Monte Carlo methods allow essentially exact calculations of ground-state and finite-temperature equilibrium properties of interacting many fermion systems. Their effectiveness, however, is severely limited by the well-known sign problem, which prevents calculations at large system sizes and low temperatures.

The recently developed ground-state and finite-temperature constrained path Monte Carlo methods allow simulations of fermion systems without any decay of sign. The elimination of the sign decay is possible by studying the behavior of the paths in the general space of Slater determinants. In doing so, we have underlined the common feature of the sign problem under different forms of quantum Monte Carlo methods. The CPMC methods are approximate. Benchmark calculations have shown that accurate results can be obtained with simple choices of $\widehat{B}_T$ or $|\psi_T\rangle$. An improved trial propagator or wave function will lead to improved results.

The CPMC methods make possible calculations under the field-theoretical formalism whose required computer time scales algebraically, rather than exponentially, with inverse temperature and system size. With the second-quantized representation, it complements the fixed-node GFMC method and the restricted path-integral Monte Carlo method [24] in real space. The algorithm automatically accounts for particle permutations and allows easy computations of both diagonal and off-diagonal expectations, as well as imaginary-time correlations. While much work is needed to study various forms of $\widehat{B}_T$ or $|\psi_T\rangle$, and to understand the subtleties of the methods because of the nonorthogonal and over-complete nature of the Slater determinant space involved, we expect the method and the concept brought forth here to see many applications, and to significantly enhance the applicability of quantum simulations in interacting fermion systems.

Acknowledgments: It is a pleasure to thank my collaborators and colleagues, especially E. Allman, J. Carlson, J.E. Gubernatis, and H. Krakauer, for collaborations and discussions. The author is grateful to the CRM and to the organizers for the opportunity to participate in the workshop. Support from the US National Science Foundation under a CAREER Award (DMR-9734041) is gratefully acknowledged. Part of the calculations were performed on computers at the NCSA (UIUC) and at Boston University. The author is a Research Corporation Cottrell Scholar.

7 References

[1] K.E. Schmidt and M.H. Kalos, in *Applications of the Monte Carlo Method in Statistical Physics*, ed. K. Binder (Springer Verlag, Heidelberg, 1984).

[2] E.Y. Loh Jr., J.E. Gubernatis, R.T. Scalettar, S.R. White, D.J. Scalapino, and R.L. Sugar, Phys. Rev. B **41**, 9301 (1990).

[3] S. Zhang, in *Quantum Monte Carlo Methods in Physics and Chemistry*, eds. M.P. Nightingale and C.J. Umrigar, NATO ASI Series (Kluwer Academic Publishers, Dordrecht, 1999).

[4] S. Zhang, J. Carlson, and J.E. Gubernatis, Phys. Rev. Lett. **78**, 4486 (1997).

[5] S. Zhang, J. Carlson, and J.E. Gubernatis, Phys. Rev. Lett. **74**, 3652 (1995); Phys. Rev. B **55**, 7464 (1997).

[6] J. Carlson, J.E. Gubernatis, G. Ortiz, and S. Zhang, Phys. Rev. B, **59**, 12788 (1999).

[7] S. Zhang, Phys. Rev. Lett. **83**, 2777 (1999); Comp. Phys. Comm. **127**, 150 (2000).

[8] M.H. Kalos, D. Levesque, and L. Verlet, Phys. Rev. **A9**, 2178 (1974); D.M. Ceperley and M.H. Kalos, in *Monte Carlo Methods in Statistical Physics*, ed. K. Binder (Springer-Verlag, Heidelberg, 1979).

[9] See, e.g., M.H. Kalos and P.A. Whitlock, *Monte Carlo Methods*, Vol. I (Wiley, New York, 1986).

[10] R. Blankenbecler, D.J. Scalapino, and R.L. Sugar, Phys. Rev. D **24**, 2278 (1981).

[11] G. Sugiyama and S.E. Koonin, Ann. Phys. (NY) **168**, 1 (1986).

[12] S.R. White, D.J. Scalapino, R.L. Sugar, E.Y. Loh Jr., J.E. Gubernatis, R.T. Scalettar, Phys. Rev. B **40**, 506 (1989).

[13] S. Sorella et al., Int. Jour. Mod. Phys. B **1** 993 (1988).

[14] See, e.g., D.R. Hamann and S.B. Fahy, Phys. Rev. B **41**, 11352 (1990).

[15] J. Hubbard, Phys. Rev. Lett. **3**, 77 (1959).

[16] J.E. Hirsch, Phys. Rev. B **31**, 4403 (1985).

[17] J.W. Negele and H. Orland, *Quantum Many-Particle Systems* (Addison-Wesley, New York, 1987), Chapter 7.

[18] S. Rombouts, K. Heyde, and N. Jachowicz, Phys. Lett. A **242**, 271 (1998).

[19] R.T. Scalettar, private communication.

[20] S.B. Fahy and D.R. Hamann, Phys. Rev. Lett. **65**, 3437 (1990); Phys. Rev. B **43**, 765 (1991).

[21] J.B. Anderson, J. Chem. Phys. **63**, 1499 (1975); **65**, 4122 (1976).

[22] J.W. Moskowitz, K.E. Schmidt, M.A. Lee, and M.H. Kalos, J. Chem. Phys. **77**, 349 (1982); P.J. Reynolds, D.M. Ceperley, B.J. Alder, and W.A. Lester, J. Chem. Phys. **77**, 5593 (1982).

[23] D.F.B. ten Haaf, H.J.M. van Bemmel, J.M.J. van Leeuwen, W. van Saarloos, and D.M. Ceperley, Phys. Rev. B **51**, 13039 (1995); H.J.M. van Bemmel, D.F.B.ten Haaf, W. van Saarloos, J.M.J. van Leeuwen, and G. An, Phys. Rev. Lett., **72**, 2442 (1994).

[24] D.M. Ceperley, Phys. Rev. Lett. **69**, 331 (1992); in *Simulation in Condensed Matter Physics and Chemistry*, eds. K. Binder and G. Ciccotti (1996).

[25] C.J. Umrigar, M.P. Nightingale, and K.J. Runge, J. Chem. Phys. **99**, 2865 (1993); M.P. Nightingale, *Quantum Monte Carlo Methods in Physics and Chemistry*, eds. M.P. Nightingale and C.J. Umrigar, NATO ASI Series (Kluwer Academic Publishers, Dordrecht, 1999).

[26] S. Zhang and M.H. Kalos, Phys. Rev. Lett. **67**, 3074 (1991); J.B. Anderson, C.A. Traynor, and B.M. Boghosian, J. Chem. Phys. **95**, 7418 (1991).

[27] S. Zhang and E.C. Allman, in *Computer Simulation Studies in Condensed Matter Physics XII*, eds. D.P. Landau, S.P. Lewis, and H.B. Schuttler (Springer Verlag, Heidelberg, Berlin, 1999).

[28] N. Furukawa and M. Imada, J. Phys. Soc. Jpn. **61**, 3331 (1992).

[29] See, e.g., M. Guerrero, J.E. Gubernatis, and S. Zhang, Phys. Rev. B **57**, 11980 (1998); J. Bonca and J. E. Gubernatis, Phys. Rev. B **58**, 6992 (1998); M. Guerrero, G. Ortiz, J.E. Gubernatis, Phys. Rev. B **59**, 1706 (1999); S. Zhang and E.C. Allman, in *Computer Simulation Studies in Condensed Matter Physics XII*, eds. D.P. Landau, S.P. Lewis, and H.B. Schuttler (Springer Verlag, Heidelberg, Berlin, 1999).

[30] O.F. Sankey, A.A. Demkov, and T. Lenosky, Phys. Rev. B **57**, 15129 (1998); S. Mazumdar, S. Ramasesha, R.T. Clay, D.K. Campbell, Phys. Rev. Lett. **82**, 1522 (1999); K.E. Schmidt and S. Fantoni, Phys. Lett. B **446**, 99 (1999).

[31] H. De Raedt and M. Frick, Phys. Reports, **231**, 109 (1993).

[32] S.R. White, Phys. Rev. Lett. **69**, 2863 (1992); Phys. Rev. B **48**, 10345 (1993).

[33] R.M. Noack, private communication.

[34] A. Parola, S. Sorella, S. Baroni, R. Car, M. Parrinello, and E. Tosatti, Physica C **162-164**, 771 (1989); A. Parola, S. Sorella, M. Parrinello, and E. Tosatti, Phys. Rev. B **43**, 6190 (1991); G. Fano, F. Ortolani, and A. Parola, Phys. Rev. B **42**, 6877 (1990).

[35] C.J. Umrigar et al., Phys. Rev. Lett. **60** 1719 (1988).

[36] N. Trivedi and D.M. Ceperley, Phys. Rev. B **41**, 4552 (1990)

Appendix A Brief Review of Configuration-Space Methods

A.1 Variational Monte Carlo

This is perhaps the simplest way of performing a quantum Monte Carlo calculation for ground-state properties. It is conceptually different from most other QMC approaches. Its theoretical basis is the variational principle. A variational trial wave function $|\Psi_T\rangle$ is first constructed with a set of parameters that can be varied. One then optimizes these parameters so as to obtain the lowest possible variational ground-state energy. The only role Monte Carlo plays is in computing the many-dimensional integrals that are necessary for the ground-state energy.

We start with writing the trial state $|\Psi_T\rangle$ in terms of the configurations of the system

$$|\Psi_T\rangle = \sum_R |R\rangle\langle R \mid \Psi_T\rangle, \tag{A.1}$$

where the coefficients $\langle R \mid \Psi_T\rangle = \Psi_T(R)$ depend on the set of variational parameters p_i. Here, $|R\rangle$ is a state in configuration space as defined in Section 2.3. For example, for M electrons of the same spin on an $N = L \times L$ two-dimensional square lattice, a configuration could be a state with electrons on lattice sites $[i_1, i_2, \ldots, i_M]$ such that $1 \leq i_k \leq N$ for $1 \leq k \leq M$

and $i_{k1} \neq i_{k2}$ for any pair of particle indices $k1 \neq k2$. The variational ground-state energy in the state $|\Psi_T\rangle$ can be expressed as

$$E_T = \frac{\langle\Psi_T|\widehat{H}|\Psi_T\rangle}{\langle\Psi_T \mid \Psi_T\rangle} = \frac{\sum_R E_L(R)\langle\Psi_T \mid R\rangle\langle R \mid \Psi_T\rangle}{\sum_R\langle\Psi_T \mid R\rangle\langle R \mid \Psi_T\rangle}, \tag{A.2}$$

where the local energy in configuration R is defined as

$$E_L(R) = \frac{\langle\Psi_T|\widehat{H}|R\rangle}{\langle\Psi_T \mid R\rangle}. \tag{A.3}$$

The many-dimensional sum in (A.2), while in general impossible to evaluate analytically, is in exactly the form of (2.7) and can therefore be evaluated by Monte Carlo. We can sample the function $f(R) = \langle\Psi_T \mid R\rangle\langle R \mid \Psi_T\rangle = |\Psi_T(R)|^2$ by the Metropolis algorithm. Given $\mathcal{M}$ independent Monte Carlo samples, we evaluate the integral for E_T by (2.9) and obtain a Monte Carlo estimate, E_{VMC}. In addition, we can easily evaluate the statistical error of E_{VMC}: $\delta E = \sigma_E/\sqrt{\mathcal{M}}$. The variance of the local energy, σ_E, is defined as

$$\sigma_E = \sqrt{\langle E^2\rangle - \langle E\rangle^2}, \tag{A.4}$$

where $\langle E\rangle = E_{\mathrm{VMC}}$ and $\langle E^2\rangle = \left(\sum_R E_L^2(R)\right)/\mathcal{M}$. It is known [35] that σ_E, which has a lower bound of 0, is a better quantity to optimize than E_{VMC}. We have often used a combination of both quantities as indicators in our search of variational parameters.

We use the Metropolis algorithm to generate a set of $\mathcal{M}$ configurations distributed according to $f(R)$. The general Metropolis algorithm is reviewed briefly in Section 3.1 under the AFQMC algorithm. For a lattice system such as the two-dimensional Hubbard model, working in configuration space allows some further simplification. For example, the kernel K for each electron can simply propose a move uniformly to one of its 4 neighboring sites. The acceptance is then given by

$$A = \min\left[\frac{|\Psi_T(R_{\mathrm{new}})|^2}{|\Psi_T(R_{\mathrm{old}})|^2}, 1\right], \tag{A.5}$$

where R_{new} and R_{old} are the new and old configurations, respectively. Clearly, the move would be rejected if the neighboring site is already occupied by an electron of like spins. For most choices of the trial wave function $|\Psi_T\rangle$, which could contain Slater determinants multiplied by two-body (or higher) correlation functions, the evaluation of (A.5) is simple because R_{new} and R_{old} differ only by one electron position. The random walk is repeated sufficient number of steps so that $\mathcal{M}$ independent walkers are obtained.

A.2 Green's Function Monte Carlo (GFMC)

The basic premise of the GFMC method [8] is the same as that of the ground-state CPMC method. That is, it involves random walks based on (2.2). The random walk formalism utilizes first-quantized representation and is in configuration space.

For a lattice system such as the Hubbard or t-J model, one starts with the operator [36]

$$\widehat{F} \equiv C - \widehat{H} \tag{A.6}$$

in place of $e^{-\Delta\tau\widehat{H}}$. The operator $\widehat{F}$ can be viewed as a first order expansion of the latter. But for discrete systems, projection with $\widehat{F}$ leads to the exact ground-state wave function as well, because the system has a spectrum that is bounded. In this case, it is often advantageous to use the operator $\widehat{F}$ in the projection, because of the sparseness of its matrix representation. The constant C is chosen to ensure that all diagonal elements are positive and the spectrum of $\widehat{F}$ is positive. Similar to ground-state AFQMC or CPMC, if an initial state $|\Psi^{(0)}\rangle$ has any overlap with the ground state $|\Psi_0\rangle$ of $\widehat{H}$, the process

$$|\Psi^{(n)}\rangle = \widehat{F}^n|\Psi^{(0)}\rangle \tag{A.7}$$

will lead to $|\Psi_0\rangle$ at large n. The Green's function Monte Carlo method realizes the above process by a Monte Carlo random walk in configuration space.

In order to improve efficiency of the random walk process, one more mathematical manipulation of (A.7) is necessary. This is done by introducing an operator $\widetilde{\widehat{F}}$ whose matrix elements are related to those of $\widehat{F}$ by a similarity transformation:

$$\widetilde{\widehat{F}}(R', R) \equiv \langle\Psi_T \mid R'\rangle\langle R'|\widehat{F}|R\rangle\frac{1}{\langle\Psi_T \mid R\rangle}. \tag{A.8}$$

The stochastic realization of (A.7) is actually with $\widetilde{\widehat{F}}$ instead of $\widehat{F}$. While mathematically equivalent, the use of $\widetilde{\widehat{F}}$ can significantly reduce the fluctuation of the Monte Carlo process if $|\Psi_T\rangle$ is a reasonable approximation of $|\Psi_0\rangle$. As we have seen with the constrained path Monte Carlo methods, this is the idea of importance sampling.

The program is then to start with a set of walkers distributed according to $\langle\Psi_T \mid R\rangle\langle R \mid \Psi_T\rangle$. This can be accomplished with a variational Monte Carlo calculation prior to GFMC. The walkers then undergo random walks in R-space. For each walker, denoted by R, a step in the random walk means randomly selecting and moving to a new position R' with probability $\widetilde{\widehat{F}}(R', R)/\sum_{R'}\widetilde{\widehat{F}}(R', R)$. Note that the sparsity in the structure of $\widehat{H}$ means that there are only $\mathcal{O}(N)$ possible R''s, whose corresponding probabilities can be computed. Because the overall normalization $\sum_{R'}\widetilde{\widehat{F}}(R', R)$ is not a constant, each walker carries a weight which fluctuates, or a branching scheme is introduced to allow the total number of walkers to fluctuate, or both.

The ground state energy is given exactly by the so-called *mixed estimate* (see (2.4))

$$E_0 = \frac{\langle \Psi_T | \widehat{H} | \Psi_0 \rangle}{\langle \Psi_T \mid \Psi_0 \rangle} = \frac{\sum_R E_L(R) \langle \Psi_T \mid R \rangle \langle R \mid \Psi_0 \rangle}{\sum_R \langle \Psi_T \mid R \rangle \langle R \mid \Psi_0 \rangle}, \tag{A.9}$$

where $E_L(R)$ is defined in (A.3). After a sufficient number of steps, the walkers are distributed according to $\langle \Psi_T \mid R \rangle \langle R \mid \Psi_0 \rangle$. E_0 can therefore be computed from such walkers as the (weighted) average of $E_L(R)$ with respect to walker positions R. We denote this Monte Carlo estimate of E_0 by E_{GFMC}.

Expectation values of operators other than $\widehat{H}$ can also be computed from the mixed estimate. However, if the operator does not commute with $\widehat{H}$, this estimate is *not* exact. This is an important distinction which requires careful analysis of the bias in the results due to $|\psi_T\rangle$. There exist ways to improve upon the mixed estimate and to possibly extract exact estimates of expectation values, but we will not discuss them here. The computation of *off-diagonal expectations* in general presents difficulty in GFMC, because it is difficult to sample two groups of random walkers $\{R'\}$ and $\{R\}$ whose overlap is well behaved statistically to allow for computation of matrix elements such as $\langle \Psi_0 \mid R' \rangle \langle R' | \widehat{O} | R \rangle \langle \Psi_0 \mid R \rangle$.

The sign problem occurs in GFMC if the particles are fermions (e.g., the Hubbard model) or any of the matrix elements in (A.8) are negative (e.g., anti-ferromagnetic coupling of quantum spins). The fixed-node method eliminates the sign decay by imposing the nodal structure of the trial wave function $|\psi_T\rangle$ on the projected wave function. For continuum systems, such as atoms or molecules, the fixed-node approximation is straightforward to implement [21, 22]. One would simply modify the transition probability in (A.8), setting all negative matrix elements to zero. At the limit of $\Delta\tau \to 0$, this ensures that the random walk process solves the Schrödinger equation under the boundary condition that $\Psi_0^{\mathrm{FN}}(R) = 0$ for all configurations R that satisfy $\Psi_T(R) = 0$, where $|\Psi_0^{\mathrm{FN}}\rangle$ is then the approximate solution for $|\Psi_0\rangle$ under the fixed-node approximation.

For discrete systems, however, the implementation of the fixed-node approximation is more subtle. Unlike in continuum systems, the random walk does *not* become continuous at the limit of $\Delta\tau \to 0$. A method was developed a few years ago [23] which successfully generalized the continuum fixed-node approximation to lattice fermion systems. In addition to setting the transition probability to zero when the matrix element is negative in (A.8), one also modifies diagonal matrix elements $\widetilde{\widetilde{F}}(R, R)$ when the corresponding configuration R is in the vicinity of the node defined by $|\Psi_T\rangle$. R is in the vicinity of the trial node if there exists at least one negative $\widetilde{\widehat{F}}(R',R)$, i.e., at least one "hop" which would take the walker across the trial node. For each such configuration R, an extra part is added to the diagonal matrix element. The extra part serves as an effective potential. It is defined as $\sum_{R'}^{-} \widetilde{\widehat{F}}(R',R)$, where the negative sign means that the sum is only over those configurations R' that lead to negative elements $\widetilde{\widehat{F}}(R',R)$. The idea of this effective potential is in fact related to the discussion in the first part of Section 4.5.

Part II

Lagrangian, Functional Integral, Renormalization Group, Conformal and Bosonization Methods

3

Renormalization Group Technique for Quasi-One-Dimensional Interacting Fermion Systems at Finite Temperature

C. Bourbonnais, B. Guay, and R. Wortis

ABSTRACT We review some aspects of the renormalization group method for interacting fermions. Special emphasis is placed on the application of scaling theory to quasi-one-dimensional systems at nonzero temperature. We begin by introducing the scaling ansatz for purely one-dimensional fermion systems and its extension when interchain coupling and dimensionality crossovers are present at finite temperature. Next, we review the application of the renormalization group technique to the one-dimensional electron gas model and clarify some peculiarities of the method at the two-loop level. The influence of interchain coupling is then included and results for the crossover phenomenology and the multiplicity of characteristic energy scales are summarized. The emergence of the Kohn–Luttinger mechanism in quasi-one-dimensional electronic structures is discussed for both superconducting and density-wave channels.

1 Introduction

Scaling ideas have exerted a far-reaching influence on our understanding of complex many-body systems. Their use in the study of critical phenomena has offered a broad and fertile field of work at the heart of which is the Wilson formulation of the renormalization group method [1, 2]. In the Wilson view, fluctuations of the order parameter at all length scales up to the correlation length are the key ingredient in explaining the existence of power law singularities that govern critical properties in accordance with scaling laws. Equally powerful is the extension of scaling concepts to anisotropic systems, namely, when small anisotropic parameters are magnified as a result of their coupling to singular fluctuations. Intermediate length scales thus emerge giving rise to changes or *crossovers* in the critical behavior [3–5]. The horizon of applications of this methodology was further widened when scaling was applied to quantum critical systems for which quantum mechanics governs fluctuations of the order parameter at the critical point [6, 7].

There has been a parallel expansion of the use of scaling tools in the description of *many-fermion* systems. This was especially true for the Kondo impurity and the one-dimensional fermion gas problems. In the latter, the reduction of spatial dimension is well known to enlarge the range of fluctuations which have a peculiar impact on the properties of the system [8, 9]. In this context the scaling theory commonly termed *multiplicative renormalization group* [10, 11] contributed significantly to the completion of a coherent microscopic picture of the 1D fermion gas system as a paradigm of a non-Fermi–Luttinger-liquid state. However, this scaling theory rests for the most part on the logarithmic structure of infrared singularities that compose 1D perturbation theory and as such departs somewhat from the standard Wilson picture. It remains closer to earlier formulations of the renormalization group in quantum field theory. Yet, fluctuations that characterize the 1D fermion gas do show a multiplicity of length scales stretching from the shortest distance, of the order of the lattice constant, up to the quantum coherence length or the de Broglie wavelength of fermions. Early attempts to reexamine the properties of low-dimensional fermion gas system along these lines were motivated by the coupled chains problem [12–16], which finds concrete applications in the physics of quasi-one-dimensional organic conductors [17, 18]. The description of interacting fermions when strong spatial anisotropy is present shares many traits with its counterpart in critical phenomena [4, 5]. Hence the Wilson method provides a powerful framework to study how deviations from perfect 1D scaling are introduced by spatial anisotropy. The corresponding length scales enter as key components of the notoriously complex description of dimensionality crossovers in which the 1D state evolves either toward the emergence of a Fermi liquid or the formation of long-range order at finite temperature.

The Wilson renormalization group approach has attracted increasing interest in several areas of the many-fermion problem [19], notably in the description of two-dimensional systems in connection with high-T_c materials [20–22], Fermi liquids [23–25], and ladder systems [26–28]. Within the bounds of this review, it would therefore be impossible to give anything more than a cursory account of the expanding literature in this field [29–31]. With the goal of keeping this paper self-contained for the nonspecialist, however, our interest will be more selective and will focus on the application of the renormalization group to the quasi-one-dimensional fermion gas at finite temperature [32]. Although many aspects of this problem have already been discussed in detail in previous reviews [15, 16], it is useful to revisit some issues while in addition discussing features of the Wilson method that deserve closer examination but which have as yet never received a detailed investigation. This is especially true in regard to the formulation of the method itself, in particular the way successive integrations over fermion states in outer momentum shells (the Kadanoff transformation) are carried out when high-order calculations are performed. We close this review with a detailed discussion of the dual nature of the Kohn–Luttinger mechanism [33] from which interchain pairing correlations find their origin in quasi-one-dimensional systems [34–37]. With the aid of the renormalization group [38, 39], we show how this mechanism may yield instabilities of the metallic state toward either unconventional superconducting or density-wave order.

In Section 2, we introduce scaling notions for fermions in low dimensions from a phenomenological standpoint, and predictions of the scaling hypothesis are compared to the solution in the noninteracting case in Section 3. The formulation of the renormalization group in the classical Wilson scheme is given in Section 4, and explicit calculations are carried out up to the two-loop level. In Section 5, the influence of interchain coupling is discussed and we go through various possibilities of dimensionality crossovers in the quasi-one-dimensional Fermi gas. In Section 6, we study how the Kohn–Luttinger mechanism for interchain pairing correlation emerges from the renormalization group flow and under what conditions it can lead to long-range order. We conclude in Section 7.

2 Scaling Ansatz for Fermions

Before embarking on the program sketched above, it will be useful first to discuss on purely phenomenological grounds how scaling notions apply to 1D and quasi-1D fermion systems. The predictions of the scaling ansatz are explicitly checked in the trivial but instructive case of the noninteracting Fermi gas.

2.1 One Dimension

Let us look at how scaling ideas apply to the description of purely one-dimensional fermionic systems at low temperature. The temperature T is the parameter that controls the quantum delocalization or the spatial coherence of each particle in one dimension. This appears as a characteristic length scale, ξ, which is precisely the de Broglie wavelength. The temperature dependence of ξ is readily determined by considering the effect of thermal excitations on fermion states within an energy shell $\delta\epsilon(k) \sim T$ ($k_B \equiv 1$, $\hbar \equiv 1$), where $\epsilon(k)$ is the energy spectrum of fermions evaluated with respect to the Fermi level. In the vicinity of the Fermi points $\pm k_{\rm F}$, one has $\delta\epsilon(k) \approx v_{\rm F}(|k| - k_{\rm F})$, where $v_{\rm F}$ is the Fermi velocity, and the coherence length can be written

$$\xi \sim v_{\mathrm{F}}/T \sim T^{-\nu}, \tag{2.1}$$

where the exponent ν is then equal to unity. As $T \to 0$, ξ goes to infinity and the system develops long-range quantum coherence. The "time" τ required for quantum delocalization up to ξ is therefore $\sim 1/T$, which simply implies that

$$\tau \sim \xi^z, \tag{2.2}$$

where $z = 1$ is the dynamical exponent of the fermion field. Since "time" and distance must be considered on the same footing, both ν and z can be taken as independent of interaction.

Let us now consider the Matsubara–Fourier transform

$$G(k, \omega_n) = \int_0^\beta \int_0^L e^{-ikx + i\omega_n \tau} G(x, \tau)\, dx\, d\tau, \tag{2.3}$$

where

$$G(x, \tau) = -\langle T_\tau \psi_\alpha(x, \tau) \psi_\alpha^\dagger(0, 0)\rangle \tag{2.4}$$

is the one-particle time-ordered correlation function expressed as a statistical average over fermion fields $\psi_\alpha^{(\dagger)}$ of spin α. Here $\omega_n = (2n+1)\pi T$ corresponds to the Matsubara frequencies. For noninteracting fermions and $T = 0$, $G(k) \sim (|k| - k_{\mathrm{F}})^{-1}$ has a simple pole singularity at $|k| \to k_{\mathrm{F}}$ and $\omega_n = 0$, indicating that single particle excitations are well defined. However, when fermions interact in one dimension, such a picture is expected to be altered. In the scaling picture, this can be illustrated by adding an anomalous power dependence on the wave vector near k_{F}, namely

$$G(k) \sim (|k| - k_{\mathrm{F}})^{-1+\theta}. \tag{2.5}$$

The absence of a quasi-particle pole introduces θ, the anomalous dimension for the Green's function ($\theta \geq 0$ where the equality occurs in the noninteracting limit). Correspondingly, the absence of a Fermi liquid component at zero temperature will affect the equal-time decay of coherence at large distance, which takes the form

$$G(x) \approx \frac{\hat{C}}{x^{\bar{d}-1+\theta}}, \tag{2.6}$$

where the effective dimension $\bar{d} = 2$ has a space and time component. A similar decay in time takes place for large τ at $x = 0$. Looking at the temperature dependence of the Green's function, the scaling hypothesis allows us to write

$$G(k, \omega_n) \approx C T^{-\bar{\gamma}} g\big(\delta k \xi, \omega_n \xi^z\big), \tag{2.7}$$

where $\delta k = |k| - k_{\mathrm{F}}$, $\bar{\gamma}$ is the thermal exponent of the single particle Green's function, and $g(x, y)$ is a scaling function. Consistency between (2.7) and (2.5) leads to the following relation between the exponents

$$\bar{\gamma} = (1 - \theta)\nu. \tag{2.8}$$

Scaling can also be applied to the free energy per unit length, the temperature dependence of which can be expressed as

$$f = -\frac{T}{L} \ln Z \approx A T^{1-\alpha}. \tag{2.9}$$

The exponent α is connected to the temperature variation of specific heat

$$\begin{aligned} C_L &= -T \frac{\partial^2 f}{\partial T^2} \\ &\approx A' T^{-\alpha}, \end{aligned} \tag{2.10}$$

where

$$\alpha = 1 - \bar{d}\nu. \tag{2.11}$$

As long as the ansatz $\nu = 1$ holds for the coherence length, one then always has $\alpha = -1$ in one spatial dimension. The specific heat is therefore linear in temperature for fermions even though, according to (2.5), the system is no longer a Fermi liquid at low energy. It should be noted that the familiar Sommerfeld result $C_V \sim T$ for a Fermi liquid (and Fermi gas) in higher dimensions indicates that the importance put on fermion states near the Fermi level one-dimensionalizes the sum over states for the evaluation of internal energy.

Scaling concepts can also be applied to the critical behavior of correlations that involve *pairs* of fermions in one dimension. Superconducting and $2k_{\mathrm{F}}$ density-wave responses are among the pair susceptibilities that may present singularities at zero temperature. Consider the Matsubara–Fourier transform of the susceptibility

$$\chi(q, \omega_m) = \int_0^\beta \int_0^L e^{-iqx + i\omega_m \tau} \chi(x, \tau)\, dx\, d\tau, \tag{2.12}$$

where $\omega_m = 2m\pi T$. The time-ordered pair correlation function is

$$\chi(x, \tau) = -\langle T_\tau O(x, \tau) O^\dagger(0, 0) \rangle, \tag{2.13}$$

in which $O^\dagger \sim \psi_\alpha^\dagger \psi_\beta^\dagger$ ($O^\dagger \sim \psi_\alpha^\dagger \psi_\beta$) are the composite operators of the superconducting (density-wave) channel, commonly called the Cooper (Peierls) channel . At $T = 0$, these can show long-range coherence leading to an algebraic decay of pair correlations with distance. Following the example of the correlation of the order parameter at a critical point, one can write a relation of the form

$$\chi(x) \approx e^{-iq_0 x} \frac{D}{x^{\bar{d} - 2 + \eta}}, \tag{2.14}$$

where $\eta \geq 0$ is an anomalous exponent that characterizes the decay of pair correlation. Here $q_0 = 0$ $(2k_{\rm F})$ is the characteristic wave vector of correlations in the Cooper (Peierls) channel. A similar power law variation with time τ can also be written at small distance. Correspondingly, in Fourier space, one has in the static limit

$$\chi(q) \approx \frac{\overline{D}}{q^{2-\eta}}, \tag{2.15}$$

where q refers to deviations with respect to an ordering wave vector q_0 in the Cooper or Peierls channel. For $q = 0$ and at finite frequency, a similar expression holds with q being replaced by the frequency. Now at finite temperature, the scaling hypothesis allows us to write

$$\chi(x,T) \approx De^{-iq_0 x} x^{-\bar{d}+2-\eta} \mathcal{X}(x/\xi), \tag{2.16}$$

where $\mathcal{X}$ is scaling function ($\mathcal{X}(0) = 1$). Here the correlation length ξ for the pair field is assumed to have the same power law dependence on temperature as the coherence length (2.1) for both fermions in a pair. In Fourier space, one can write for the divergence of the susceptibility in a given channel

$$\chi(q,\omega_m,T) \approx \overline{D}T^{-\gamma}\overline{\mathcal{X}}(q\xi,\omega_m\xi^z). \tag{2.17}$$

Consistency between (2.17) and (2.15) in the static case leads to the Fisher relation

$$\gamma = (2-\eta)\nu \tag{2.18}$$

between the exponents of the static susceptibility, the correlation function and the correlation length.

2.2 Anisotropic Scaling and Crossover Phenomena

Let us now move from one-dimensional systems to quasi-one–dimensional systems by which we mean weakly coupled chains. A dimensionality crossover in low-dimensional fermion systems is induced by a small coupling between chains. The nature of this crossover will depend on the kind of coupling involved. Weakly coupled chains correspond to so-called quasi-one-dimensional anisotropy, a situation realized in practice for electronic materials such as the organic conductors [17,18,40], lattice of spin chains, etc. In ordinary critical phenomena, scaling concepts were extended to anisotropic systems and have been successful in describing the new length scales introduced by anisotropy [3–5]; their use in the study of anisotropic fermion systems appears therefore quite natural. For example, deconfinement of quantum coherence for single particles can give rise to the restoration of a Fermi liquid component below some characteristic low energy scale. In addition, the deconfinement of pair correlations in more than one spatial

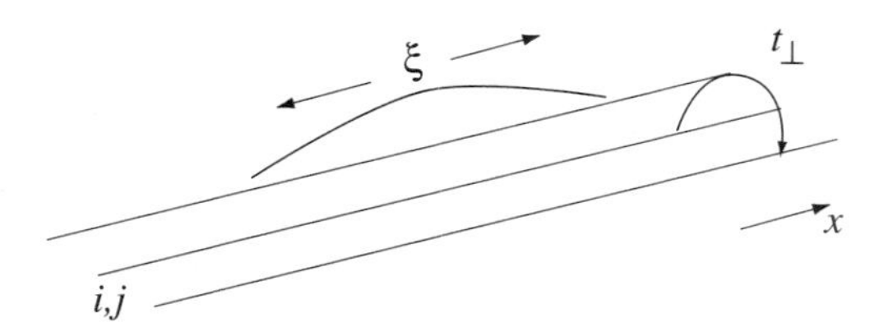

FIGURE 3.1. One-fermion interchain hopping. ξ is the fermion coherence length along the chains.

dimension can lead to the emergence of true long-range order at finite temperature [41].

Let us consider first the case of interchain single-particle hopping (Fig. 3.1), whose amplitude $t_\perp$ is small compared to the Fermi energy of isolated chains ($t_\perp \ll E_{\mathrm{F}}$). Applying the extended scaling ansatz to the single fermion Green's function at $|k| = k_{\mathrm{F}}$, and $\omega_{n=0} = \pi T$, [4] one can write

$$G(T, k_\perp) \approx D T^{-\bar{\gamma}} X_G(B t_\perp(k_\perp)/T^{\phi_x}). \tag{2.19}$$

$t_\perp(k_\perp) = t_\perp \cos k_\perp$ and $k_\perp$ is the transverse momentum of the Fermion spectrum (here the inter-chain distance $d_\perp$ has been set equal to unity). $\phi_x > 0$ is the crossover exponent which governs the magnification of the single-particle perturbation $t_\perp$ as ξ increases along the chains (Fig. 3.1). $X_G(y)$ is called a crossover scaling function and it determines transitory aspects of the crossover and the temperature range over which it is achieved. Here $X_G(0) = 1$. When $y \ll 1$, the coherence of the fermionic system is confined along the chains and the physics is essentially dominated by the one-dimensional expression (2.7). Otherwise for $y \gg 1$, a different temperature dependence is expected to arise. The temperature scale for the single-particle crossover is thus determined by the condition $y \sim 1$, that is

$$T_x \sim t_\perp^{1/\phi_x}. \tag{2.20}$$

Well below T_x and on the full Fermi surface $\mathbf{k} = \mathbf{k}_{\mathrm{F}}$, one has

$$G(T) \sim z(t_\perp) T^{-\dot{\bar{\gamma}}}, \tag{2.21}$$

where $z(t_\perp)$ is the quasi-particle weight and $\dot{\bar{\gamma}} = 1$ is the Fermi liquid exponent that restores the canonical dimension of the Green's function. In order to match (2.19) and (2.21) well below T_x, one must have [4]

$$z(t_\perp) = z_\infty t_\perp^{-(\bar{\gamma} - \dot{\bar{\gamma}})/\phi_x}, \tag{2.22}$$

where z_∞ is a nonuniversal constant. Applying scaling to the dynamical-coherent-part of the single-particle Green's function, the recovery of Fermi liquid behavior yields

$$G(\mathbf{k}, \omega_n) \approx \frac{z(t_\perp)}{[i\omega_n - E(\mathbf{k})]^{\dot{\bar{\gamma}}}} \tag{2.23}$$

with $\dot{\bar{\gamma}} = 1$ again.

Interchain coupling will affect the temperature dependence of pair correlations as well, which can undergo a dimensionality crossover and in turn an instability toward the formation of long-range order either in the Cooper or the Peierls channel at finite temperature. This possibility is of course bound to the number of components of the order parameter and the spatial dimensionality of the system [41]. The scaling behavior of pair correlations is therefore analogous to the one found in ordinary anisotropic critical phenomena. Properties of the system close to the true critical point at T_c can then be described by the usual reduced temperature

$$\dot{t} = \frac{T - T_c}{T_c}. \tag{2.24}$$

The static pair susceptibility at some ordering wave vector $\mathbf{Q}_0 \equiv (q_0, q_{\perp 0})$ will then diverge as $\dot{t} \to 0$ according to

$$\chi(T) \approx \dot{C}(g_\perp)\dot{t}^{-\dot{\gamma}}, \tag{2.25}$$

where $\dot{\gamma}$ is the susceptibility exponent.

Applying the extended scaling hypothesis to the static susceptibility in the superconducting or the density-wave channel, one can write

$$\chi(T) \approx \bar{D}T^{-\gamma} X(Bg_\perp / T^{\phi_{x^2}}), \tag{2.26}$$

where $X(y)$ is a crossover scaling function ($X(0) = 1$), $g_\perp$ is the interchain coupling for pairs of fermions and ϕ_{x^2} is a *two*-particle crossover exponent [15, 42]. Following the same arguments as in the single-particle case, the change in the exponents will occur when $y \sim 1$, which allows us to introduce the crossover temperature for pair correlations

$$T_{x^2} \sim g_\perp^{1/\phi_{x^2}}. \tag{2.27}$$

It is worth noting that the interchain coupling $g_\perp$ for pairs of particles differs from $t_\perp$, and the two-particle crossover can occur above its single particle counterpart T_x, in which case the restoration of Fermi liquid behavior is absent. Instead only pair correlations deconfine, possibly leading to long-range order.

If the scaling hypothesis holds, the expressions (2.25) and (2.14) have to match near T_c, which implies

$$\dot{C}(g_\perp) = C_\infty g_\perp^{-(\gamma - \dot{\gamma})/\phi_{x^2}}, \tag{2.28}$$

where C_∞ is a nonuniversal constant. Another prediction of the scaling hypothesis concerns the shift in T_c from zero temperature [4]

$$T_c(g_\perp) \sim T_x \sim g_\perp^{1/\phi_{x^2}} \tag{2.29}$$

When the system undergoes its two-particle dimensionality crossover, the temperature dependence of singularities near T_c is governed by the correlation length

$$\dot{\vec{\xi}} = \dot{\vec{\xi}}_0 \dot{t}^{-\dot{\nu}}, \tag{2.30}$$

where $\dot{\vec{\xi}}_0$ is the coherence length at $T \sim T_x$, which fixes a short anisotropic length scale for the onset of critical fluctuations described by the exponent $\dot{\nu}$. At T_c, one has the critical form

$$\chi(\mathbf{q}) \approx \frac{\dot{D}}{\left(\dot{\xi}_{0\|}^2 q^2 + \dot{\xi}_{0\perp}^2 q_\perp^2\right)^{(2-\dot{\eta})/2}}, \tag{2.31}$$

which according to (2.25) leads to the Fisher relation

$$\dot{\gamma} = (2 - \dot{\eta})\dot{\nu}. \tag{2.32}$$

3 Free Fermion Limit

Before facing the task of calculating the exponents and various scaling quantities of interacting fermion models, it is instructive first to analyze the properties of the free fermion gas model. We will consider first the purely 1D system and then introduce interchain single-particle hopping and see how they actually satisfy the above scaling relations.

3.1 One Dimension

Let us consider a one-dimensional system of free fermions coupled to source pair fields h for which the partition function is given by a functional integral over the anticommuting (Grassmann) fields ψ (see [19, 43] for reviews on Grassmann fields)

$$\begin{aligned} Z_0[h^*, h] &= \iint \mathcal{D}\psi^* \mathcal{D}\psi e^{S_0[\psi^*,\psi] + S_h[\psi^*,\psi,h^*,h]} \\ &= \iint \mathcal{D}\psi^* \mathcal{D}\psi \exp\Big\{ \sum_{\tilde{k},\alpha} [G_p^0(\tilde{k})]^{-1} \psi^*_{p,\alpha}(\tilde{k}) \psi_{p,\alpha}(\tilde{k}) \\ &\qquad + \sum_{\mu,\tilde{q}} [O^*_\mu(\tilde{q}) h_\mu(\tilde{q}) + \text{c.c}] \Big\}. \end{aligned} \tag{3.1}$$

Here S_0 is the action of the system in the absence of fields and

$$G_p^0(\tilde{k}) = -\langle \psi(\tilde{k})\psi^*(\tilde{k})\rangle_0 \\ = [i\omega_n - \epsilon_p(k)]^{-1} \tag{3.2}$$

is the free fermion propagator in the Fourier-Matsubara space where $\tilde{k} = (k, \omega_n)$. The fermion spectrum is linearized with respect to the Fermi points $\pm k_{\mathrm{F}}$, namely,

$$\epsilon_p(k) = v_{\mathrm{F}}(pk - k_{\mathrm{F}}), \tag{3.3}$$

where $p = \pm$ refers to right (+) and left (−) going fermions and v_{F} is the Fermi velocity. A natural cutoff on the bandwidth can be imposed on this continuum model by restricting the summations on the wave vector k of branch p in the interval $pk_{\mathrm{F}} + k_0 > k > pk_{\mathrm{F}} - k_0$. The cutoff $k_0 \sim 1/a$, where a is the lattice constant on the band wave vector, then leads to a finite bandwidth $E_0 = 2v_{\mathrm{F}}k_0 \equiv 2E_{\mathrm{F}}$ which we will take equal to twice the Fermi energy $E_{\mathrm{F}} = v_{\mathrm{F}}k_{\mathrm{F}}$.

In one dimension, the fermion gas is unstable to the formation of pair correlations in both Cooper and Peierls channels. This shows up as infrared singularities for pair susceptibilities, which can be calculated by adding S_h to the action, which linearly couples source fields to fermion pairs. In the Peierls channel ($\mu = \mu_P$), the particle-hole fields

$$O_{\mu_P=0}(\tilde{q}) = \sqrt{\frac{T}{L}} \sum_{\tilde{k},\alpha} \psi^*_{-,\alpha}(\tilde{k} - \tilde{q})\psi_{+,\alpha}(\tilde{k}) \tag{3.4}$$

and

$$O_{\mu_P\neq0}(\tilde{q}) = \sqrt{\frac{T}{L}} \sum_{\tilde{k},\alpha,\beta} \psi^*_{-,\alpha}(\tilde{k} - \tilde{q})\sigma^{\alpha\beta}_{\mu_P}\psi_{+,\beta}(\tilde{k}) \tag{3.5}$$

correspond to charge-density-wave (CDW: $\mu_P = 0$) and spin-density-wave (SDW: $\mu_P \neq 0$) correlations respectively near the wave vector $2k_{\mathrm{F}}$, where $\tilde{q} = (2k_{\mathrm{F}} + q, \omega_m)$, $\sigma_{\mu_P=1,2,3}$ are the Pauli matrices and L is the length of the system. In the Cooper channel ($\mu = \mu_C$), the particle-particle fields

$$O_{\mu_C=0}(\tilde{q}) = \sqrt{\frac{T}{L}} \sum_{\tilde{k},\alpha} \alpha\psi_{-,-\alpha}(-\tilde{k} + \tilde{q})\psi_{+,\alpha}(\tilde{k}) \tag{3.6}$$

and

$$O_{\mu_C\neq0}(\tilde{q}) = \sqrt{\frac{T}{L}} \sum_{\tilde{k},\alpha,\beta} \psi_{-,-\alpha}(-\tilde{k} + \tilde{q})\sigma^{\alpha\beta}_{\mu}\psi_{+,\beta}(\tilde{k}), \tag{3.7}$$

correspond to singlet (SS: $\mu_C = 0$) and triplet (TS: $\mu_C \neq 0$,) superconducting channels respectively, where $\tilde{q} = (q, \omega_m)$. The susceptibilities in

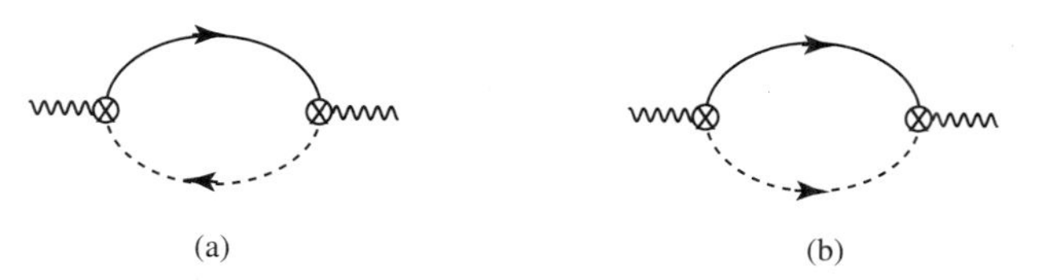

FIGURE 3.2. Diagrams for the bare susceptibilities in the Peierls (a) and Cooper (b) channels. A continuous (dashed) line correspond to a particle or hole near $+k_{\mathrm{F}}$ $(-k_{\mathrm{F}})$.

zero h_μ in both channels are defined by

$$\begin{aligned}\chi_\mu(\tilde{q}) &= -\frac{1}{Z_0}\frac{\delta^2 Z_0[h^*,h]}{\delta h_\mu^*(\tilde{q})\delta h_\mu(\tilde{q})}\bigg|_{h=0} \\ &= -\langle O_\mu(\tilde{q})O_\mu^*(\tilde{q})\rangle_0,\end{aligned} \tag{3.8}$$

and obtained from the calculation of the statistical averages

$$\chi_\mu(\tilde{q}) = -Z_0^{-1}\iint \mathcal{D}\psi^*\mathcal{D}\psi O_\mu(\tilde{q})O_\mu^*(\tilde{q})e^{S_0[\psi^*,\psi]}, \tag{3.9}$$

in the Peierls and Cooper channels. Thus in the Peierls channel, the dynamic susceptibility (Fig. 3.2a) close to $2k_{\mathrm{F}}$ is given by the standard result

$$\begin{aligned}\chi_{\mu_P}(\tilde{q}) &= 2\frac{T}{L}\sum_{\tilde{k}} G_-^0(k-2k_{\mathrm{F}}-q,\omega_n-\omega_m)G_+^0(k,\omega_n) \\ &= -N(0)\bigg\{\ln\frac{1.13E_{\mathrm{F}}}{T}+\psi\Big(\frac{1}{2}\Big) \\ &\qquad -\frac{1}{2}\Big[\psi\Big(\frac{1}{2}+\frac{iv_{\mathrm{F}}q-\omega_m}{4\pi T}\Big)+\mathrm{c.c}\Big]\bigg\}\end{aligned} \tag{3.10}$$

for all density-wave components μ at low temperature. Here $N(0)=1/(\pi v_{\mathrm{F}})$ is the density of states per spin at the Fermi level and $\psi(x)$ is the digamma function. In the evaluation of the Peierls bubble, the electron-hole symmetry

$$\epsilon_+(k) = -\epsilon_-(k-2k_{\mathrm{F}}), \tag{3.11}$$

of the fermion spectrum—called nesting—has been used in the evaluation of the Peierls bubble. The elementary susceptibility in the Cooper channel (Fig. 3.2b) has a similar behavior; that is

$$\begin{aligned}\chi_{\mu_C}(\tilde{q}) &= -2\frac{T}{L}\sum_{\tilde{k}} G_-^0(-k+q,-\omega_n+\omega_m)G_+^0(k,\omega_n) \\ &= -N(0)\bigg\{\ln\frac{1.13E_{\mathrm{F}}}{T}+\psi\Big(\frac{1}{2}\Big) \\ &\qquad -\frac{1}{2}\Big[\psi\Big(\frac{1}{2}+\frac{iv_{\mathrm{F}}q-\omega_m}{4\pi T}\Big)+\mathrm{c.c}\Big]\bigg\}\end{aligned} \tag{3.12}$$

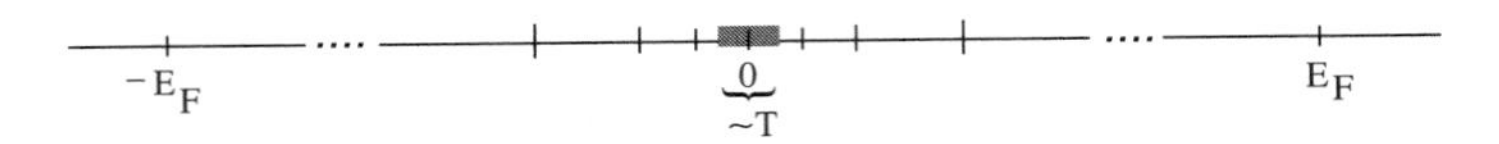

FIGURE 3.3. Energy scales for the 1D fermion problem. Between the Fermi (E_F) and thermal energy (T), there is no characteristic energy scale.

for which the inversion symmetry (which is independent of the spatial dimension)

$$\epsilon_+(k) = \epsilon_-(-k) \tag{3.13}$$

of the spectrum has been used.

In the static limit, both channels show logarithmic singularities of the form $\chi_\mu(T) \sim \ln(E_F/T)$, indicating that pair correlations have no characteristic energy scale between E_F and T. All intermediate scales between E_F and T turn out to give similar contributions to the integrals (3.10) and (3.12), which are roughly of the form $\int_T^{E_F} d\epsilon_+/\epsilon_+$ at $q = 0$ and $\omega_m = 0$ as shown in Fig. 3.3, a feature that can be equated with scaling [1].

This logarithmic divergence can also be seen as a power law with a vanishing exponent; that is

$$\chi_\mu(T) = \lim_{\gamma \to 0^+} -N(0)\gamma^{-1}\left[\left(\frac{T}{1.13E_F}\right)^{-\gamma} - 1\right]. \tag{3.14}$$

According to the relation (2.18), this should imply $\eta \to 2$, namely, a logarithmic singularity in q. This is indeed consistent with the asymptotic behavior of $\chi_\mu(q)$ at $T = 0$ and $\omega_m = 0$, which is found to be

$$\chi_\mu(q) = N(0) \ln \frac{|v_F q|}{2E_F}, \tag{3.15}$$

in agreement with (2.15).

The origin of scale invariance can be further sharpened if one looks at the decay of the fermion coherence at large distance. Consider the Green's function

$$\begin{aligned} G_p^0(x,\tau) &= \frac{T}{L} \sum_{k,\omega_n} G_p^0(k,\omega) e^{ikx - i\omega_n \tau} \\ &= \frac{e^{ipk_F x}}{2\pi i} \frac{1}{\xi \sinh[(x + ipv_F\tau)/\xi]} \end{aligned} \tag{3.16}$$

where

$$\xi = \frac{v_F}{\pi T} \tag{3.17}$$

is a characteristic length scale for the quantum coherence of each fermion and in fact the de Broglie quantum length. At low temperature for $a \ll x \ll \xi$, one finds the power law decay of coherence as a function of distance

$$G_p^0(x) \sim 1/x, \tag{3.18}$$

which signals homogeneity or scaling. Reverting to (2.6), we verify that $\theta = 0$ in the free fermion limit. For $x \gg \xi$,

$$G_p^0(x) \sim e^{-x/\xi} \tag{3.19}$$

there is no homogeneity and the coherence is exponentially damped with distance. The absence of energy scale between $E_{\rm F}$ and T thus corresponds in real space to the absence of length scale between ξ and a. Rewriting the elementary the Cooper and Peierls susceptibilities as

$$\chi_\mu(T) \approx -N(0) \ln \frac{\xi}{a}, \tag{3.20}$$

one then sees that the quantum spatial coherence of particles becomes an essential ingredient in building up logarithmic singularities in pair correlations of the Cooper and Peierls channels. This can be directly verified by the decay of pair correlations in space and time. From the definition (2.13) and the use of Wick theorem for free fermions, one obtains

$$\begin{aligned} \chi_\mu(x,\tau) &= \mp 2 G_-^0(\mp x, \mp\tau) G_+^0(-x,-\tau) \\ &= e^{-iq_0 x} \frac{1}{2\pi^2\xi^2} \frac{1}{\sinh[(x + iv_{\rm F}\tau)/\xi] \sinh[(x - iv_{\rm F}\tau)/\xi]}, \end{aligned} \tag{3.21}$$

where the upper (lower) sign corresponds to the Cooper (Peierls) case. For $a \ll x \ll \xi$, one finds

$$\chi_\mu(x) \propto \frac{e^{-iq_0 x}}{x^2}, \tag{3.22}$$

in agreement with (2.14), in which $\bar{d} = 2$ and $\eta = 2$. At large distance, when $x \gg \xi$, one has

$$\chi_\mu(x) \propto e^{-iq_0 x} e^{-2x/\xi}, \tag{3.23}$$

where the effective coherence length for pairs is half the one of a single fermion.

3.2 Interchain Coupling

A very simple illustration of a dimensionality crossover for the single-particle coherence can be readily given in the free fermion case by adding an interchain hopping term to the the free action, which becomes

$$\begin{aligned} S_0[\psi^*,\psi] &= S_0[\psi^*,\psi]_{1D} + \sum_{\tilde{k},p,\alpha} \sum_{\langle i,j \rangle} t_\perp \psi^*_{p,\alpha,i}(\tilde{k}) \psi_{p,\alpha,j}(\tilde{k}) \\ &= \sum_{\mathbf{k},\omega_n} \sum_{p,\alpha} [G_p^0(\mathbf{k},\omega_n)]^{-1} \psi^*_{p,\alpha}(\mathbf{k},\omega_n) \psi_{p,\alpha}(\mathbf{k},\omega_n) \end{aligned} \tag{3.24}$$

where $t_\perp$ is the single-particle hopping integral between nearest-neighbor chains i and j. Considering a linear array of $N_\perp$ chains, the free propagator that parameterizes S_0 in Fourier is

$$G_p^0(\mathbf{k}, \omega_n) = \big(i\omega_n - E_p(\mathbf{k})\big)^{-1}, \tag{3.25}$$

where

$$E_p(\mathbf{k}) = \epsilon_p(k) - 2t_\perp \cos k_\perp \tag{3.26}$$

is the full fermion spectrum. The propagator can be trivially written in the extended scaling scaling form (2.19) with $\bar{\gamma} = 1$, where the scaling function takes the simple form

$$X_G(y) = (1 + y)^{-1}, \tag{3.27}$$

with $y = Bt_\perp \cos(k_\perp)/T$ and $B = (i\pi/2)^{-1}$. This implies $\phi_{x^1} = 1$ for the single-particle crossover exponent. This value of ϕ_{x^1} is found in the anisotropic free field theory of critical phenomena [4]. The Fourier transform leads to single-particle fermion correlation function

$$G_p^0(x, n_\perp, \tau) = G_p^0(x, \tau) e^{-i\pi n_\perp} J_{n_\perp}(x/\xi_x), \tag{3.28}$$

where $n_\perp$ is the transverse separation expressed in number of chains. The crossover scaling function is then embodied in the Bessel function $J_{n_\perp}(x/\xi_x)$ of order $n_\perp$, which takes sizable values when the intrachain distance x reaches values of the order of the characteristic length scale $\xi_x = v_\mathrm{F}/(2t_\perp)$ needed for coherent hopping from one chain to another (Fig. 3.1). For $x \sim \xi$, the crossover condition is $\xi \sim \xi_x$, and this leads once again to $T_{x^1} \sim t_\perp$, namely, $\phi_{x^1} = 1$.

4 The Kadanoff–Wilson Renormalization Group

4.1 One-Dimensional Case

When the motion of fermions is entirely confined to one spatial dimension, particles cannot avoid each other and hence the influence of interactions is particularly important. The absence of length scale up to ξ found in the free fermion gas problem for both single and pair correlations will carry over in the presence of interactions, which couple different energy scales. The low-energy behavior of the system in which we are interested will be influenced by all the higher energy states up to the highest energy at the band edge. Renormalization group ideas can be used to tackle the problem of the cascade of energy or length scales that characterizes interacting fermions. This will be done in the framework of the fermion gas model in which the direct interaction between fermions is parametrized by a small set of distinct scattering processes close to the Fermi level (Fig. 3.4) [9]. This is justified given the particular importance shown by the Cooper and Peierls

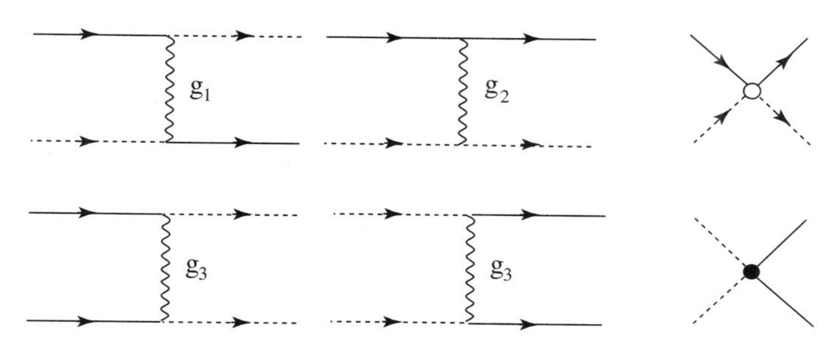

FIGURE 3.4. Backward (g_1), forward (g_2), and Umklapp (g_3) couplings of the 1-D fermion gas model and the corresponding diagrams. The open (full) circle corresponds to the generic vertex part for backward and forward (Umklapp) scatterings.

infrared singularities for electronic states near the Fermi level. We will consider the backward (g_1) and forward (g_2) scattering processes for which particles on opposite sides of the Fermi level exchange momentum near $2k_{\mathrm{F}}$ and zero, respectively. When the band is half-filled, $4k_{\mathrm{F}}$ coincides with a reciprocal lattice vector, making possible Umklapp processes, denoted by g_3, in which two particles can be transferred from one side of the Fermi level to the other. Thus, adding the interaction part S_I to the action, the partition function in the absence of source fields takes the form

$$
\begin{aligned}
Z &= \iint \mathcal{D}\psi^* \mathcal{D}\psi \; e^{S_0[\psi^*,\psi]+S_I[\psi^*,\psi]} \\
&= \iint \mathcal{D}\psi^* \mathcal{D}\psi \exp\Big\{ \sum_{\tilde{k},\alpha} [G_p^0(\tilde{k})]^{-1} \psi^*_{p,\alpha}(\tilde{k}) \psi_{p,\alpha}(\tilde{k}) \\
&\quad + \pi v_{\mathrm{F}} \frac{T}{L} \sum_{\{p,\tilde{k},\tilde{q},\alpha\}} g_{\{\alpha\}} \psi^*_{+,\alpha_1}(\tilde{k}_1+\tilde{q}) \psi^*_{-,\alpha_2}(\tilde{k}_2-\tilde{q}) \psi_{-,\alpha_3}(\tilde{k}_2) \psi_{+,\alpha_4}(\tilde{k}_1) \\
&\quad - \pi v_{\mathrm{F}} \frac{T}{2L} \sum_{\{p,\tilde{k},\tilde{q},\alpha\}} g_3 \psi^*_{p,\alpha}(\tilde{k}_1+p\tilde{q}) \psi^*_{-p,\alpha'}(\tilde{k}_1-p\tilde{q}+p\tilde{G}) \psi_{-p,\alpha'}(\tilde{k}_2) \psi_{p,\alpha}(\tilde{k}_1) \Big\},
\end{aligned}
\tag{4.1}
$$

where

$$
g_{\{\alpha\}} = g_1 \delta_{\alpha_1\alpha_3} \delta_{\alpha_2\alpha_4} - g_2 \delta_{\alpha_1\alpha_4} \delta_{\alpha_2\alpha_3} \tag{4.2}
$$

and $\tilde{G} = (4k_{\mathrm{F}}, 0)$ is the reciprocal lattice vector. The parameter space of the action is then

$$
\mu_S = (G_p^0, g_1, g_2, g_3), \tag{4.3}
$$

where all the couplings are expressed in units of πv_{F}.

The idea behind the Kadanoff–Wilson renormalization group is the transformation or renormalization of the action S following successive partial integration of $\bar{\psi}$'s of momentum located in the energy shell $\frac{1}{2}E_0(\ell)d\ell$ on both sides of the Fermi level, where $E_0(\ell) = E_0 e^{-\ell}$ is the effective bandwidth at step ℓ of integration and $d\ell \ll 1$. The transformation of S from ℓ to $\ell + d\ell$ that keeps the partition function invariant is usually written as

$$\begin{aligned} Z &= e^{A(\ell)} \iint_< \mathcal{D}\psi^* \mathcal{D}\psi e^{S[\psi^*,\psi]_<} \iint \mathcal{D}\bar{\psi}^* \mathcal{D}\bar{\psi} e^{S[\psi^*,\psi,\bar{\psi}^*,\bar{\psi}]_{d\ell}} \\ &= e^{A(\ell+d\ell)} \iint_< \mathcal{D}\psi^* \mathcal{D}\psi e^{S[\psi^*,\psi]_{\ell+d\ell}}, \end{aligned} \tag{4.4}$$

where $A(\ell)$ corresponds to the free energy density at the step ℓ. The integration measure in the outer shell is

$$\mathcal{D}\bar{\psi}^* \mathcal{D}\bar{\psi} = \prod_{p,\alpha,\{\tilde{k}\}'} d\bar{\psi}^*_{p,\alpha}(\tilde{k})\, d\bar{\psi}_{p,\alpha}(\tilde{k}). \tag{4.5}$$

and $\{\tilde{k}\}' = \{k\}'\{\omega_n\}$. Here $\{k\}'$ corresponds to the momentum outer shells

$$k_0 e^{-\ell} + k_{\rm F} > k > k_0 e^{-\ell-d\ell} + k_{\rm F} \qquad (-k_0 e^{-\ell-d\ell} - k_{\rm F} > k > -k_0 e^{-\ell} - k_{\rm F})$$

above the Fermi level and

$$-k_0 e^{-\ell-d\ell} + k_{\rm F} > k > -k_0 e^{-\ell} + k_{\rm F} \qquad (-k_0 e^{-\ell} - k_{\rm F} > k > -k_0 e^{-\ell-d\ell} - k_{\rm F})$$

below for right (left) moving fermions, while $\{\omega_n\}$ covers all the Matsubara frequencies. The remaining inner shell ($<$) fermion degrees of freedom are kept fixed. The parameters of the action denoted as μ_S transform according to

$$R_{d\ell}\mu_S(\ell) = \mu_S(\ell + d\ell), \tag{4.6}$$

where

$$\mu_S(\ell) = \big(z(\ell)G_p^0, z_1(\ell)g_1, z_2(\ell)g_2, z_3(\ell)g_3\big) \tag{4.7}$$

contains the factors z and $z_{i=1,2,3}$ that renormalize the parameter space of the action at ℓ.[1] The flow of parameters is conducted down to $\ell_T \equiv \ln E_{\rm F}/T$ corresponding to the highest value of ℓ at which scaling applies as discussed in Section 2. This leads in turn to the temperature dependence of the parameter space $\mu_S(T)$. It should be mentioned that the Kadanoff transformation will generate new terms which were not present in the action at $\ell = 0$. In principle, the relevance of these can be seen explicitly from a rescaling of energy and field "density" after each partial integration. Thus, performing an energy change by the factor $s = e^{d\ell} > 1$, one has

$$\begin{aligned} \epsilon'_p &= s\epsilon_p \\ \omega'_n &= s\omega_n. \end{aligned} \tag{4.8}$$

[1] The notation used here for the renormalization factors of μ_S differs from the one of [15,16], from which the correspondance $z \leftrightarrow z_1^{-1}$, $z_1 \leftrightarrow z_2$, $z_2 \leftrightarrow z_3$, and $z_3 \leftrightarrow z_4$ can be established.

The fermion fields in the action will then transform according to

$$\psi'^{(*)} = z^{-1/2} s^{-1/2} \psi^{(*)}, \tag{4.9}$$

which in its turn implies that the rescaling of interaction parameters is given by

$$g_i' = g_i z_i z^2, \quad \text{for } i = 1, 2, 3. \tag{4.10}$$

This contains only anomalous renormalization factors, indicating that the interaction parameters are all marginal. However, if interactions between three, four, or more particles were added to the *bare* action, the amplitude of these would decay as $1/s$ to some positive power, and would therefore be irrelevant in the RG sense. A different conclusion would be reached, however, if these many-body interactions were generated in the flow of renormalization. In this case, the amplitudes of interactions becomes scale dependent and such couplings can remain marginal or even become relevant (see Sections 4.3 and 5.1). Besides this use, rescaling for our fermion problem is not an essential step of the RG transformation. We will consider the renormalization of μ_S with respect to ℓ as a transformation that describes an effective system with a reduced band width $E_0(\ell)$ or a magnified length scale $v_F/E_0(\ell)$.

In practice, the partial integration (4.4) is most easily performed with the aid of diagrams. We first decompose the $\tilde{k}$ sums in the action into outer and inner shells momentum variables

$$\sum_{\{\tilde{k}\}} = \sum_{\{\tilde{k}\}'} + \sum_{\{\tilde{k}\}_<} . \tag{4.11}$$

This allows us to write

$$S[\psi^*, \psi] = S[\psi^*, \psi]_< + S[\psi^*, \psi, \bar{\psi}^*, \bar{\psi}] \tag{4.12}$$

where $S[\psi^*, \psi]_<$ of the action with all the ψ's in the inner shell, whereas

$$S[\psi^*, \psi, \bar{\psi}^*, \bar{\psi}] = S_0[\bar{\psi}^*, \bar{\psi}] + \sum_{i=1}^{4} S_{I,i}[\psi^*, \psi, \bar{\psi}^*, \bar{\psi}] \tag{4.13}$$

consists in a free part in the outer momentum shell and an interacting part as a sum of terms, $S_{I,i}$, having $i = 1, \ldots, 4$ $\bar{\psi}$'s in the outer momentum shell. Their integration following the partial trace operation (4.4) is performed perturbatively with respect to $S_0[\bar{\psi}^*, \bar{\psi}]$. Making use of the linked cluster theorem, the outer shell integration becomes

$$
\begin{aligned}
Z &= e^{A(\ell)} \iint_< \mathcal{D}\psi^* \, \mathcal{D}\psi \, e^{S[\psi^*,\psi]_<} \iint \mathcal{D}\bar{\psi}^* \, \mathcal{D}\bar{\psi} \, e^{S_0[\bar{\psi}^*,\bar{\psi}]} \\
&\qquad \times \exp\Big\{ \sum_{i=1}^{4} S_{I,i}[\psi^*,\psi,\bar{\psi}^*,\bar{\psi}] \Big\} \\
&= e^{A(\ell)} \iint_< \mathcal{D}\psi^* \, \mathcal{D}\psi \, \exp\Big\{ S[\psi^*,\psi]_< \\
&\qquad + \sum_{n=1}^{\infty} \frac{1}{n!} \Big\langle \Big(\sum_i S_{I,i}[\psi^*,\psi,\bar{\psi}^*,\bar{\psi}] \Big)^n \Big\rangle_{\bar{0},c} \Big\}
\end{aligned} \tag{4.14}
$$

where

$$
\langle \ldots \rangle_{\bar{0},c} = Z_{\bar{0}}^{-1} \iint \mathcal{D}\bar{\psi}^* \, \mathcal{D}\bar{\psi} \, (\ldots) e^{S_0[\bar{\psi}^*,\bar{\psi}]} \tag{4.15}
$$

is a free fermion average corresponding to a connected diagram evaluated in the outer momentum shell and

$$
Z_{\bar{0}} = \iint \mathcal{D}\bar{\psi}^* \, \mathcal{D}\bar{\psi} \, e^{S_0[\bar{\psi}^*,\bar{\psi}]}, \tag{4.16}
$$

is the outer shell contribution to the free partition function.

4.2 One-Loop Results

The KW method may be used to obtain low-order results presented in previous review papers [15, 16], and no particular technical difficulties arise. The following two subsections present material which has already been developed using the parquet summation [8, 9] and the first-order multiplicative renormalization group [10, 11, 44]. Nevertheless, it is useful to supply the basic properties of the logarithmic theory of the 1D fermion gas model using this method.

4.2.1 Incommensurate Band Filling

For band filling that is incommensurate, there is no possibility of Umklapp scattering so that one can drop g_3 in the action. The renormalization of backward and forward scattering amplitudes at the one-loop level are obtained from the $n = 2$ outer-shell averages $\frac{1}{2}\langle (S_{I,2})^2 \rangle$, where

$$
\begin{aligned}
S_{I,2}[\psi^*,\psi,\bar{\psi}^*,\bar{\psi}] &= S^C_{I,2} + S^P_{I,2} + S^L_{I,2} \\
&\Longleftrightarrow (\bar{\psi}^*_+ \bar{\psi}^*_- \psi_- \psi_+ + \text{c.c}) + (\bar{\psi}^*_+ \psi^*_- \bar{\psi}_- \psi_+ + \text{c.c}) \\
&\qquad + (\bar{\psi}^*_+ \psi^*_- \psi_- \bar{\psi}_+ + \psi^*_+ \bar{\psi}^*_- \bar{\psi}_- \psi_+),
\end{aligned} \tag{4.17}
$$

the three terms of which correspond to putting two particles or two holes in the outer momentum shell (the Cooper channel), one particle and one hole on opposite branches (the Peierls channel), and finally putting a particle and a hole on the same branch (the Landau channel). The latter does not lead to logarithmic contributions at the one-loop level and can be ignored for the moment. The flow of the scattering amplitudes becomes $g_{i=1,2}(\ell + d\ell) = z_{i=1,2}(d\ell)g_{i=1,2}(\ell)$ where the renormalization factors at the one-loop level are

$$\begin{aligned} z_1(d\ell) &= 1 - 2g_1(\ell)I_P(d\ell) + 2g_2(\ell)[I_P(d\ell) + I_C(d\ell)] \\ z_2(d\ell) &= 1 + g_2^{-1}(\ell)g_1^2(\ell)I_C(d\ell) + g_2(\ell)[I_P(d\ell) + I_C(d\ell)]. \end{aligned} \tag{4.18}$$

The outer shell integration of the Peierls channel at $2k_{\mathrm{F}}$ and zero external frequency enables the particle and hole to be simultaneously in the outer-shell. The result is

$$\begin{aligned} I_P(d\ell) &= -\pi v_{\mathrm{F}} \frac{T}{L} \sum_{\{\tilde{k}\}'} G_+^0(k,\omega_n) G_-^0(k - 2k_{\mathrm{F}}, \omega_n) \\ &= \left\{ \int_{-E_0(\ell)/2}^{-E_0(\ell+d\ell)/2} + \int_{E_0(\ell+d\ell)/2}^{E_0(\ell)/2} \right\} \frac{\tanh\left(\frac{1}{4}\beta E_0(\ell)\right)}{2E_0(\ell)} \, dE_0(\ell) \\ &\simeq \tfrac{1}{2} d\ell. \end{aligned} \tag{4.19}$$

Similarly in the Cooper channel at zero pair momentum and external frequency, we have

$$\begin{aligned} I_C(d\ell) &= \pi v_{\mathrm{F}} \frac{T}{L} \sum_{\{\tilde{k}\}'} G_+^0(k,\omega_n) G_-^0(-k, -\omega_n) \\ &= -I_P(d\ell). \end{aligned} \tag{4.20}$$

Both contributions are logarithmic and their opposite signs lead to important cancellations (interference). The diagrams which remain after the cancellation are shown in Fig. 3.5.

With an obvious rearrangement, the flow equations may be written

$$\begin{aligned} \frac{dg_1}{d\ell} &= -g_1^2 \\ \frac{d}{d\ell}(2g_2 - g_1) &= 0. \end{aligned} \tag{4.21}$$

The solution for $g_1(\ell)$ is found at once to be

$$g_1(\ell) = \frac{g_1}{1 + g_1 \ell}, \tag{4.22}$$

showing that backward scattering is marginally irrelevant (relevant) if the bare coupling g_1 is repulsive (attractive). As for the combination $2g_2 - g_1$, it remains marginal, an invariance that reflects the conservation of particles on each branch [45, 46]. Another feature of these equations is the fact that

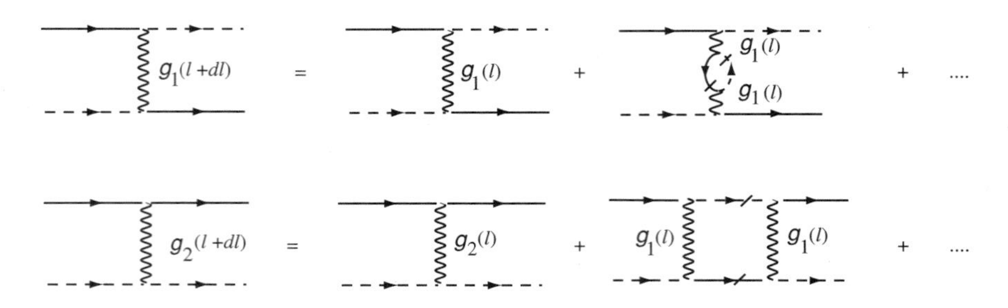

FIGURE 3.5. Recursion relations for backward (g_1) and forward (g_2) scattering at the one-loop level. The slashed lines refer to a particle or a hole in the outer shell.

the flow of g_1 is entirely uncoupled from the one of $2g_2 - g_1$. This feature, which follows from the interference between the Cooper and the Peierls channels, gives rise to an important property of 1D interacting fermion systems, namely, the separation of long wavelength spin and charge degrees of freedom. This is rendered manifest by rewriting the interacting part of the action as follows [9]:

$$S_I[\psi^*, \psi]_\ell = -\pi v_{\mathrm{F}}(2g_2 - g_1) \sum_{p,\tilde{q}} \rho_p(\tilde{q})\rho_{-p}(-\tilde{q}) + \pi v_{\mathrm{F}} g_1(\ell) \sum_{p,\tilde{q}} \mathbf{S}_p(\tilde{q}) \cdot \mathbf{S}_{-p}(-\tilde{q}), \quad (4.23)$$

where the long-wave length particle-density and spin-density fields of branch p are defined by

$$\begin{aligned} \rho_p(\tilde{q}) &= \frac{1}{2}\sqrt{\frac{T}{L}} \sum_{\alpha,\{\tilde{k}\}_<} \psi^*_{p,\alpha}(\tilde{k}+\tilde{q})\psi_{p,\alpha}(\tilde{k}) \\ \mathbf{S}_p(\tilde{q}) &= \frac{1}{2}\sqrt{\frac{T}{L}} \sum_{\alpha,\{\tilde{k}\}_<} \psi^*_{p,\alpha}(\tilde{k}+\tilde{q})\vec{\sigma}^{\alpha\beta}\,\psi_{p,\beta}(\tilde{k}). \end{aligned} \quad (4.24)$$

The spin-charge separation is preserved at higher order and is a key property of a Luttinger liquid in one dimension [47]. In the repulsive sector $g_1 \geq 0$, the spin coupling goes to zero, while the forward scattering $g_2^*(\ell \to \infty) \to g_2 - g_1/2$ flows to a nonuniversal value on the Tomanaga–Luttinger line of fixed points $g_1^* = 0$ (Fig. 3.6).

For an attractive interaction $g_1 < 0$, an instability occurs in the the spin sector at ℓ_σ defined by $g_1\ell_\sigma = -1$. Identifying ℓ with $\ell_T = \ln E_{\mathrm{F}}/T$, the temperature dependence of the Peierls or Cooper loop, the characteristic temperature scale for strong coupling is

$$T_\sigma = E_{\mathrm{F}} e^{-1/|g_1|} \equiv \tfrac{1}{2}\Delta_\sigma, \quad (4.25)$$

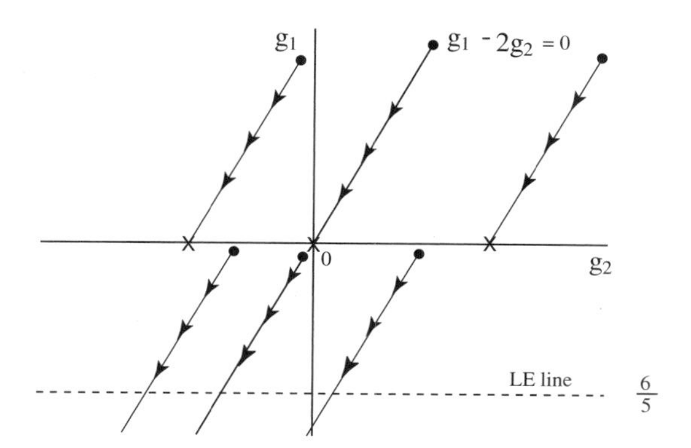

FIGURE 3.6. Flow of coupling constants in the $g_1 g_2$ plane at the one-loop level. In the repulsive region $g_1 \geq 0$, the system has a line of fixed points on the Tomonaga–Luttinger line $g_1^* = 0$. In the attractive sector $g_1 < 0$, the flows is to strong coupling. The dashed line corresponds to the Luther–Emery line at $g_1 = -\frac{6}{5}$.

where Δ_σ is defined as the spin gap. According to (4.23), backward scattering can be equated with an exchange interaction between right- and left-going spin densities which is antiferromagnetic in character when g_1 is attractive. Therefore strong attractive coupling signals the presence of a spin gap Δ_σ. Although the divergence in (4.22) is an artifact of the one-loop approximation, the flow of $g_1(\ell)$ to strong coupling is nevertheless preserved at higher order. This is shown by the fact that the system inevitably crosses the so-called Luther–Emery line at $g_1(\ell_{LE}) = -\frac{6}{5}$ (Fig. 3.6), where an exact solution using the technique of bosonization confirms the existence of a spin gap [48, 49].

4.2.2 Umklapp Scattering at Half-Filling

At half-filling, there is one fermion per site so that $4k_{\mathrm{F}} = 2\pi/a$ coincides with a reciprocal lattice vector G enabling Umklapp scattering processes denoted by g_3 in Fig. 3.4. The presence of g_3 processes reinforces the importance of the $2k_{\mathrm{F}}$ density-wave channel compared to the superconducting one. Flows will be modified accordingly. Thus at the one-loop level, the outer shell corrections coming from $\frac{1}{2}\langle (S_{I,2})^2 \rangle_{\bar{0},c}$ will include additional contributions

$$S_{I,2} \iff S_{I,2}\big|_{g_3=0} + (\, \bar{\psi}_+^* \psi_+^* \bar{\psi}_- \psi_- + \text{perm.} + \text{c.c}) \\ + (\, \bar{\psi}_+^* \bar{\psi}_+^* \psi_- \psi_- + \psi_+^* \psi_+^* \bar{\psi}_- \bar{\psi}_- + \text{c.c}\,) \quad (4.26)$$

involving outer shell fields in the Peierls channel and pairs of particles or holes on the same branch. At the one-loop level only the former gives rise to logarithmic corrections, whereas the latter will be involved in higher order corrections (Section 4.3.2). The former contributions, along with (4.18), will lead to the recursion formulas $g_{i=1,2,3}(\ell+d\ell) = z_{i=1,2,3}(d\ell)g_{i=1,2,3}(\ell)$, which can be written in form

$$\begin{aligned} g_1(\ell+d\ell) &= g_1(\ell) - 2g_1^2(\ell)I_P(d\ell), \\ (2g_2-g_1)(\ell+d\ell) &= (2g_2-g_1)(\ell) + 2g_3^2(\ell)I_P(d\ell), \\ g_3(\ell+d\ell) &= g_3(\ell) + 2(2g_2-g_1)(\ell)g_3(\ell)I_P(d\ell). \end{aligned} \tag{4.27}$$

The flow of the coupling constants are then governed by

$$\begin{aligned} \frac{dg_1}{d\ell} &= -g_1^2, \\ \frac{d}{d\ell}(2g_2-g_1) &= g_3^2, \\ \frac{dg_3}{d\ell} &= g_3(2g_2-g_1), \end{aligned} \tag{4.28}$$

which corresponds to the diagrammatic parquet summation of Dzyaloshinskii and Larkin [9]. By combining the last two equations, the flow is found to follow the hyperbolas described by the renormalization invariant $C = [(2g_2-g_1)(\ell)]^2 - g_3^2(\ell)$. Umklapp scattering is then entirely uncoupled from g_1 and will only affect the charge sector. When $g_1 - 2g_2 > |g_3|$, $g_3(\ell)$ scales to zero and is marginally irrelevant, while $2g_2^* - g_1^*$ is nonuniversal. The excitation of charge degrees of freedom are then gapless in this case. The attractive Hubbard model in which $g_{1,2,3} = U < 0$ falls in this category.

Otherwise, the flows of g_3 and $2g_2 - g_1$ evolve to strong coupling and both couplings are marginally relevant. There is a singularity at ℓ_ρ corresponding to the temperature scale

$$T_\rho = E_{\mathrm{F}} e^{-1/\sqrt{|C|}} \equiv \tfrac{1}{2}\Delta_\rho. \tag{4.29}$$

Following the example of the incommensurate case, the couplings inevitably cross the Luther–Emery line at $g_1 - 2g_2 = -\frac{6}{5}$, at which point the problem

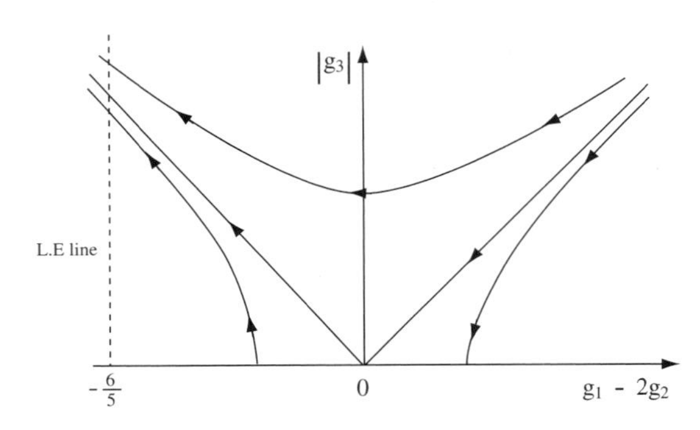

FIGURE 3.7. The flow of couplings in the charge sector. The condition $g_1 - 2g_2 > |g_3|$ corresponds to weak coupling where Umklapp scattering is a marginally irrelevant coupling. Otherwise, both $g_1 - 2g_2$ and g_3 are marginally relevant. The Luther–Emery line is located at $g_1 - 2g_2 = -\frac{6}{5}$.

can be solved exactly and the existence of a charge gap Δ_ρ is confirmed [48]. The special case of the repulsive Hubbard limit at half-filling where $g_{1,2,3} = U > 0$ falls on the separatrix $2g_2 - g_1 = |g_3|$ (Fig. 3.7). In this case the solution of (4.28) leads to a singularity at $T_\rho = E_F e^{-1/|g_3|}$. A charge gap Δ_ρ is consistent with the characteristics of a 1D Mott insulator found in the Lieb–Wu exact solution [9, 50, 51].

4.3 Two-Loop Results

At the one-loop level, corrections to the action parameters μ_S resulting from the partial trace are confined to the scattering amplitudes $g_{i=1,2,3}(\ell)$ and aside from a renormalization of the chemical potential, the flow leaves the single particle properties unchanged. At the two-loop level, outer shell logarithmic contributions will affect both the one-particle self-energy and the four-point vertices. This yields an essential feature of the 1D electron gas model which is the absence of quasi-particle states.

4.3.1 One-Particle Self-Energy

Let us consider first the one-particle part. It is easy to see that the term $\frac{1}{2}\langle(S_{I,3})^2\rangle_{\bar{0},c}$ at $n = 2$ has the form of a one-particle self-energy correction. However, if we strictly follow the classical Wilson scheme, the constraint of having *all* internal lines in the outer shell leads to a vanishing contribution. It becomes clear then that one-particle corrections can only be logarithmic at the two-loop level if the internal lines refer to different momentum shells. In the classical Wilson scheme, this possibility actually results from a cascade of contractions on the three-particle interaction (Fig. 3.8). Such interactions, although absent in the initial action, are generated along the RG flow and enlarge the parameter space.

In order to see how these interactions are generated from the partial trace (4.14) and under what conditions they give rise to logarithmic corrections at the two-loop level, it is useful to label the steps of the outer shell integration by the index j, for which $E_0(\ell_j) = E_0 e^{-(j-2)d\ell}$ is the scaled bandwidth at $j \geq 3$. Thus at each step $j - 2$, the contraction $\frac{1}{2}\langle S_{I,1}^2\rangle_{\bar{0},c}$ yields a three-particle term in the action (Fig. 3.8a), which is denoted $\delta S_\lambda^{(j)}$. Its explicit expression for the forward and backward scattering is given in Eq. (A.1) of Section A.1. The analogous expression $\delta S_{\lambda u}^{(j)}$ for Umklapp with $\lambda_u^\pm$ as the scattering amplitudes is given in (A.6)) of Section A.2. Owing to the presence of the internal line, the rescaling of energy and fields shows that the amplitudes $\lambda^\pm$ and $\lambda_u^\pm$ are marginal [24]. At the next step of partial integration $j - 1$, two fermion fields of these three-particle terms can in turn be contracted in $\langle\delta S_{\lambda,2}^{(j)}\rangle_{\bar{0},c}$, from which we will retain an effective two-particle interaction with two internal lines tied to the adjacent shells $j-2$ and $j-1$ (Fig. 3.8b and (A.2)). These contributions are nonlogarithmic and differ from those already retained in the one-loop

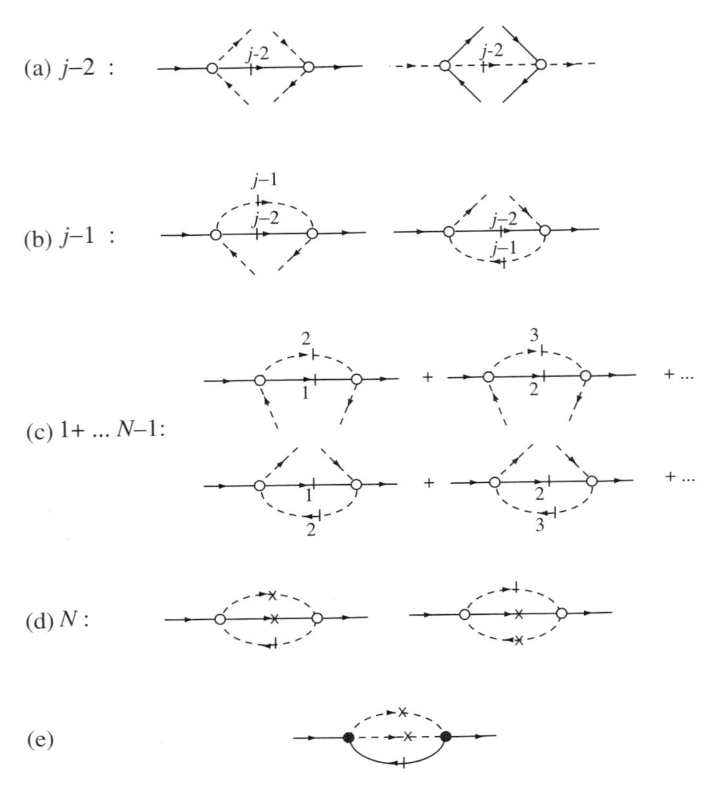

FIGURE 3.8. The cascade of contractions contributing to logarithmic one-particle self-energy corrections in the local approximation for the RG flow: (a) backward and forward scattering contributions to the three-particle interactions; λ^+ (left) and λ^- (right) from $\frac{1}{2}\langle S^2_{I,1}\rangle_{\bar{0},c}$ at the step $(j-2)$ where $j \geq 3$; (b) second contraction in the outer shell at the step $j-1$ with two internal lines on successive energy shells (only shown for λ^+); (c) the summation of diagrams given in (b) up to $j = N-1$; (d) outer shell one-particle self-energy correction at the step N ($Nd\ell = e^{\ell+d\ell}$) for the top inner shell state k of the $+k_{\rm F}$ branch in the local approximation—here a crossed (slashed) line refers to fermions in the interval $[0,\ell]$ ($[\ell,\ell+d\ell]$); (e) the result of a similar cascade of contractions for Umklapp scattering.

calculation in Section 4.2 for which both internal lines were put in the same outer shell. The repetition of these contractions is carried out by a sum on j up to $N-1$ (Fig. 3.8c), which yields $\delta S'_\lambda \equiv \sum_j^{N-1}\langle \delta S^{(j)}_{\lambda,2}\rangle_{\bar{0},c}$ ($\delta S'_{\lambda,u} \equiv \sum_j^{N-1}\langle \delta S^{(j)}_{\lambda,u,2}\rangle_{\bar{0},c}$ for Umklapp) as shown in (A.2) (Eq. A.7 for Umklapp). At the next step, N, of partial integration, two more fields are in turn contracted, which leads to a sum of one-particle self-energy corrections for a k state at the top of the inner shell.[2] Since each term of the sum refers to couplings at different steps j, the correction is nonlocal.

In the following, we will adopt a *local* scheme of approximation, which consists in neglecting the slow (logarithmic) variation of the coupling constants in the intermediate integral over momentum transfer q in (A.5) and evaluate them at ℓ (Fig. 3.8d). It is worth noting that the local approximation becomes exact for couplings or combination of couplings that are invariant with respect to ℓ. This turns out to be the case for the Tomanaga–Luttinger model when only g_2 coupling is present ($g_1 = g_3 = 0$), or for the combination $2g_2 - g_1$ when $g_3 = 0$. Neglecting small scaling deviations, the end result given by (A.5) (Eq. A.8 for Umklapp) is logarithmic and contributes to a $1/z$ renormalization factor of $[G_p^0(k, \omega_n)]^{-1}$ for k located at the top of the inner shell. The flow equation for $z(\ell)$ is

$$\frac{d}{d\ell}\ln z = -\frac{1}{16}[(2g_2 - g_1)^2 + 3g_1^2 + 2g_3^2]. \tag{4.30}$$

The solution as a function of ℓ can be written as a product $z = z_\sigma z_\rho$ of renormalization factors, in which

$$z_\sigma(\ell) = \exp\left\{-\frac{3}{16}\int_0^\ell g_1^2 d\ell'\right\} \tag{4.31}$$

for the spin part and

$$z_\rho(\ell) = \exp\left\{-\frac{1}{16}\int_0^\ell [(2g_2 - g_1)^2 + 2g_3^2]d\ell'\right\} \tag{4.32}$$

for the charge. Prior to analyzing the impact of these equations on one-particle properties, to work consistently at two-loop order, we must proceed to the evaluation of the two-loop contribution to the coupling constant flow.

4.3.2 Four-Point Vertices

To obtain the two-loop corrections to four-point vertices within a classical KW scheme, we come up against difficulties similar to those found for the one-particle self-energy. We can indeed verify that contractions of the form $\langle S_{I,3}^2 S_{I,2}^L\rangle_{\bar{0},c}$ at $n = 3$, which would lead to two-loop corrections for the four-point vertices, are actually vanishingly small if all internal lines are in the same momentum shell. Here again nonzero contributions can only be found if the internal lines refer to different shells, which implies that the contractions must be made starting from interactions that involve more than two particles (Fig. 3.9) [15]. Here we skip the technical details

[2] One-particle self-energy corrections for deeper inner shell states k can be obtained by considering at the level of Fig. 3.8b contractions of the three-particle interaction with internal lines separated by more than one outer shell.

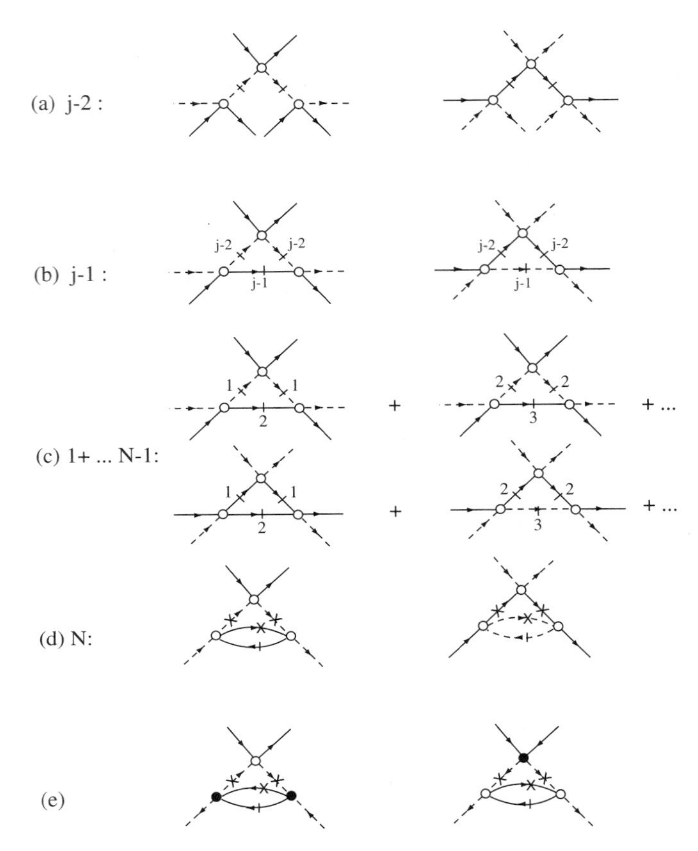

FIGURE 3.9. The cascade of contractions contributing to the four-point vertices in the local approximation of the RG flow: (a) backward and forward scattering contributions to the generation of four-particle interactions from $\frac{1}{2}\langle S^L_{I,2} S_{I,1} S_{I,1}\rangle_{\bar{0},c}$ at the step $j-2$ ($j \geq 3$); (b) second contraction in the outer shell at the step $j-1$; (c) the summation of diagrams given in (b) up to $j = N-1$ as effective three-particle interactions; (d) two-loop outer shell correction to four-point vertices in the local approximation at the step $Nd\ell \to \ell + d\ell$. Here a crossed (dashed) line refers to fermion lines in the interval $[0,\ell]$ ($[\ell, \ell+d\ell]$); (e) the result of a similar cascade of contractions for Umklapp scattering.

of the calculation and only sketch the main steps, which run parallel to those of the one-particle self-energy. The cascade of contractions starts from the four-particle interaction resulting from $3\langle S^L_{I,2} S_{I,1} S_{I,1}\rangle_{\bar{0},c}/3!$ at $n = 3$, which involves the contribution $S^L_{I,2}$ of the Landau channel. The corresponding diagrams are shown in Fig. 3.9a at the step $j-2$. For the required contractions to be possible, the momentum transfer must be zero. At the next step, two more fields are contracted and a line is put in the adjacent shell $j-1$ (Fig. 3.9b). As a function of j, these corrections add up as shown in Fig. 3.9c to lead to effective three-particle interactions at the step $N-1$. At the last step, N, of the cascade, one arrives at a two-loop correction for the four-point vertices (Fig. 3.9d). Following the example of one-particle self-energy, these corrections are nonlocal in the coupling constants. In the local approximation scheme, one finds an outer shell contribution that is $\mathcal{O}(g^3 d\ell)$ (Fig. 3.9d). The two-loop contributions involving Umklapp scattering are shown in Fig. 3.9e. Finally, the outer shell contributions to the renormalization factors $z_{1,2,3}$ (neglecting corrections to logarithmic scaling at ℓ) are given by

$$\begin{aligned} z_1^{(2)}(d\ell) &= \tfrac{1}{4}\Big(g_3^2(\ell) + 2g_2(\ell)\big(g_2(\ell) - g_1(\ell)\big)\Big) d\ell, \\ z_2^{(2)}(d\ell) &= -\tfrac{1}{4} g_2^{-1}(\ell)\Big(g_1^3(\ell) - 2g_1^2(\ell)g_2(\ell) + 2g_2^2(\ell)g_1(\ell) \\ &\qquad - 2g_2^2(\ell) + g_3^2(\ell)\big(g_2(\ell) - g_1(\ell)\big)\Big) d\ell, \\ z_3^{(2)}(d\ell) &= \tfrac{1}{4}\big(g_1^2(\ell) + 2g_1(\ell)g_2(\ell) - 2g_2^2(\ell)\big). \end{aligned} \tag{4.33}$$

Now, combining these with the one-loop results of (4.28) and performing the rescaling of the fields, one has $z^{-1/2}\psi^{(*)} \to \psi^{(*)}$

$$\mu_S = (G_p^0, z_1 z^2 g_1, z_2 z^2 g_2, z_3 z^2 g_3, \ldots), \tag{4.34}$$

which renders explicit the renormalization of the couplings g_i. Neglecting scaling deviations at small ℓ, the corresponding scaling equations are

$$\begin{aligned} \frac{dg_1}{d\ell} &= -g_1^2 - \tfrac{1}{2} g_1^3, \\ \frac{d}{d\ell}(2g_2 - g_1) &= g_3^2[1 - \tfrac{1}{2}(2g_2 - g_1)], \\ \frac{dg_3}{d\ell} &= g_3(2g_2 - g_1)[1 - \tfrac{1}{4}(2g_2 - g_1)] - \tfrac{1}{4} g_3^3, \end{aligned} \tag{4.35}$$

which reproduces the old results of the multiplicative renormalization group method [11, 44]. We see that the spin coupling g_1 is still uncoupled to the charge couplings $2g_2 - g_1$ and g_3. Furthermore, the one-loop singularities, previously found at $T_{\nu=\sigma,\rho}$, are removed and replaced by strong coupling fixed points for the couplings as $\ell \to \infty$. For the interesting cases already analyzed at the one loop level, we have first the attractive one where $g_1 < 0$ and $g_1 - 2g_2 > |g_3|$, for which $g_1^* \to -2$, $g_3^* \to 0$, while $g_1^* - 2g_2^*$ is strongly attractive but nonuniversal; second, in the repulsive case when $g_1 > 0$ and $g_1 - 2g_2 < |g_3|$, for which Umklapp is marginally relevant with $g_3^* \to 2$, $g_2^* \to 1$ and $g_1^* \to 0$. Since in both cases, the couplings cross the Luther–Emery line, the scale for the gap Δ_ν remains of the order of the one-loop temperature scale T_ν in both sectors.

The integration of (4.30) at large $\ell \gg \ell_\nu$ then leads to the power law decay

$$z(\ell) \sim (E_0(\ell)/\Delta_\nu)^{\theta^*}. \tag{4.36}$$

The exponent evaluated at the fixed point in the presence of a single gap can be separated into charge (θ_ρ) and spin (θ_σ) components:

$$\begin{aligned}\theta^* &= \theta^*_\rho(2g^*_2 - g^*_1, g^*_3) + \theta^*_\sigma(g^*_1) \\ &\simeq \tfrac{3}{4}.\end{aligned} \tag{4.37}$$

The case when there is no Umklapp and $g_1 > 0$ corresponds to the Tomonaga-Luttinger model at large ℓ. The couplings remain weak and one finds the power law decay

$$z(\ell) = D_\sigma(\ell)\big(E_0(\ell)/E_0\big)^{\theta_\rho}, \tag{4.38}$$

with

$$\theta_\rho = \frac{1}{16}(2g_2 - g_1)^2, \tag{4.39}$$

and

$$D_\sigma(\ell) = \exp\bigg(-\frac{3}{16}\int_0^\ell g_1^2(\ell')\,d\ell'\bigg). \tag{4.40}$$

Here the exponent θ_ρ is nonuniversal [11]. A contribution from the spin degrees of freedom appears through the weak transient $D_\sigma(\ell)$. The physics of this gapless case is that of the Luttinger liquid.

The factor z is the quasi-particle weight at Fermi level. In all cases, it vanishes at $\ell \to \infty$ or $T \to 0$ indicating the absence of a Fermi liquid in one dimension. Actually the same factor also coincides with the single-particle density of states at the Fermi level

$$N^*(\ell) = N(0)z(\ell), \tag{4.41}$$

which vanishes in the same conditions.

The connection between the one-particle Green's function computed at $\ell = 0$ and that computed at nonzero ℓ can be written in the form

$$G_p(\tilde{k}, \mu_S) = z(\ell)G_p(\tilde{k}, \mu_S(\ell)). \tag{4.42}$$

When $\ell \to \infty$, the number of degrees of freedom to be integrated out goes to zero, thus $G_p(\tilde{k}, \mu_S(\ell)) \to G^0_p(k \to k_{\mathrm{F}}, \omega_n)$ and as a function of $T \to 0$, one finds

$$G_p(k_{\mathrm{F}}, \omega_{n=0}, \mu_S) \approx CT^{-\bar{\gamma}}, \tag{4.43}$$

where, consistent with the scaling ansatz (2.7), $\bar{\gamma} = 1 - \theta$ and $C = 1/(i\pi E^\theta_{\mathrm{F}})$. If we identify the vanishing energy $\frac{1}{2}E_0(\ell)$ in $z(\ell)$ with $v_{\mathrm{F}}(pk - k_{\mathrm{F}})$ close to k_{F}, we have at $T = 0$

$$G_p(k, \mu_S) \propto |pk - k_{\mathrm{F}}|^{-1+\theta}, \tag{4.44}$$

which agrees with the scaling expressions (2.5)–(2.6) showing the absence of a Fermi liquid.

4.3.3 Remarks

At this point, casting a glance back at the RG approach described above, a few remarks can be made in connection with the alternative formulation presented in [15, 16]. In the classical Wilsonian scheme presented here, a momentum cutoff k_0 (or bandwidth E_0) is imposed on the spectrum at the start and all contractions done at a given step of the RG refer to fermion states of the outer shell tied to that particular step. In this way, we have seen that the two-loop level calculations require a cascade of outer shell contractions starting from many-particle interactions that were not present in the bare action. The cascade therefore causes nonlocality of the flow which depends on the values of coupling constants obtained at previous steps. Thus obtaining logarithmic scaling at higher than one-loop forces us to resort to a local approximation.

In the formulation of [15,16], the Kadanoff transformation at high order is performed differently. In effect, a cutoff k_0 is not imposed on the range of momentum for the spectrum in the free part of the action S_0 but it only bounds the k-summations of the interaction term S_I. The latter summations, of which there two in both of the interaction terms (as in (4.1)), are involved in the ultraviolet regularization of all singular diagrams, while the remaining summation on the momentum transfer q is treated independently. In this way, the interaction term $S_I[\psi^*, \psi, \bar{\psi}^*, \bar{\psi}]$ with fields to be integrated out in the Kadanoff transformation is defined differently than in (4.13). It has no $S_{I,1}$ term with a single field in the outer shell, while $S_{I,3}$ and $S_{I,4}$ contain a momentum transfer summation corresponding to the high-energy interval $|v_F q| > E_0(\ell)/2$. The generation of new many-particle terms *via* $S_{I,1}$ is therefore absent, and the evaluation of $\frac{1}{2}\langle(S_{I,3})^2\rangle_{\bar{0},c}$ for the one-particle self-energy correction directly yields (aside from some differences in the range of momentum transfer integration and in the scaling deviations) the end result of the local approximation of Section 4.3.1 at the two-loop level. A similar conclusion holds for the four-point vertices when the calculation of $\frac{1}{2}\langle S^2_{I,3} S^L_{I,2}\rangle_{\bar{0},c}$ is performed. Despite its simplicity, however, this alternative formulation is less transparent as to the origin of logarithmic scaling in the 1D fermion gas model at high order. Moreover, the presence of a summation over a finite interval of momentum transfer for each Kadanoff transformation removes from the fermion spectrum its natural cutoff k_0 and stretches the standard rules of the Wilson procedure.

4.4 Response Functions

We now calculate the response of the system to the formation of pair correlations in the Peierls and Cooper channels. In the Peierls channel, density-wave correlations with charge or spin buildup centered either on sites or bonds differ at half-filling due to the presence of Umklapp scattering. Actually, because $4k_F$ is a reciprocal lattice vector, density waves at $+2k_F$ and

$-2k_{\rm F}$ turn out to be equivalent. In a continuum model, one can thus define the following symmetric and antisymmetric combinations of pair fields

$$O_{\mu_P^\pm}(\tilde{q}) = \tfrac{1}{2}\big(O^*_{\mu_P}(\tilde{q}) \pm O_{\mu_P}(\tilde{q})\big), \tag{4.45}$$

corresponding to site (+) and bond (−) (standing) density-wave correlations.

We consider the linear coupling of these, as well as those previously introduced for the Cooper channel in (3.6)–(3.7), to source fields h_μ; the interacting part of the action S_I then becomes

$$S_I[\psi^*, \psi, h^*, h] = S_I[\psi^*, \psi] + S_h[\psi^*, \psi], \tag{4.46}$$

where

$$S_h[\psi^*, \psi] = \sum_{\mu,\tilde{q}} \big(h_\mu(\tilde{q}) O^*_\mu(\tilde{q}) + \text{c.c}\big). \tag{4.47}$$

Following the substitution of the above S_I in the action (4.12), the KW transformation will generate corrections that are linear in $h^{(*)}_\mu$ and also terms with higher powers. In the linear response theory, only linear and quadratic terms are kept so that the action at ℓ takes the form

$$\begin{aligned} S[\psi^*, \psi, h^*, h]_\ell = S[\psi^*, \psi]_\ell + \sum_{\mu,\tilde{q}} \big\{ & z^h_\mu(\ell) h^*_\mu(\tilde{q}) O_\mu(\tilde{q}) + \text{c.c.} \\ & - \chi_\mu(\ell) h^*_\mu(\tilde{q}) h_\mu(\tilde{q}) \big\}, \end{aligned} \tag{4.48}$$

where $z^h_\mu(\ell)$ is the renormalization factor for the pair vertex part in the channel μ (Fig. 3.10a). The term which is quadratic in the source fields corresponds to the response function (Fig. 3.10b)

$$\chi_\mu(\ell) = -(\pi v_{\rm F})^{-1} \int_0^\ell \bar{\chi}_\mu(\ell') d\ell', \tag{4.49}$$

where

$$\bar{\chi}_\mu(\ell) = \big(z^h_\mu(\ell) z(\ell)\big)^2 \tag{4.50}$$

is known as the auxiliary susceptibility of the channel μ [9, 11], which is decomposed here into a product of pair vertex and self-energy renormalizations. In the presence of source fields, the parameter space of the action at the step ℓ then becomes

$$\mu_S(\ell) = (zG^0_p, z_1 g_1, z_2 g_2, z_3 g_3, z^h_\mu, \chi_\mu, \ldots). \tag{4.51}$$

At the one-loop level $z = 1$ and corrections to z^h_μ come from the outer shell averages $\langle S_h S_{I,2} \rangle_{\bar{0},c}$ for $n = 2$. Their evaluation is analogous to the one given in Section 4.2 and leads to the flow equation (Fig. 3.10)

$$\frac{d}{d\ell} \ln z^h_\mu = \tfrac{1}{2} g_\mu(\ell), \tag{4.52}$$

where $g_{\mu_P^\pm=0} = g_2 \mp g_3 - 2g_1$ and $g_{\mu_P^\pm\neq 0}(\ell) = g_2 \pm g_3$ correspond to the combinations for $2k_F$ CDW and SDW site and bond correlations, respectively, whereas $g_{\mu_C=0} = -g_1 - g_2$ and $g_{\mu_C\neq 0} = g_1 - g_2$ refer to couplings for SS and TS pair correlations. At the two-loop level, the one-particle self-energy corrections contribute and from (4.30), (4.50), and (4.52), a flow equation for the auxiliary susceptibility can be written:

$$\frac{d}{d\ell} \ln \bar{\chi}_\mu = g_\mu - \tfrac{1}{8}[(2g_2 - g_1)^2 + 3g_1^2 + 2g_3^2], \tag{4.53}$$

which coincides with the well known results of the multiplicative renormalization group [11,44]. In conditions that lead to strong coupling, the use of the two-loop fixed point values found in (4.35) yields the asymptotic power law form in temperature

$$\chi_\mu(T) \sim \left(\frac{T}{\Delta_\nu}\right)^{-\gamma_\mu^*}, \quad \text{for } \nu = \sigma, \rho. \tag{4.54}$$

In the repulsive sector $g_1 > 0$ and at half-filling, there is a charge gap ($\Delta_\rho \neq 0$). The bond CDW, usually called bond-order-wave (BOW), and site SDW correlations show a power law singularity with exponent $\gamma_{\mathrm{BOW}}^* = \gamma_{\mathrm{SDW}}^* = \frac{3}{2}$, as in the old calculations of the multiplicative renormalization group [44]. In the attractive sector, when $g_1 < 0$, and Umklapp scattering is irrelevant, the gap is in the spins ($\Delta_\sigma \neq 0$) and both SS and incommensurate CDW are singular with the fixed point exponent $\gamma_{\mathrm{SS,CDW}}^* = \frac{3}{2}$ [11]. (Here the SS response has a larger amplitude.) However, these critical indices obtained from a two-loop RG expansion can only be considered as qualitative estimates resulting from fixed points, which are themselves artefacts of the two-loop RG. Three-loop calculations have been achieved in the framework of multiplicative renormalization group for $g_3 = 0$, and γ_{SS} and γ_{CDW} turn out to be close to unity. The bosonization technique on the

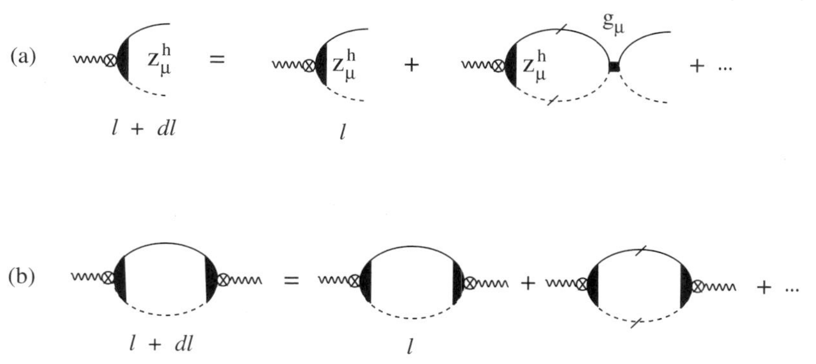

FIGURE 3.10. Diagrams of the flow equations of the pair vertex part z_μ^h (a) and susceptibility χ_μ (b) in the channel μ.

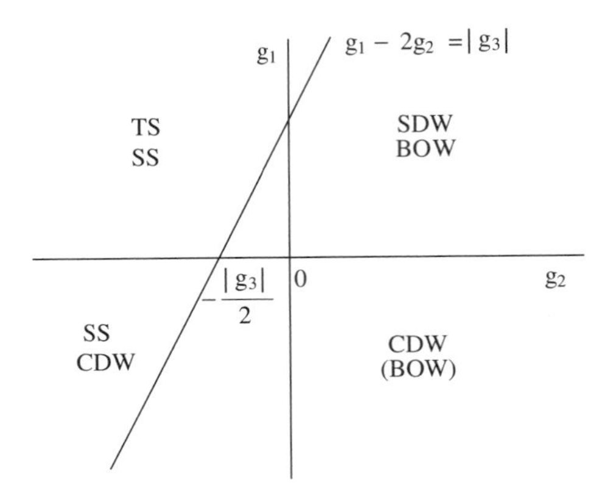

FIGURE 3.11. Phase diagram of the 1D interacting Fermi gas in the presence of Umklapp scattering.

Luther–Emery line [48, 49, 52], allows an exact determination of these indices at $\ell \gg \ell_\nu$ which are all equal to unity.

In the gapless regime for both spin and charge when $g_1 > 0$ and $g_1 - 2g_2 > |g_3|$, the $\mu = \mathrm{TS}$ and $\mu = \mathrm{SS}$ responses are singular

$$\chi_\mu(T) \sim \left(\frac{T}{E_\mathrm{F}}\right)^{-\gamma_\mu^*}. \tag{4.55}$$

Here the exponents $\gamma_{\mathrm{TS}}^* = \gamma_{\mathrm{SS}}^* = -g_2^* - g_2^{*2}/2$ are equal and take nonuniversal values on the Tomanaga–Luttinger line where only the attractive g_2^* coupling remains at $\ell \to \infty$ (here the amplitude of the TS response is larger). Finally, there is a sector of coupling constants, that is when $g_1 < 0$ and $g_1 - 2g_2 < |g_3|$, where a gap develops in both the spin and charge degrees of freedom. In this sector, BOW (site CDW) is singular for $g_3 > 0$ ($g_3 < 0$) with an exponent larger than 2 [44], which indicates once again that the KW perturbative RG becomes inaccurate in the strong coupling sector. The instabilities of the 1D interacting Fermi systems are summarized in Fig. 3.11.

Before closing this section we would like to analyze the scaling properties of the two-particle response function. To do so, it is useful to restore the wave vector dependence of χ_μ by evaluating the upper bound of the loop integral in (4.49) at

$$\ell \to \ell(q,T) = \ln\frac{1.13E_\mathrm{F}}{T} + \psi\left(\frac{1}{2}\right) - \frac{1}{2}\left[\psi\left(\frac{1}{2} + i\frac{v_\mathrm{F}q}{4\pi T}\right) + \mathrm{c.c}\right], \tag{4.56}$$

which coincides with the complete analytical structure of the free Peierls and Cooper loop in (3.10) and (3.12) in the static limit, respectively. At low temperature when $|v_\mathrm{F}q| \gg 4\pi T \to 0$, we can use (3.15) and obtain

$$\chi_\mu(q) \propto q^{-\gamma_\mu} \quad \text{or} \quad \chi_\mu(x) \propto x^{\gamma_\mu - 2}, \tag{4.57}$$

which are consistent with the scaling relation $2 - \eta_\mu = \gamma_\mu$ at $\nu = 1$ (Eq. 2.18).

5 Interchain Coupling: One-Particle Hopping

Let us now consider the effect of a small interchain hopping (Fig. 3.1) in the framework of the KW renormalization group. The bare propagator of the action at $\ell = 0$ is then replaced by $G_p^0(\mathbf{k}, \omega_n)$ given in (3.25). The parameter space for the action at $\ell = 0$ then becomes

$$\mu_S = (G_p^0(k, \omega_n), t_\perp, g_1, g_2, g_3).$$

Now, applying the rescaling transformations (4.8)–(4.9) in the absence of interaction, the rescaling of 1D energy leads to the transformation

$$t'_\perp = s t_\perp, \tag{5.1}$$

which means that the canonical dimension of $t_\perp$ is unity and it is therefore a *relevant* perturbation. An estimate of the temperature at which a dimensionality crossover in the coherence of single-particle motion occurs can be obtained by replacing s with the ratio ξ/a and setting $t'_\perp \sim E_\mathrm{F}$. We find $T_{x^1} \sim t_\perp$ in agreement with the results of Section 3.2. How this is modified in the presence of interactions will depend on the anomalous dimension θ which the fermion field acquires under 1D renormalization. When the partial trace is performed at low values of ℓ, one-particle self-energy corrections are essentially 1D in character and are governed by (4.30). At step ℓ, the effective 2D propagator then takes the form

$$z(\ell) G_p^0(\mathbf{k}, \omega_n) = z(\ell)[G_p^{0-1}(\tilde{k}) + 2t_\perp z(\ell)\cos k_\perp]^{-1}. \tag{5.2}$$

Therefore the power law decay of the quasi-particle weight and the density of states (cf. (4.41)) along the chains leads to a renormalized hopping amplitude $z t_\perp$ which decreases as a function of ℓ. The condition for the single particle crossover temperature T_{x^1} is then renormalized, becoming $z(T) t_\perp \xi/a \sim E_\mathrm{F}$, which implies

$$\begin{aligned} T_{x^1} &\sim t_\perp \left(\frac{t_\perp}{E_\mathrm{F}}\right)^{(1-\theta)/\theta} \\ &\propto t_\perp^{1/\phi_{x^1}}, \end{aligned} \tag{5.3}$$

where $\phi_{x^1} = 1 - \theta$ is the single particle crossover exponent [15, 53]. In this simple picture, well below T_{x^1}, the system behaves as a Fermi liquid. The effective bare propagator (5.2) for $E_p(\mathbf{k}) < \frac{1}{2} E_0(\ell_{x^1})$, where $\ell_{x^1} = \ln E_\mathrm{F}/T_{x^1}$, has the Fermi liquid form (2.7) with a weight that is consistent

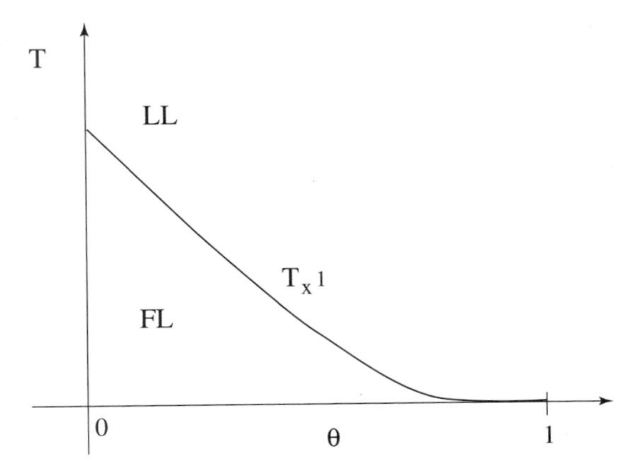

FIGURE 3.12. Crossover temperature variation as a function of the one-particle self-energy exponent θ.

with the extended scaling expression (2.22) with $\bar{\gamma} = 1 - \theta$, $\dot{\bar{\gamma}} = 1$ and the nonuniversal constant $z_\infty = E_{\mathrm{F}}^{\theta/(\theta-1)}$.

As long as $\theta < 1$, $t_\perp$ remains a relevant variable and T_{x^1} is finite. For $\theta = 1$ and $\theta > 1$, however, it becomes marginal and irrelevant respectively. In these latter cases, transverse band motion has no chance to develop and fermions are confined along the chains at all temperatures. In the KW framework, conditions for 1D single-particle confinement are likely to be realized in the presence of a gap since in this case we have seen that $\theta \sim 1$ [54–56]. Single-particle confinement is also present in the absence of any gaps, as in the Tomanaga–Luttinger model, where $g_1 = g_3 = 0$ and $\theta \sim 1$ when $|g_2|$ is close to unity [48, 49] and other interacting fermion models [57, 58]. The dependence of the single-particle crossover temperature on the interaction strength, parametrized by the exponent θ, is given in Fig. 3.12.

A point worth stressing here is that the expression (5.2) for the effective propagator is equivalent to an RPA approach for the interchain motion and is therefore not exact [59–62]. It becomes exact when interchain single-particle hopping has an infinite range [59]. Another point worth stressing is the fact that the temperature interval over which the single-particle dimensionality crossover takes place is not small. In effect, as already noticed in Section 2.1, the critical 1D temperature domain stretches from E_{F} to T making the crossover as a function of temperature gradual.

5.1 Interchain Pair Hopping and Long-Range Order

Although transverse coherent band motion is absent above T_{x^1}, a finite probability for fermions to make a jump on neighboring chains still survives within the thermal coherence time $\tau \sim 1/T$. When this "virtual" motion is combined with electron-electron interactions along the stacks, it generates an effective interchain hopping for *pairs* of particles that are created in all 1D channels of correlations (Peierls, Cooper, and Landau) [14, 15, 37]. When coupled with singular pair correlations along the chains, as may occur in the Cooper or Peierls channels, effective transverse two-particle interaction may arise and lead to the occurrence of long-range order above T_{x^1}, even when fermions are still confined along the chains either by thermal fluctuations or by the presence of a gap. In other words, interchain pair processes generate a different kind of dimensionality crossover, which in turn introduces a new fixed point connected to long-range order. It should be noticed, however, that long-range order is not possible in two dimensions if the order parameter has more than one component [41]. In such a case, a finite coupling in a third direction is required.

To see how these new possibilities emerge from the renormalization group method described above, the quasi-1D propagator (5.2) is used in the evaluation of the one-loop outer shell corrections $\frac{1}{2}\langle S_{I,2}^2\rangle_{\bar{0},c}$ and the perturbative influence of $t_\perp$ in both Peierls and Cooper channels is shown in Fig. 3.13. The first term is independent of $t_\perp$ and reproduces the 1D one-loop corrections. The second diagram expresses the generation at each outer shell integration of an elementary contribution to interchain hopping of a pair of particles in the Peierls or Cooper channel. This outer-shell-generating term is given by

$$S_\perp^0[\psi^*,\psi] = \frac{d\ell}{2} z^{-2}(\ell)\pi v_{\mathrm{F}} \sum_\mu \sum_{\widetilde{Q}} f_\mu(\mathbf{Q},\ell) O_\mu^*(\widetilde{Q}) O_\mu(\widetilde{Q}), \tag{5.4}$$

where $\mathbf{Q} = (q_0, q_\perp), \widetilde{Q} = (q, q_\perp, \omega_m)$ and

$$f_\mu(\mathbf{Q},\ell) = \pm 2 \left[\frac{g_\mu(\ell) z(\ell) t_\perp}{E_0(\ell)}\right]^2 \cos q_\perp . \tag{5.5}$$

The upper (lower) sign refers to the Peierls (Cooper) channel.

This term was not present in the bare action and must be added to $\mu_S(\ell)$. Furthermore, it will be coupled to other terms of S giving rise to additional corrections and hence to an effective interchain two-particle action of the form

$$S_\perp[\psi^*,\psi] = -\tfrac{1}{2} z^{-2}(\ell) \sum_\mu \sum_{\widetilde{Q}} V_\mu(\mathbf{Q},\ell) O_\mu^*(\widetilde{Q}) O_\mu(\widetilde{Q}), \tag{5.6}$$

where $V_\mu(\mathbf{Q},\ell)$ is the pair tunneling amplitude. As a function of ℓ, the parameter space of the action then becomes

$$\mu_S(\ell) = (zG_p^0, zt_\perp, z_1 g_1, z_2 g_2, z_3 g_3, V_\mu, \dots)$$

Among the most interesting pair tunneling processes in the Peierls channel, $V_{\mu_P^+ \neq 0}$ corresponds to an interchain kinetic exchange and promote site

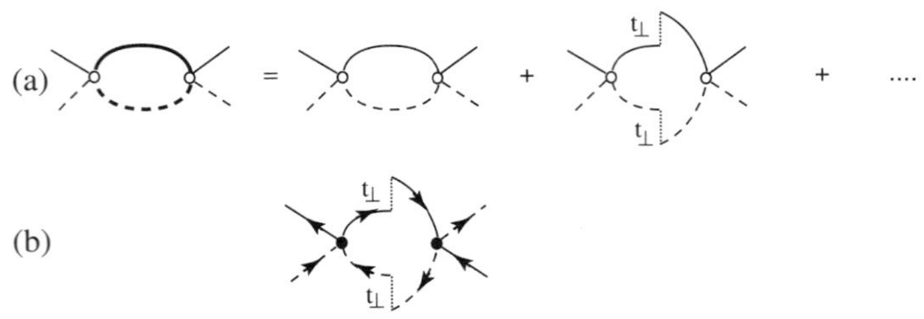

FIGURE 3.13. a) Generation of interchain pair hopping from one-loop backward and forward vertex corrections in the Cooper and Peierls channels at each outer shell integration. Here the thick (thin) continuous and dashed lines refer to $\pm k_{\mathrm{F}}$ 2D (1D) propagator. b) The Umklapp contribution to interchain pair hopping in the Peierls channel.

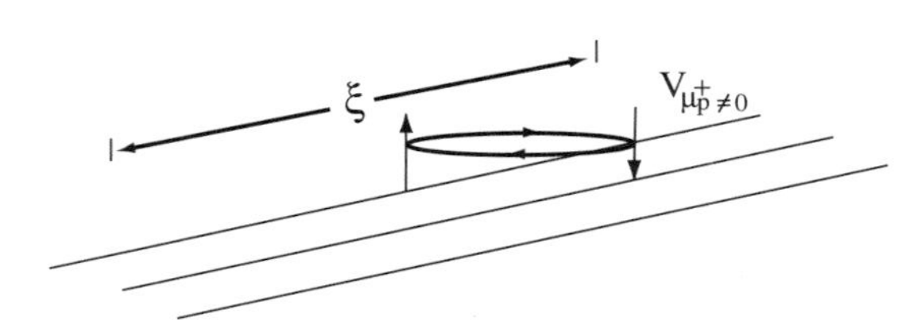

FIGURE 3.14. Antiferromagnetic interchain exchange $V_{\mu_P^+}$ occurring between two spins within the coherence length ξ. In the presence of a charge gap, the exchange takes place in the interval $\xi \to \xi_\rho \sim v_{\mathrm{F}}/\Delta_\rho$.

SDW or AF ordering (Fig. 3.14), while $V_{\mu_P^-=0}$ is an interchain BOW coupling which is involved in spin-Peierls ordering [16]. Both favor the transverse propagation of order at $\mathbf{Q}_0 = (2k_{\mathrm{F}}, \pi)$, which actually coincides with the best nesting vector of the whole *warped* Fermi surface. In the Cooper channel, V_{μ_C} describes the interchain Josephson coupling for SS and TS Cooper pair hopping. According to Fig. 3.11, these terms can only couple to singular intrachain correlations when Umklapp scattering is irrelevant. Long-range order in the Cooper channel is uniform and occurs at the wave vector $\mathbf{Q}_0 = (0, 0)$.

Long-range order will occur at the critical point determined by $\ell_c = \ln(E_{\mathrm{F}}/T_{c,\mu})$ where $V_\mu(\mathbf{Q}_0, \ell_c) \to -\infty$. A reasonable way to obtain an approximate $T_{c,\mu}$ is a transverse RPA approach (Fig. 3.15), which consists of treating interchain coupling in the mean-field approximation while the intrachain correlations are taken into account rigorously using (4.53) or the exact asymptotic form of the pair susceptibility when it is known. Thus substituting $S_I \to S_I + S_\perp$ in (4.14), one gets the one-loop outer-shell contribution $\frac{1}{2}\langle S_{I,2}^2 + 2S_{I,2}S_{\perp,2} + S_{\perp,2}^2\rangle_{\bar{0}}$. The first term leads to the generating term (5.4) to second order in $zt_\perp$, whereas the second term introduces

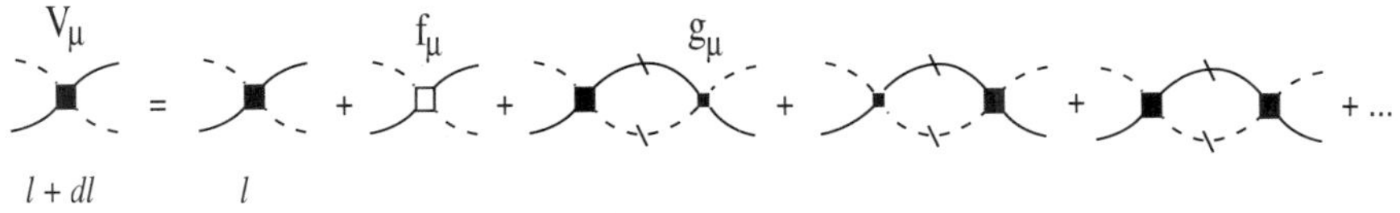

FIGURE 3.15. Recursion relation for the interchain pair-hopping term in the transverse RPA treatment (Eq. 5.7).

pair vertex corrections to the composite field $O_\mu^{(*)}$ which can then be expressed in terms of $\bar{\chi}_\mu$ given in (4.53). Finally, the third term is the RPA contribution that enables the free propagation of pairs in the transverse direction and hence the possibility of long-range order at finite temperature (Fig. 3.15). In all cases other than the generating term, the outer-shell integration in the RPA approach is made by neglecting $t_\perp$, with the result

$$\frac{d}{d\ell} V_\mu(\mathbf{Q}, \ell) = f_\mu + V_\mu \frac{d}{d\ell} \ln \bar{\chi}_\mu - \tfrac{1}{2}(V_\mu)^2 \tag{5.7}$$

for the flow equation of the interchain pair hopping. The solution of this inhomogeneous equation at ℓ has the simple pole structure

$$V_\mu(\mathbf{Q}, \ell) = \frac{\overline{V}_\mu(\mathbf{Q}, \ell)}{1 - \frac{1}{2}\big(\bar{\chi}_\mu(\ell)\big)^{-1} \overline{V}_\mu(\mathbf{Q}, \ell)\chi_\mu(\ell)}, \tag{5.8}$$

where

$$\overline{V}_\mu(\mathbf{Q}, \ell) = \bar{\chi}_\mu(\ell) \int_0^\ell d\ell' \, f_\mu(\mathbf{Q}, \ell')\big(\bar{\chi}_\mu(\ell')\big)^{-1} \\ \times \big[1 - \tfrac{1}{2}\big(\bar{\chi}_\mu(\ell')\big)^{-1}\overline{V}_\mu(\mathbf{Q}, \ell')\chi_\mu(\ell')\big]^2. \tag{5.9}$$

5.1.1 Critical Temperature in the Random Phase Approximation

Owing to the structure of $\overline{V}_\mu$, a general analytic solution for $T_{c,\mu}$ is impossible. However, an approximate expression for the critical temperature can be readily obtained, by first neglecting transients in the power laws of $\bar{\chi}_\mu \sim e^{\gamma_\mu \ell}$, $z \sim e^{-\theta\ell}$, and the couplings $g_\mu(\ell)$, and by setting to unity the term in square brackets in the integrand of (5.9). Thus, far from $T_{c,\mu}$ where the RPA term can be neglected, one can write

$$V_\mu(\mathbf{Q}_0, \ell) \approx 2\frac{\langle g_\mu^2 \rangle t_\perp^2}{E_0^2}\big((2 - 2\theta - \gamma_\mu)^{-1}(e^{\ell(2-2\theta-\gamma_\mu)} - 1\big), \tag{5.10}$$

where $\langle g_\mu^2 \rangle$ is an average of the coupling constants over the interval $[0, \ell]$. In the presence of a gap in the charge or spin sector, the above scaling form at $\ell \gg \ell_{\nu=\rho,\sigma} = \ln(E_{\mathrm{F}}/T_{\sigma,\rho})$ is governed by the strong coupling condition $2 - 2\theta^* - \gamma_\mu^* < 0$, while for $\ell \ll \ell_{\nu=\rho,\sigma}$, the couplings are weak and

$2-2\theta-\gamma_\mu>0$. In strong coupling when $\ell|2-2\theta^*-\gamma_\mu^*|\gg 1$, the expression (5.10) actually coincides with $\overline{V}_\mu$ so that its substitution in (5.8) leads to a non-self-consistent pole at

$$T_{c,\mu}\approx T_\nu\left[\frac{g_{\perp\mu}}{\delta_\mu\gamma_\mu^*}\left(1+\frac{g_{\perp\mu}}{\delta_\mu\gamma_\mu^*}\right)^{-1}\right]^{1/\gamma_\mu^*}, \tag{5.11}$$

where

$$g_{\perp\mu}=\frac{\langle g_\mu^2\rangle z^2(\ell_\nu)}{2-2\theta-\gamma_\mu}\frac{t_\perp^2}{\Delta_\nu^2} \tag{5.12}$$

is an effective interchain pair-hopping term at the step ℓ_ν. This form is reminiscent of a perturbative expansion in terms of the small parameter $zt_\perp/\Delta_\nu$ indicative of interchain hopping of bound pairs [63–65]. The positive constant

$$\delta_\mu=1+g_{\perp\mu}(\Delta_\nu/E_0)^{\gamma_\mu}\chi_\mu(\Delta_\nu)<1 \tag{5.13}$$

corresponds to the integral of (5.9) over the interval $[0,\ell_\nu]$ at high energy. These results in the presence of a gap allow us to write

$$T_{c,\mu}\sim T_{x^2}\sim g_{\perp\mu}^{1/\phi_{x^2}}, \tag{5.14}$$

which is compatible with (2.27) from the extended scaling ansatz. T_{x^2} is thus a temperature scale for the dimensionality crossover for the correlation of *pairs* of particles induced by $g_{\perp\mu}$, and $\phi_{x^2}=\gamma_\mu^*$ is the two-particle exponent [15]. T_{x^2} may also be equated with the onset of critical (Gaussian) fluctuations. A point that is worth noticing here concerns the variation of the critical temperature with the gap Δ_ν. Actually, for most interesting situations, the flow of coupling constants will cross the Luther–Emery line and the use of the expression (4.51) with $\gamma^*=1$ will lead to the dependence $T_{c,\mu}\approx z^2(\ell_\nu)t_\perp^2/\Delta_\nu$ in which $T_{c,\mu}$ increases as the amplitude of the gap decreases (Fig. 3.16). In the absence of a gap, the interchain pair-hopping term is still present and it can lead to long-range ordering. Sufficiently far from the critical point and for $\ell(2-2\theta-\gamma_\mu)\gg 1$, one finds

$$\begin{aligned}V_\mu(\mathbf{Q}_0,\ell)&=\overline{V}_\mu(\mathbf{Q}_0,\ell)\\&\approx-2g_{\perp\mu}(2-2\theta-\gamma_\mu)^{-1}(E_0(\ell)/E_0)^{2\theta-2}\end{aligned} \tag{5.15}$$

where in weak coupling $g_{\perp\mu}=\langle g_\mu^2\rangle t_\perp^2/E_0^2$. Substituting this expression for $\overline{V}_\mu$ in the RPA expression (5.8), the critical point will occur at

$$T_{c,\mu}\sim E_{\mathrm{F}}\big(g_{\perp\mu}/[\gamma_\mu(2-2\theta-\gamma_\mu)]\big)^{1/(2(1-\theta))}. \tag{5.16}$$

Therefore, in weak coupling, $T_{c,\mu}\sim T_{x^2}\sim(g_{\perp\mu})^{1/\phi_{x^2}}$, and one can define the two-particle crossover exponent $\phi_{x^2}=2(1-\theta)$ which is twice ϕ_{x^1} (plus anomalous corrections coming from the g_μ). The critical temperature $T_{c,\mu}\approx\langle g_\mu^2\rangle^{1/2}zt_\perp$ will then decrease when interaction decreases. This

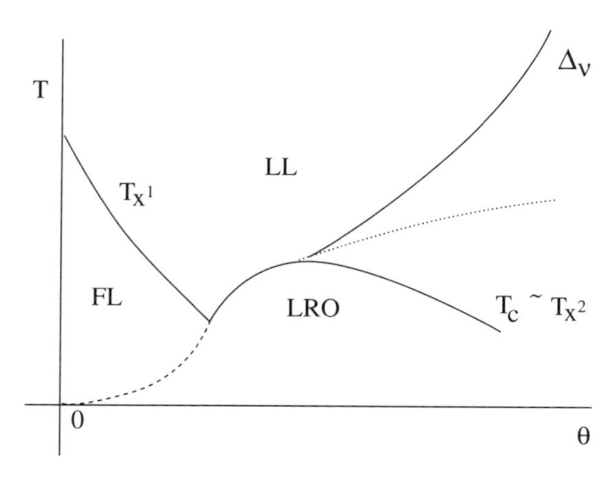

FIGURE 3.16. Schematic diagram of the temperature (T) and interaction strength (θ) dependence of the single-particle and pair dimensionality crossovers from the Luttinger liquid (LL) or Luther–Emery liquid ($T < \Delta_\nu$) towards a Fermi liquid (FL) or long-range ordering (LRO). The dashed (dotted) line corresponds to the onset of LRO from a Fermi liquid (gapless LL in strong coupling).

result is meaningful as long as $T_{x^2} > T_{x^1}$, that is, when the two-particle dimensionality crossover that marks the onset of Gaussian critical fluctuations still falls in a temperature region where the transverse single particle motion is incoherent. The interval of coupling constants where this occurs is relatively narrow, however, since when there is no gap $t_\perp$ is a relevant parameter and the calculated T_{x^1} increases with decreasing interactions. When the strong and weak coupling regimes are looked at together, a maximum of $T_{c,\mu}$ is therefore expected at the boundary (Fig. 3.16).

Finally, it should be stressed that whenever the strong coupling condition $2 - 2\theta - \gamma < 0$ is achieved in the absence of any gaps as in the case for the Tomanaga–Luttinger model, there is no maximum of $T_{c,\mu}$. Instead, it is found to increase monotonically with the interaction strength (see the dotted line of Fig. 3.16) [59, 64].

5.1.2 Response Functions

The calculation of the response functions in the presence of interchain pair hopping can be done following the procedure given in Section 4.4. At the RPA level for the pair hopping, one must add the outer-shell contribution $\langle S_h S_{\perp,2}\rangle_{\bar{0},c}$ to the scaling equation for $\bar{\chi}_\mu$ in the 1D regime (see (4.53)), to get

$$\frac{d}{d\ell} \ln \bar{\chi}_\mu(\mathbf{Q}, \ell) = \frac{d}{d\ell} \ln \bar{\chi}_\mu(\ell) - V_\mu. \tag{5.17}$$

Using (5.8), the integration of this equation leads to

$$\bar{\chi}_\mu(\mathbf{Q},\ell) \approx \frac{\bar{\chi}_\mu(\ell)}{\left[1 - \frac{1}{2}\left(\bar{\chi}_\mu(\ell)\right)^{-1}\overline{V}_\mu(\mathbf{Q},\ell)\chi_\mu(\ell)\right]^2}, \tag{5.18}$$

which presents a singularity that is quadratic in $(T - T_{c,\mu})^{-1}$ at $\mathbf{Q} = \mathbf{Q}_0$. The exponent $\dot{\gamma}$ (cf. (2.25)) for the auxiliary susceptibility is therefore twice the value expected for χ_μ in the transverse RPA approach. For the cases of two-particle crossovers in Section 5.1.1, this expression at $\mathbf{Q}_0$ can also be written in an extended scaling form similar to (2.26)

$$\bar{\chi}_\mu(\mathbf{Q}_0,\ell) \approx \bar{\chi}_\mu(\mathbf{Q}_0,\ell) X(Bg_{\perp\mu}/T^{\phi_{x^2}}) \tag{5.19}$$

where $X(y) = 1/(1-y)^2$ is the RPA crossover scaling function for the auxiliary susceptibility.

In the present transverse RPA scheme, the small longitudinal q dependence can be incorporated in $\bar{\chi}_\mu$ through the substitution (4.56), while the $q_\perp$ dependence will come from an expansion of $\overline{V}_\mu$ in (5.9) around the transverse ordering wavevector $q_{\perp 0}$ in the channel considered. The $\mathbf{Q}$ dependence of $\bar{\chi}_\mu$ will take the form

$$\bar{\chi}_\mu(\mathbf{Q},T) \propto (\dot{t} + \dot{\xi}_0^2 q^2 + \dot{\xi}_{\perp 0}^2 q_\perp^2)^{-2} \tag{5.20}$$

where $\dot{\xi}_0 \sim v_{\mathrm{F}}/T_c$ is the longitudinal coherence length; the transverse coherence length $\dot{\xi}_{\perp 0} < 1$ is smaller than the interchain distance indicating strong anisotropy. The growth of the correlation length is governed by the usual Gaussian expression $\vec{\dot{\xi}} = \vec{\dot{\xi}}_0 \dot{t}^{-1/2}$, which according to (2.30), leads to the correlation exponent $\dot{\nu} = \frac{1}{2}$. Similar agreement with scaling is found for the response function, for which the above RPA treatment leads to the familiar results $\dot{\gamma} = 1$ and $\dot{\eta} = 0$.

5.2 Long-Range Order in the Deconfined Region

5.2.1 Critical Temperatures in the Ladder Approximation

When the calculated value of T_{x^2} becomes smaller than the single-particle deconfinement temperature T_{x^1}, the two-particle crossover is irrelevant and the expression (5.16) for $T_{c,\mu}$ becomes incorrect. In effect, thermal fluctuations have energies smaller than the characteristic energy related to the warping of the Fermi surface, which then becomes coherent. The $t_\perp$ expansion of Fig. 3.13 stops being convergent and the flow equation (5.7) must be modified accordingly. Now the effective low-energy spectrum

$$E_p(\mathbf{k}) = \epsilon_p(k) - 2zt_\perp \cos k_\perp \tag{5.21}$$

is characterized by the electron-hole symmetry (nesting) $E_+(\mathbf{k} + \mathbf{Q}_0) = -E_-(\mathbf{k})$ and logarithmic corrections will also be present below $E_0(\ell_{x^1})$

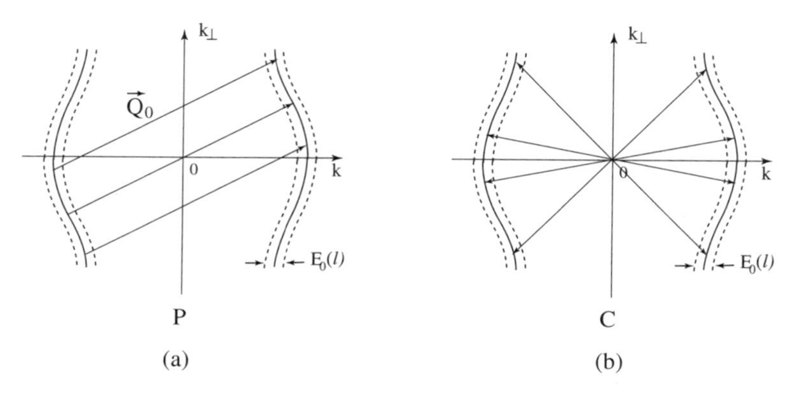

FIGURE 3.17. (a) Particle-hole pairing in the Peierls channel at $\mathbf{Q}_0 = (2k_\mathrm{F}, \pi)$. (b) Particle-particle (or hole-hole) pairing in the Cooper channel. Constant energy surfaces at $\pm\frac{1}{2}E_0(\ell > \ell_{x^1})$ from the Fermi surface are shown as dashed lines.

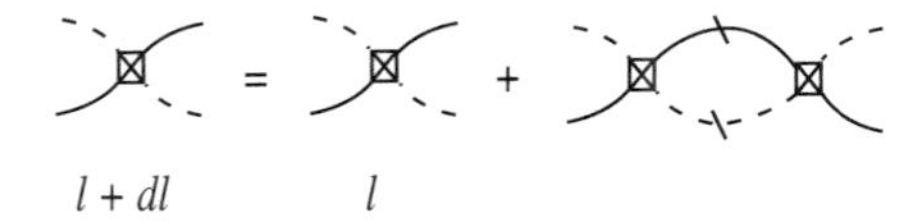

FIGURE 3.18. Recursion relation in the ladder approximation for the effective coupling $\mathcal{G}_\mu \equiv V_\mu - g_\mu$ (crossed square) in the channel μ below the temperature of the single-particle dimensionality crossover T_{x^1}.

(Fig. 3.17a). Similarly in the Cooper channel, the inversion property of the spectrum $E_+(\mathbf{k}) = E_-(-\mathbf{k})$ also leads to a logarithmic singularity in the same energy range (Fig. 3.17b). We will follow a simple two-cutoff scaling scheme, [15, 63, 66] in which the flows of V_μ (Eq. 5.7) and g_μ (Eq. 4.35) are stopped at ℓ_{x^1} and their values are used as boundary conditions for the outer shell integration below $E_0(\ell_{x^1})$ which is done with respect to constant warped energy surfaces (Fig. 3.17). The outer shell integration will be done at the one-loop level in the so-called ladder approximation, meaning that the interference between the Cooper and Peierls channels will be neglected. This decoupling between the two channels at $T \ll T_{x^1}$ is actually not a bad approximation for $T_{c,\mu}$ when the couplings are attractive and favor superconductivity or when the nesting properties of the warped Fermi surface are good and a density-wave instability is expected. The approximation becomes less justified, however, if, for example, nesting deviations are present in the spectrum and the instability of the Peierls channel is weakened. In this case, it has been shown that the residual interference between the two channels gives rise to nontrivial influence on pairing which becomes nonuniform along the Fermi surface—e.g., unconventional superconductivity for repulsive interactions and anisotropy of the SDW gap [67].

In the ladder approximation, the evaluation of $\frac{1}{2}\langle S_{\perp,2}^2\rangle_{\bar{0},c}$ at $\mathbf{Q}_0$ in each channel allows us to write

$$\frac{dV_\mu}{d\ell} = V_\mu g_\mu - \tfrac{1}{2}(V_\mu)^2. \tag{5.22}$$

This expression is incomplete, however, because $\frac{1}{2}\langle S_{I,2}^2\rangle_{\bar{0},c}$ evaluated at $\mathbf{Q}_0$, gives a logarithmic outer shell correction of the form $\frac{1}{2}g_\mu^2 d\ell$, which leads to the flow equation

$$\frac{dg_\mu}{d\ell} = \tfrac{1}{2}(g_\mu)^2. \tag{5.23}$$

In the action, the corresponding combination of couplings is proportional to $O_\mu^* O_\mu$, and will then add to V_μ. Therefore it is useful to define $\mathcal{G}_\mu \equiv V_\mu - g_\mu$ for which one obtains the flow equation of Fig. 3.18:

$$\frac{d}{d\ell}\mathcal{G}_\mu = -\tfrac{1}{2}(\mathcal{G}_\mu)^2. \tag{5.24}$$

The relevant parameter space for the action for $E_0(\ell) < E_0(\ell_{x^1})$ in the ladder approximation becomes

$$\mu_S(\ell > \ell_{x^1}) = (G_p^0(\mathbf{k},\omega_n), \mathcal{G}_\mu). \tag{5.25}$$

The solution of (5.24) has the following simple pole structure

$$\mathcal{G}_\mu(\ell) = \frac{\mathcal{G}_\mu(\ell_{x^1})}{1 + \frac{1}{2}\mathcal{G}_\mu(\ell_{x^1})(\ell - \ell_{x^1})}, \tag{5.26}$$

where $\mathcal{G}_\mu(\ell_{x^1})$ is the boundary condition deduced from the 1D scaling equations (5.7) and (4.35) at ℓ_{x^1}. The presence of an instability of the Fermi liquid towards the condensation of pairs in the channel μ occurs at the critical temperature

$$T_{c,\mu} = T_{x^1}\ \exp\bigl(-2/|\mathcal{G}_\mu(\ell_{x^1})|\bigr). \tag{5.27}$$

This BCS type of instability occurs when the effective coupling constant is attractive ($\mathcal{G}_\mu < 0$). The phase diagram coincides with the one found in 1D when there is perfect nesting (see Fig. 3.11). The critical temperature in each sector is associated with those correlations having the largest amplitude. The variation of the critical temperature as a function of the interaction strength is shown by the dashed line in Fig. 3.16. We also notice that the BCS expression (5.27) satisfies the extended scaling hypothesis (2.29), namely, $T_{c,\mu} \propto T_{x^1} \propto t_\perp^{1/\phi_{x^1}}$.

5.2.2 Response Functions in the Deconfined Region

By adding the coupling of fermion pairs to source fields, the calculation of response functions in the channel μ below T_{x^1} can readily be done. In the ladder approximation, the outer shell contraction $\frac{1}{2}\langle S_h S_{I,2}\rangle_{\bar{0},c}$, which amounts to making the substitution $g_\mu \to \mathcal{G}_\mu$ in Fig. 3.10, leads at once to the flow equation for the pair vertex part:

$$\frac{d}{d\ell}\ln z_\mu^h = -\tfrac{1}{2}\mathcal{G}_\mu. \tag{5.28}$$

Using the ladder result (5.26), one gets for the auxiliary susceptibility $\bar{\chi}_\mu = (z_\mu^h)^2$ at $\ell \geq \ell_{x^1}$

$$\bar{\chi}_\mu(\mathbf{Q}_0,\ell) = \frac{\bar{\chi}_\mu(\mathbf{Q}_0,\ell_{x^1})}{[1+\frac{1}{2}\mathcal{G}_\mu(\ell_{x^1})(\ell-\ell_{x^1})]^2}, \tag{5.29}$$

which presents a power law singularity that is quadratic in $(T-T_{c,\mu})^{-1}$ as the temperature approaches $T_{c,\mu}$ from above. The integration of this expression following the definition given in (4.49) causes no difficulties and one finds for the response function

$$\chi_\mu(\mathbf{Q}_0,\ell) = \chi_\mu(\mathbf{Q}_0,\ell_{x^1}) - (\pi v_{\mathrm{F}})^{-1}\frac{\bar{\chi}_\mu(\mathbf{Q}_0,\ell_{x^1})(\ell-\ell_{x^1})}{1+\frac{1}{2}\mathcal{G}_\mu(\ell_{x^1})(\ell-\ell_{x^1})}. \tag{5.30}$$

The singularity appears in the second term and is linear in $(T-T_{c,\mu})^{-1}$, which is typical of a ladder approximation. The critical exponent defined in (2.25) is $\dot{\gamma}=1$ and corresponds to the prediction of mean-field theory. As for the amplitude of the singular behavior, it satisfies the extended scaling constraint shown in (2.28). It is worth noting that these scaling properties are also satisfied by the auxiliary susceptibility.

In going beyond the ladder approximation, the RG technique for fermions becomes less reliable in the treatment of fluctuation effects beyond the Gaussian level which are usually found sufficiently close the critical point – at least in three dimensions. Standard approaches of critical phenomena, such as the statistical mechanics of the order parameter from a Ginzburg–Landau free-energy functional, can be used sufficiently close to $T_{c,\mu}$.

6 Kohn–Luttinger Mechanism in Quasi-One-Dimensional Metals

6.1 *Generation of Interchain Pairing Channels*

The outer shell partial integration described in the RPA treatment of Section 5.1 for $T > T_{x^1}$, namely, in the presence of interchain pair hopping, was restricted to the channel μ, corresponding to the most singular correlations of the 1D problem (Fig. 3.11). However, looking more closely at the possible contractions of fermion fields in $S_{\perp,2}$, a new pairing possibility emerges, as shown in Fig. 3.19. This consists of putting the two outer-shell

FIGURE 3.19. Possible outer-shell contractions for interchain pair hopping: intrachain channel μ (left) and interchain channel $\bar{\mu}$ (right). The black square corresponds to the pair-hopping amplitude $V_{\mu ij}$ between chains i and j.

fermion fields on separate chains i and j, which can then be equated with *interchain pairing*. The attraction between the two particles in the pair is mediated by the combination of pair hopping and intrachain correlations. Thus interchain pairing is similar to the Kohn–Luttinger mechanism for superconductivity by which Friedel charge oscillations can lead to an effective attraction between two electrons and in turn to an instability of the 3D metal at a finite but extremely low temperature [33]. In quasi-1D systems, however, the K-L mechanism has a dual nature given that interchain pair hopping can exist not only in the Peierls channel but also in the Cooper channel. Therefore for a Josephson coupling, the interchain outer-shell decomposition of Fig. 3.19 leads to an attraction between a particle and hole on different chains. In this case the attraction is mediated by the exchange of Cooper pairs.

6.1.1 Coupling Constants

In the following we will restrict our description of the onset of interchain pairing correlations within the KW RG to the incommensurate case where only g_1 and g_2 scattering processes are present. Thus in the temperature domain above T_{x^1}, which corresponds to $\ell < \ell_{x^1}$, we have seen in Section 5.1 that the outershell contraction in which the fermion lines are diagonal in the chain indices yields the RPA equation (5.7). As pointed out above, another possible contraction would be to put the two outer shell fermion fields on distinct chains.

This is best illustrated by first rewriting the pair-hopping term (5.6) in terms of chains indices. Then putting two fields in the outershell, we consider the sum of two contributions

$$
\begin{aligned}
S_{\perp,2} + S'_{\perp,2} \\
&= -\frac{1}{2}\pi v_{\mathrm{F}} z^{-2}(\ell) \sum_{\mu,\tilde{q}} \sum_{i,j} V_{\mu,ij}(\ell) \big(\bar{O}^*_{\mu,i}(\tilde{q}) O_{\mu,j}(\tilde{q}) + \mathrm{c.c}\big) \\
&\quad + \frac{1}{2}\pi v_{\mathrm{F}} z^{-2}(\ell) \sum_{\bar{\mu},\tilde{q}} \sum_{i,j} U_{\bar{\mu},ij}(\ell) \big(\bar{O}^*_{\bar{\mu},ij}(\tilde{q}) O_{\bar{\mu},ji}(\tilde{q}) + \mathrm{c.c}\big), \quad (6.1)
\end{aligned}
$$

the first of which leads to (5.7) at the one-loop level, while the second is connected to the interchain channel denoted $\bar{\mu}$. The interchain composite fields are defined by

$$O_{\bar{\mu}_C,ij}(\tilde{q}) = (T/L)^{1/2} \sum_{\tilde{k}} \alpha \psi_{-,-\alpha,i}(-\tilde{k}) \sigma^{\alpha\beta}_{\bar{\mu}_C} \psi_{+,\beta,j}(\tilde{k}+\tilde{q}) \tag{6.2}$$

for the singlet (ISS: $\bar{\mu}_C = 0$) and triplet (ITS: $\bar{\mu}_C = 1,2,3$) interchain Cooper channel $\bar{\mu}_C$, and by

$$O_{\bar{\mu}_P,ij}(\tilde{q}) = (T/L)^{1/2} \sum_{\tilde{k}} \psi^*_{-,\alpha,i}(\tilde{k}-\tilde{q}) \sigma^{\alpha\beta}_{\bar{\mu}_P} \psi_{+,\beta,j}(\tilde{k}) \tag{6.3}$$

for the charge-density-wave (ICDW: $\bar{\mu}_P = 0$) and spin-density-wave (ISDW: $\bar{\mu}_P = 1,2,3$) pair fields of the interchain Peierls channel $\bar{\mu}_P$. This introduces the combinations of couplings

$$U_{\bar{\mu},ij} = \sum_{\mu} c^{\bar{\mu}}_{\mu} V_{\mu,ij}. \tag{6.4}$$

The constants $c^{\bar{\mu}}_{\mu}$ are obtained from

$$\begin{aligned} U_{\bar{\mu}_{C(P)}=0,ij} &= \tfrac{3}{2} V_{\mu_{P(C)}\neq 0,ij} - \tfrac{1}{2} V_{\mu_{P(C)}=0,ij} && \text{ISS (ICDW)} \\ U_{\bar{\mu}_{C(P)}\neq 0,ij} &= \tfrac{1}{2} V_{\mu_{P(C)}\neq 0,ij} + \tfrac{1}{2} V_{\mu_{P(C)}=0,ij} && \text{ITS (ISDW).} \end{aligned} \tag{6.5}$$

Therefore at the one-loop level in addition to the RPA term, one obtains the term

$$\tfrac{1}{2}\langle (S'_{\perp,2})^2 \rangle_{\bar{0},c} = \pi v_F z^{-2}(\ell) \frac{d\ell}{4} \sum_{\bar{\mu},\tilde{q}} \sum_{i,j} U^2_{\bar{\mu},ij}(\ell) O^*_{\bar{\mu},ij}(\tilde{q}) O_{\bar{\mu},ij}(\tilde{q}), \tag{6.6}$$

which is also logarithmic. Owing to the dependence on the chain indices of the last composite field on the right-hand side of this equation, this term cannot be rewritten in terms of the intrachain channels and therefore stands as a new coupling which we will write $\delta\mathcal{S}_\perp$. It must be added to the parameter space of the action. To do so, it is convenient to recast (6.6) in terms of interchain backward and forward scattering amplitudes:

$$\begin{aligned} \delta\mathcal{S}_\perp = -\frac{T}{L}\pi v_F z^{-2}(\ell)\, d\ell \sum_{\{\alpha\},\tilde{q}} \sum_{i,j} \big(f_{2,ij}(\ell) \delta_{\alpha_1\alpha_4}\delta_{\alpha_2\alpha_3} - f_{1,ij}(\ell)\delta_{\alpha_1\alpha_3}\delta_{\alpha_2\alpha_4} \big) \\ \times \psi^*_{+,\alpha_1,i}(\tilde{k}_1+\tilde{q}) \psi^*_{-,\alpha_2,j}(\tilde{k}_2-\tilde{q}) \psi_{-,\alpha_3,j}(\tilde{k}_2) \psi_{+,\alpha_4,i}(\tilde{k}_1). \end{aligned} \tag{6.7}$$

The outer-shell backscattering amplitude $f_{1,ij}$ then corresponds to intrachain momentum transfer near $2k_F$ between the particles and a change in the chain indices during the process, whereas the forward generating amplitude $f_{2,ij}$ corresponds to small intrachain momentum with no change in the chain index (Fig. 3.20). The explicit expressions are

$$\begin{aligned} f_{1,ij} &= -\tfrac{1}{2}[V^2_{\mu_P\neq 0,ij} - V_{\mu_P\neq 0,ij}V_{\mu_P=0,ij}] \\ &\qquad + \tfrac{1}{2}[V^2_{\mu_C\neq 0,ij} - V_{\mu_C\neq 0,ij}V_{\mu_C=0,ij}] \\ 2f_{2,ij} - f_{1,ij} &= \tfrac{1}{4}[3V^2_{\mu_C\neq 0,ij} + V^2_{\mu_C=0,ij}] \\ &\qquad + \tfrac{1}{4}[3V^2_{\mu_C\neq 0,ij} + V^2_{\mu_P=0,ij}]. \end{aligned} \tag{6.8}$$

Since the pair-hopping amplitudes V_μ are themselves generated as a function of ℓ, it is clear that the $f's$ will be very small except when the V_μ approach the strong coupling regime. Successive KW transformations will lead to the renormalization of the above interchain backward- and forward-scattering amplitudes which we denote $g^\perp_{1,ij}$ and $g^\perp_{2,ij}$, respectively. These define the part of the action denoted $\mathcal{S}_\perp$, which when expressed in terms of uniform charge and spin fields introduced in (4.24), takes the rotationally invariant form

$$\begin{aligned} \mathcal{S}_\perp[\psi^*,\psi]_\ell \\ = -\pi v_{\mathrm{F}} z^{-2}(\ell)\sum_{p,\tilde{q}}\sum_{ij}\Big((2g^\perp_{2,ij}(\ell) - g^\perp_{1,ij}(\ell))\rho_{p,i}(\tilde{q})\rho_{-p,j}(-\tilde{q}) \\ - g^\perp_{1,ij}(\ell)\mathbf{S}_{p,i}(\tilde{q})\cdot\mathbf{S}_{-p,j}(-\tilde{q})\Big) \end{aligned} \tag{6.9}$$

This expression clearly shows that interchain pair-hopping generates a coupling between uniform charge and spin excitations of different chains. The flow equations of these couplings will now be obtained from $\frac{1}{2}\langle\mathcal{S}^2_{\perp,2}\rangle_{\bar{0},c}$ at the one-loop level. Aside from the scale-dependent-generating terms, the contributions consist of the same type of interfering Cooper and Peierls diagrams found in the purely one-dimensional problem (Fig. 3.5). These outer-shell logarithmic contributions can be evaluated at once, resulting in the following flow equations for the interchain spin and charge variables:

$$\begin{aligned} \frac{d}{d\ell}g^\perp_{1,ij} &= f_{1,ij} - (g^\perp_{1,ij})^2 \\ \frac{d}{d\ell}2g^\perp_{2,ij} - g^\perp_{1,ij} &= 2f^\perp_{2,ij} - f^\perp_{1,ij}. \end{aligned} \tag{6.10}$$

It is interesting to note that, following the example of the purely one-dimensional couplings, the interference between the interchain and Cooper channels causes these equations to be decoupled. Hence the spin and charge degrees of freedom on different chains are not mixed through $g^\perp_1$ and $g^\perp_2$.

The solution for the charge coupling is straightforward and one obtains a parquet-type solution for the spin part:

$$g^\perp_{1,ij}(\ell) = \frac{\bar{g}^\perp_{1,ij}(\ell)}{1+\bar{g}^\perp_{1,ij}(\ell)\ell}, \tag{6.11}$$

where

$$\bar{g}^\perp_{1,ij}(\ell) = \int_0^\ell d\ell' f_{1,ij}(\ell')[1+\bar{g}^\perp_{1,ij}(\ell')]^2. \tag{6.12}$$

FIGURE 3.20. Perpendicular backward-scattering ($g_1^\perp$) and forward-scattering ($g_2^\perp$) terms generated by interchain pair tunneling.

According to (6.8), $f_{1,ij}(\ell')$ and therefore $\bar{g}^\perp_{1,ij}(\ell)$ are negative in the repulsive sector ($g_1 > 0$ and $2g_2 - g_1 > 0$) where SDW correlations are singular (Fig. 3.11); $g^\perp_{1,ij}(\ell)$ has then the possibility to flow to the strong attractive coupling sector indicating the existence of a spin gap. Given the place held by $g^\perp_{1,ij}(\ell)$ in the hierarchy of generated couplings of the RG transformation, however, this can only occur if the the SDW pair tunneling or exchange term $V_{\mu_P\neq 0,ij}$ first reaches the strong coupling domain – a possibility realized as the system enters the Gaussian critical domain near T_{x^2}. Since the above flow equations (6.10) hold for all pairs of chains ij for $\ell < \ell_{x^1}$, the amplitude of the spin gap goes to zero as the separation $|i-j| \to \infty$. It is worth noting that the spin gap for nearest-neighbor chains is reminiscent of the one occurring in the two-chain problem [68, 69] .

6.1.2 Interchain Response Functions

To establish the nature of correlations introduced by interchain pairing, it is convenient to express outershell decomposition of $S_\perp + \mathcal{S}_\perp$ in the corresponding channel

$$S_{\perp,2} + \mathcal{S}_{\perp,2}\Big|_{\bar{\mu}^\pm} = \frac{1}{2}\pi v_F z^{-2} \sum_{\tilde{q},\bar{\mu}^\pm} \sum_{ij} g^\perp_{\bar{\mu}^\pm,ij}(\ell)\bar{\mathcal{O}}^*_{\bar{\mu}^\pm,ij}(\tilde{q})\mathcal{O}_{\bar{\mu}^\pm,ij}(\tilde{q}) + \text{c.c}, \quad (6.13)$$

where we have introduced the combinations of couplings

$$\begin{aligned} g^\perp_{\text{ISS}^\pm,ij} &= -g^\perp_{1,ij} - g^\perp_{2,ij} \pm U_{\text{ISS},ij} \\ g^\perp_{\text{ITS}^\pm,ij} &= g^\perp_{1,ij} - g^\perp_{2,ij} \pm U_{\text{ITS},ij} \end{aligned} \quad (6.14)$$

for symmetric (+) and antisymmetric (−) interchain superconductivity and their analog expressions

$$\begin{aligned} g^\perp_{\text{ICDW}^\pm,ij} &= g^\perp_{2,ij} - 2g^\perp_{1,ij} \pm U_{\text{ICDW},ij} \\ g^\perp_{\text{ISDW}^\pm,ij} &= g^\perp_{2,ij} \pm U_{\text{ISDW},ij} \end{aligned} \quad (6.15)$$

for interchain density-wave. The corresponding pair fields are given by

$$\mathcal{O}_{\bar{\mu}^\pm,ij} = \tfrac{1}{2}\big(O_{\bar{\mu},ij} \pm O_{\bar{\mu},ji}\big) \quad (6.16)$$

for symmetric (+) and antisymmetric (−) composite fields.

We add the coupling to source fields $S_h = \sum O^*_{\bar{\mu}^\pm,ij} h_{\bar{\mu}^\pm,ij}$+c.c.. The one-loop corrections to the pair vertex part are obtained from $\langle\!\langle (S_{\perp,2} + \mathcal{S}_{\perp,2}) S_h \rangle\!\rangle_{\bar{0},c}$ (which is equivalent to making the substitution $g_\mu \to g^\perp_{\text{ISS}^\pm,ij}$ in Fig. 3.10), and yields the flow equations pair vertex renormalization factor $z_{\bar{\mu}^\pm,ij}$ and hence that of the auxiliary susceptibility $\bar{\chi}_{\bar{\mu}^\pm,ij} = z^2_{\bar{\mu}^\pm,ij}$ (with $z = 1$):

$$\frac{d}{d\ell} \ln \bar{\chi}_{\bar{\mu}^\pm,ij} = g^\perp_{\bar{\mu}^\pm,ij}. \tag{6.17}$$

Obtaining an expression for $\bar{\chi}_{\bar{\mu}^\pm,ij}$ in closed form for arbitrary chains i and j is not possible. However, in the case where intrachain channels have developed only short-range order and $U_{\bar{\mu},ij}$ is only sizable for nearest-neighbor chains, one can extract the dominant trend of interchain auxiliary suceptibilities in the various sectors of the $g_1 g_2$ plane. Thus by taking in Fourier space $V_\mu(\mathbf{Q},\ell) \approx (-1)^{\delta_{\mu\mu_c}} |V_\mu(\ell)| \cos q_\perp$ for the pair tunneling term (5.8), one gets $V_{\mu,ij}(\ell) = \frac{1}{2}(-1)^{\delta_{\mu\mu_c}} |V_\mu(\ell)|$ where $|V_\mu(\ell)| = |V_\mu(\mathbf{Q}_0,\ell)|$ is the maximum amplitude of $V_\mu(\mathbf{Q},\ell)$. This approximation holds sufficiently far from $T_{c,\mu}$, namely, for $\frac{1}{2}\overline{V}_\mu \chi_\mu / \bar{\chi}_\mu$ not too close to unity. The nearest-neighbor ij auxiliary susceptibility then becomes

$$\begin{aligned}\bar{\chi}_{\bar{\mu}^\pm,ij}(\ell) &= X_{\bar{\mu},ij}(\ell) \exp\left\{ \pm \frac{1}{2} \sum_\mu \int_0^\ell c^{\bar{\mu}}_\mu |V_\mu(\ell')|\, d\ell' \right\} \\ &\approx X_{\bar{\mu},ij}(\ell) \prod_\mu \left[1 - \tfrac{1}{2}\left(\bar{\chi}_\mu(\ell)\right)^{-1} \overline{V}_\mu(\ell)\chi_\mu(\ell)\right]^{\mp c^{\bar{\mu}}_\mu},\end{aligned} \tag{6.18}$$

where the values of the channel index μ and the exponent $c^{\bar{\mu}}_\mu$ are given in (6.4)–(6.5). The transient amplitude is given by

$$X_{\bar{\mu},ij}(\ell) = \exp\left\{ \int_0^\ell \beta_{\bar{\mu}}\, d\ell' \right\}, \tag{6.19}$$

where for superconductivity $\beta_{\text{ISS}} = -g^\perp_{1,ij} - g^\perp_{2,ij}$ and $\beta_{\text{ITS}} = g^\perp_{1,ij} - g^\perp_{2,ij}$, while for density-wave $\beta_{\text{ICDW}} = g^\perp_{2,ij} - 2g^\perp_{1,ij}$ and $\beta_{\text{ISDW}} = g^\perp_{2,ij}$. The onset of interchain correlations, even though they are small, will be essentially determined by the most singular correlations of the intrachain channels μ (Fig. 3.21) and the exponent $c^{\bar{\mu}}_\mu$. At finite temperature, we can write

$$\bar{\chi}_{\bar{\mu}^\pm,ij}(T) \approx X_{\bar{\mu},ij}(T) \prod_\mu \left[1 - \tfrac{1}{2}(\bar{\chi}_\mu(T))^{-1} \overline{V}_\mu(T)\chi_\mu(T)\right]^{\mp c^{\bar{\mu}}_\mu}. \tag{6.20}$$

In the repulsive sector where $g_1 > 0$, $2g_2 - g_1 > 0$, SDW correlations are singular (region I, Fig. 3.21) and lead to an enhancement of symmetric (+) interchain singlet superconducting (ISS$^+$) correlations with $c^{\bar{\mu}_C=0}_{\mu_P\neq 0} = \frac{3}{2}$. It should be noted here that, in this sector, symmetric interchain triplet

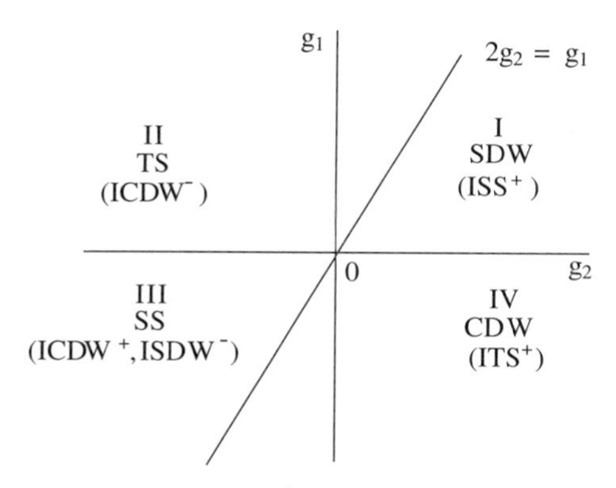

FIGURE 3.21. Phase diagram of the quasi-one-dimensional electron gas with interchain hopping and no Umklapp scattering. Dominant correlations of intrachain (interchain) pairing channels are given.

(ITS$^+$) correlations are also enhanced by SDW correlations with $c^{\bar{\mu}_C \neq 0}_{\mu_P \neq 0} = \frac{1}{2}$. The pairing on different chains results from the exchange of spin fluctuations. On the T-L line at $g_1 = 0$ and $g_2 > 0$, SDW and CDW correlations are equally singular leading to an enhancement of both ISS$^+$ and ITS$^+$ correlations with a net exponent of unity.

In region II of Fig. 3.21, where $g_1 > 0$, $2g_2 - g_1 < 0$, the TS channel is the most singularly enhanced leading to interchain charge-density-wave (ICDW$^-$) correlations with $c^{\bar{\mu}_P = 0}_{\mu_P \neq 0} = \frac{3}{2}$. Here the pairing results from a three-component (triplet) Cooper pair exchange between the chains. In the same region, an enhancement, although smaller, of antisymmetric interchain spin-density-wave correlations (ISDW$^-$) is also present with $c^{\bar{\mu}_P \neq 0}_{\mu_P \neq 0} = \frac{1}{2}$. On the T-L line at $g_2 < 0$, however, SS and TS are equally singular so both contribute to give the same enhancement for ISDW$^-$ and ICDW$^-$. If one moves to the attractive sector (region III, Fig. 3.21), where $g_1 < 0$, $2g_2 - g_1 < 0$ and SS is the most singular channel along the chains, one finds that symmetric ICDW$^+$ and antisymmetric ISDW$^-$ channels have similar enhancements $c^{\bar{\mu}_P = 0}_{\mu_C = 0} = c^{\bar{\mu}_P \neq 0}_{\mu_C = 0} = \frac{1}{2}$. Finally in the region IV, where $g_1 < 0$, $2g_2 - g_1 > 0$ and CDW correlations are the most singular, symmetric ITS$^+$ correlations are enhanced in the interchain triplet channel with the exponent $c^{\bar{\mu}_C \neq 0}_{\mu_P = 0} = \frac{1}{2}$.

6.2 *Possibility of Long-Range Order in the Interchain Pairing Channels*

If the initial parameters of the model are such that the system can reach the energy or temperature domain T_{x^1} without developing long-range order, the pair-hopping amplitude V_μ is still in the weak coupling regime. We have seen in Section 5.2.1, however, that infrared singularities in both the Peierls and Cooper channels persist above ℓ_{x^1} so that long-range order in the intrachain channel μ can occur if $\mathcal{G}_\mu < 0$. This will be so in the Peierls channel when deviations from perfect nesting of the warped Fermi surface do not exceed a given threshold (Fig. 3.17a), while in the Cooper channel if a magnetic field is below a critical value. Therefore, sufficiently large nesting deviations or magnetic fieldsS must be introduced in the model in order for long-range order in the *interchain* channels to be possible. We will now illustrate how such a possibility can be realized in the repulsive case when nesting deviations are introduced in the model. We will thus focus our attention in the region I of the phase diagram where symmetric interchain singlet superconductivity (ISS$^+$) can be stabilized (Fig. 3.21). A similar description holds for the Peierls interchain channel in the presence of a sufficiently strong magnetic field (regions II and III in Fig. 3.21).

Frustration of nesting is commonly introduced by considering the possibility of particles hopping to second nearest-neighbor chains, i.e. by adding the term $-2t_{\perp 2}\cos 2k_\perp$ to the spectrum at $\ell = 0$, where $t_{\perp 2}$ is the amplitude of the hopping ($t_{\perp 2} \ll t_\perp$). The relevant parameter space of the action at the crossover value ℓ_{x^1} becomes $\mu_S(\ell_{x^1}) = (G_p^0(\mathbf{k},\omega_n), \mathcal{G}_\mu, g^\perp_{\bar{\mu}\pm})$, where

$$G_p^0(\mathbf{k},\omega_n) = \frac{z}{i\omega_n - \epsilon_p(k) + 2t^*_\perp \cos k_\perp + 2t^*_{\perp 2}\cos 2k_\perp} \tag{6.21}$$

is the effective propagator in which z is the quasi-particle weight at ℓ_{x^1} and $t^*_{\perp(2)} \equiv z t_{\perp(2)}$. If one cranks up the amplitude of $t^*_{\perp 2}$, nesting of the Fermi surface at $\mathbf{Q}_0 = (2k_\mathrm{F}, \pi)$ is quickly suppressed so that the $T_{\mathrm{c,SDW}}$ in the region I of the phase diagram decreases rapidly and eventually vanishes above the critical $t^{*c}_{\perp 2}$ [70]. As for the infrared singularity of the Cooper channel, however, it is still present and can then lead to an instability of the normal state.

We will adopt here a simple ladder RG description of this instability, which neglects the interference between the—weakened—Peierls channel and the Cooper one. This approximation should be qualitatively valid for $t^*_{\perp 2} \gg t^{*c}_{\perp 2}$ (a more quantitative description including the interference effect can be found in [67]). We will further assume for simplicity that the enhancement of $\mathcal{G}_\mu$ above ℓ_{x^1} is negligible and the transverse short-range order of the chains is entirely contained in the ij dependence of $g^\perp_{\mathrm{ISS}^+,ij}$ in (6.14) and (6.5), which is restricted to nearest-neighbor chains. In this way, the repulsive coupling $\mathcal{G}_\mu$ is entirely local and does not compete with the interchain part: fermions on different chains are said to avoid intrachain repulsion [34, 36]. In transverse Fourier space, the outer shell decomposition

(6.13) in the channel ISS^+ becomes

$$S_{\perp,2} + \mathcal{S}_{\perp,2}\Big|_{\mathrm{ISS}^+}$$
$$= \frac{1}{2}\pi v_{\mathrm{F}} z^{-2} \sum_{\tilde{q}} \sum_{k_\perp,k'_\perp} g_{\mathrm{SS}}^{\perp}(\ell)\, \eta_{k_\perp} \bar{O}_{\mathrm{SS}}^{*}(\tilde{q},k_\perp) \eta_{k'_\perp} O_{\mathrm{SS}}(\tilde{q},k'_\perp) + \mathrm{c.c} \quad (6.22)$$

in which we have defined $O_{\mathrm{SS}}(\tilde{q}) = \sum_{k_\perp} O_{\mathrm{SS}}(\tilde{q},k_\perp)$ and where $\eta_{k_\perp} = \cos k_\perp$ is the Fourier factor that leads to a global antisymmetric order parameter for interchain singlet superconductivity. The combination of couplings at ℓ

$$g_{\mathrm{ISS}^+}^{\perp}(\ell) \equiv -g_1^{\perp}(\ell) - g_2^{\perp}(\ell) + U_{\mathrm{ISS}}(\ell) \quad (6.23)$$

is obtained from the amplitude of the transverse couplings at ℓ. In the ladder approximation, the outer shell integration $\frac{1}{2}\langle(S_{\perp,2}+\mathcal{S}_{\perp,2})^2\rangle_{\bar{0},c}$ is done in this channel. At zero external pair momentum, this leads to the flow equation

$$\frac{d}{d\ell} g_{\mathrm{ISS}^+}^{\perp} = \frac{1}{4}(g_{\mathrm{ISS}^+}^{\perp})^2, \quad (6.24)$$

which is easily integrated to give the simple pole expression

$$g_{\mathrm{ISS}^+}^{\perp}(\ell) = \frac{g_{\mathrm{ISS}^+}^{\perp}(\ell_{x^1})}{1 - \frac{1}{4} g_{\mathrm{ISS}^+}^{\perp}(\ell_{x^1})(\ell - \ell_{x^1})}. \quad (6.25)$$

The critical temperature for the superconducting instability is then $T_c = T_{x^1}\exp[-4/g_{\mathrm{ISS}^+}^{\perp}(\ell_{x^1})]$ where $g_{\mathrm{ISS}^+}^{\perp}(\ell_{x^1}) > 0$ corresponds to a net attraction in the interchain superconducting channel. From (6.22), the singlet order parameter is of the form

$$\Delta(k_\perp) = \Delta_0 \eta_{k_\perp}, \quad (6.26)$$

which has nodes on the Fermi surface.

The calculation of the interchain superconducting response causes no particular difficulty. In effect, by adding the coupling to a source field ($S_h = \sum O_{\bar{\mu}^+}^{*}(\tilde{q},k_\perp)\eta_{k_\perp} h_{\bar{\mu}^+} + \mathrm{c.c.}$), one obtains the flow equation for the auxiliary susceptibility

$$\frac{d}{d\ell} \ln \bar{\chi}_{\mathrm{ISS}^+} = \frac{1}{2} g_{\mathrm{ISS}^+}^{\perp}. \quad (6.27)$$

Using (6.25), this is integrated to give the double pole singularity

$$\bar{\chi}_{\mathrm{ISS}^+}(\ell) = \frac{\bar{\chi}_{\mathrm{ISS}^+}(\ell_{x^1})}{[1 - \frac{1}{4} g_{\mathrm{ISS}^+}^{\perp}(\ell_{x^1})(\ell - \ell_{x^1})]^2}, \quad (6.28)$$

where the boundary condition $\bar{\chi}_{\mathrm{ISS}^+}(\ell_{x^1})$ is given by the solution of (6.17) at ℓ_{x^1}. The loop integration of the above expression following the definition

(4.49), leads to the response function

$$\chi_{\mathrm{ISS}^+}(\ell) = \chi_{\mathrm{ISS}^+}(\ell_{x^1}) - (2\pi v_{\mathrm{F}})^{-1} \frac{\bar{\chi}_{\mathrm{ISS}^+}(\ell_{x^1})(\ell - \ell_{x^1})}{1 - \frac{1}{4} g_{\mathrm{ISS}^+}^{\perp}(\ell_{x^1})(\ell - \ell_{x^1})}, \tag{6.29}$$

which has the expected simple pole structure of the ladder approximation.

7 Summary and Concluding Remarks

We have reexamined the scaling tools that are used in the description of quasi-one-dimensional interacting fermion systems. On phenomenological grounds, we have emphasized that the scaling ansatz can provide a qualitative understanding of the existence of power law behavior of various quantities in one dimension while its extension for nonzero interchain coupling has allowed us to describe the different forms of dimensionality crossover in the quasi-one-dimensional case. A key feature of this phenomenology is the existence of fluctuations of single fermion and pair degrees of freedom on many length or energy scales, the origin of which can be microscopically understood using the renormalization group language. We found it useful to reconsider the formulation of this methodology and we have seen that the adoption of a classical Wilson procedure for the Kadanoff transformation has a molding influence on the RG flow beyond the one-loop level. Thus the constraint of a sharp natural cutoff k_0 on the fermion spectrum which keeps the loop momenta in the outer energy shells inevitably leads to a nonlocal flow of the parameters that define the action. Nonlocality results from the inclusion of many-particle marginal interactions in the outer shell integration. These are not present in the bare action but are generated along the RG flow. The resulting structure of outer shell integration finds an interesting parallel with the functional description of the renormalization group equations proposed by Polchinski in the framework of the ϕ^4 theory [71], and used by Zanchi and Schulz in the context of two-dimensional interacting fermion systems [20].

It is only when the marginally relevant or irrelevant couplings are treated in the local approximation that purely logarithmic scaling is recovered. Although this has only been verified at the two-loop loop level and in the one-dimensional case, similar conclusions are expected to hold at higher order and in higher dimensions as well. The procedure described in this review casts light on a previous alternative formulation of the RG which is known to offer a more direct route to logarithmic scaling at high order but which on the other hand requires the relaxation of some of the standard rules of the Wilson method. It also highlights how the difficulties linked to nonlocality of the flow equations are essentially hidden in older versions of the renormalization group.

The RG method is widely recognized as a tool well suited to the study of the influence of interchain hopping on the stability of the Luttinger and Luther–Emery liquid states in one dimension. We have reviewed the conditions under which the coupling between stacks can make these non-Fermi liquid states unstable. A possible instability occurs when the Luttinger liquid yields to the formation of a Fermi liquid. The recovery of quasi-particle excitations, here called a dimensionality crossover for the fermion quantum coherence, squares well with the picture provided by the extended scaling ansatz. A second instability emerges when the kinematics of interchain single-particle hopping is combined with intrachain pair correlations. Effective pair-hopping processes between chains then form and their coupling to singular correlations along the chains lead to a distinct (two-particle) dimensionality crossover. This is physically meaningful as long as the corresponding temperature scale is above the one predicted for single particles. Such a situation is favored by cranking up interactions, which on the one hand increases pair correlations, and on the other hand reduces single particle coherence. The temperature scale for two-particle dimensionality crossover marks the onset of long-range order, which accords well with the extended scaling ansatz. For sufficiently weak interactions, however, single-particle deconfinement occurs first and the possibility for long-range order remains as an instability of the Fermi liquid. In this case, the ladder level of the RG gives a BCS character to the transition.

The RG method can also be useful in studying how the Kohn–Luttinger mechanism for unconventional pairing works in the quasi-one-dimensional case. We have seen that this mechanism opens up channels of correlation which pair particles on different stacks as a result of pair-hopping processes between chains. Since these are present in both the Peierls and Cooper channels, the Kohn–Luttinger mechanism has a dual nature. Thus it was shown not only that the exchange of density-wave correlations can lead to interchain Cooper pairing but also that interchain Peierls pairing between a particle and a hole is also possible when both particles exchange Cooper pairs. Long-range order in these unconventional channels can only be achieved from the Fermi liquid state in which enough nesting deviations or magnetic field are present to frustrate the formation of long-range order in the the primary channels of correlation along the chains.

The results presented in this review by no means exhaust the range of possibilities supplied by the RG method. We will end this article by giving two specific examples. The first concerns the issue of single-particle deconfinement that leads to the recovery of a Fermi liquid. This question is of great interest especially in connection with the normal state properties obeserved in concrete realizations of quasi-one-dimensional materials like the organic conductors [18, 62]. We have seen that the critical domain of the primary (Luttinger liquid) fixed point in a quasi-one-dimensional system essentially covers the whole temperature range of interest. This favors in turn a slow and gradual crossover to a Fermi liquid, which means in practice that transitory effects are likely to be important in the description of these materials. Although crossover scaling effects have been underestimated in the treatment given above, they are not beyond the reach of the RG method.

Another promising avenue is the application of the RG method to interacting fermions on a lattice [72]. The Kadanoff transformation can actually be generalized to fermions with the full tight binding spectrum. Transients to the continuum limit can then be considered in a quantitative way. In the case of the fermion representation of the one-dimensional spin chain, for example, this has been shown to be an essential ingredient in the description of the thermodynamics.

Acknowledgments: C.B. thanks L.G. Caron, N. Dupuis, A.-M. Tremblay, and D. Sénéchal for numerous discussions and R. Shankar and G. Kotliar for interesting comments about the RG method at high order. This work is supported by the Natural Sciences and Engineering Research Council of Canada (NSERC), le Fonds pour la Formation de Chercheurs et l'Aide à la Recherche du Gouvernement du Québec (FCAR), and by the "superconductivity program" of the Institut Canadien de Recherches Avancées (CIAR).

8 References

[1] K.G. Wilson, Rev. Mod. Phys. **47**, 773 (1975).

[2] K.G. Wilson, Rev. Mod. Phys. **55**, 583 (1983).

[3] E. Reidel and F. Wegner, Z. Physik **225**, 195 (1969).

[4] P. Pfeuty, D. Jasnow, and M. Fisher, Phys. Rev. B **10**, 2088 (1974).

[5] M.K. Grover, Phys. Lett. **44A**, 253 (1973).

[6] J. Hertz, Phys. Rev. B **14**, 1165 (1976).

[7] S. Sachdev, *Quantum Phase Transitions* (Cambridge Univ. Press, Cambridge, U.K., 1999).

[8] Y.A. Bychkov, L.P. Gorkov, and I. Dzyaloshinskii, Sov. Phys. JETP **23**, 489 (1966).

[9] I.E. Dzyaloshinskii and A.I. Larkin, Sov. Phys. JETP **34**, 422 (1972).

[10] N. Menyhard and J. Solyom, J. Low Temp. Phys. **12**, 529 (1973).

[11] J. Solyom, Adv. Phys. **28**, 201 (1979).

[12] C. Bourbonnais, Ph.D. thesis, Université de Sherbrooke, 1985.

[13] C. Bourbonnais, Mol. Cryst. Liq. Cryst. **119**, 11 (1985).

[14] C. Bourbonnais and L.G. Caron, Physica **143B**, 450 (1986).

[15] C. Bourbonnais and L.G. Caron, Int. J. Mod. Phys. B **5**, 1033 (1991).

[16] C. Bourbonnais, in *Les Houches, Session LVI (1991), Strongly Interacting Fermions and High-T_c Superconductivity*, eds. B. Doucot and J. Zinn-Justin (Elsevier Science, Amsterdam, 1995), p. 307.

[17] D. Jérome and H. Schulz, Adv. in Physics **31**, 299 (1982).

[18] C. Bourbonnais and D. Jérome, in *Advances in Synthetic Metals, Twenty Years of Progress in Science and Technology*, eds. P. Bernier, S. Lefrant, and G. Bidan (Elsevier, New York, 1999), pp. 206–261, arXiv:cond-mat/9903101.

[19] R. Shankar, Rev. Mod. Phys. **66**, 129 (1994).

[20] D. Zanchi and H.J. Schulz, Phys. Rev. B **61**, 13 609 (2000).

[21] C.J. Halborth and W. Metzner, Phys. Rev. B **85**, 5164 (2000).

[22] C. Honerkamp, M. Salmhofer, N. Furukawa, and T.M. Rice, Phys. Rev. B **63**, 35109 (2001).

[23] G. Chitov and D. Sénéchal, Phys. Rev. B **52**, 129 (1995).

[24] N. Dupuis, Eur. Phys. J. B **3**, 315 (1998).

[25] N. Dupuis and G. Chitov, Phys. Rev. B **54**, 3040 (1996).

[26] M. Fabrizio, Phys. Rev. B **48**, 15838 (1993).

[27] H.-H. Lin, L. Balents, and M.P.A. Fisher, Phys. Rev. B **56**, 6569 (1997).

[28] J. Kishine and K. Yonemitsu, J. Phys. Soc. of Jpn. **67**, 1714 (1998).

[29] D. Zanchi, Europhys. Lett. **55**, 376 (2001).

[30] D.B. B.Binz and B. Doucot, arXiv:cond-mat/0104424 (unpublished).

[31] M. Salmhofer, *Renormalization, An Introduction* (Springer, Berlin, Heidelberg, 1999).

[32] This review will not discuss the real-space renormalization group approach based on the bosonic degrees of freedom description of the one-dimensional electron gas and ladder systems [49, 52, 68, 69]. See also D. Sénéchal, this volume; S.T. Chui and P.A. Lee, Phys. Rev. Lett. **35**, 325 (1975); T. Giamarchi and H. Schulz, J. Phys. (Paris) **49**, 819 (1988); V. Yakovenko, JETP Lett., **56**, 510 (1992); N. Nerseyan, A. Luther and F. V. Kusmartsev, Physics Letters **A 176**, 363 (1993); D. V. Khveshchenko and T. M. Rice, Phys. Rev. B **50**, 252 (1994); U. Ledermann, K. Le Hur and T. M. Rice, Phys. Rev **62**, 16383 (2000); P. Donohue, M. Tsuchiizu, T. Giamarchi and Y. Suzumura, Phys. Rev. B **63**, 45121 (2001) and references there cited.

[33] W. Kohn and J.M. Luttinger, Phys. Rev. Lett. **15**, 524 (1965).

[34] V.J. Emery, Synthetic Metals **13**, 21 (1986).

[35] L.G. Caron and C. Bourbonnais, Physica **143B**, 453 (1986).

[36] M.T. Béal-Monod, C. Bourbonnais, and V.J. Emery, Phys. Rev. B **34**, 7716 (1986).

[37] C. Bourbonnais and L.G. Caron, Europhys. Lett. **5**, 209 (1988).

[38] B. Guay and C. Bourbonnais, Synthetic Metals **103**, 2180 (1999).

[39] B. Guay, Ph.D. thesis, Université de Sherbrooke, 1997.

[40] T. Ishiguro and K. Yamaji, *Organic Superconductors*, Vol. 88 of *Springer-Verlag Series in Solid-State Science* (Springer-Verlag Heidelberg, 1990).

[41] N.D. Mermin and H. Wagner, Phys. Rev. Lett. **17**, 1133 (1966).

[42] C. Bourbonnais, J. Phys. I (France) **3**, 143 (1993).

[43] J.W. Negele and H. Orland, *Quantum Many Particle Systems*, Vol. 68 of *Frontiers in Physics Series* (Addison-Wesley, New York, 1988).

[44] M. Kimura, Prog. Theor. Phys. **63**, 955 (1975).

[45] M. Fowler, Solid State Commun. **18**, 241 (1976).

[46] E.H. Rezayi, J. Solyom, and J. Sak, Phys. Rev. B **23**, 1342 (1981).

[47] F.D.M. Haldane, J. Phys. C **14**, 2585 (1981).

[48] A. Luther and V.J. Emery, Phys. Rev. Lett. **33**, 589 (1974).

[49] V.J. Emery, in *Highly Conducting One-Dimensional Solids*, eds. J.T. Devreese, R.E. Evrard, and V.E. van Doren (Plenum Press, New York, 1979), p. 247.

[50] E. Lieb and F.Y. Wu, Phys. Rev. Lett. **20**, 1445 (1968).

[51] A. I. Larkin and J. Sak, Phys. Rev. Lett **39**, 1025 (1977).

[52] J. Voit, Rep. Prog. Phys. **58**, 977 (1995).

[53] C. Bourbonnais et al., J. Phys. (Paris) Lett. **45**, L755 (1984).

[54] M. Brech, J. Voit, and H. Buttner, Europhys. Lett. **12**, 289 (1990).

[55] J. Voit, Eur. Phys. J. B **5**, 505 (1998).

[56] E.W. Carlson, D. Orgad, S.A. Kivelson, and V.J. Emery, Phys. Rev. B **62**, 3422 (2000).

[57] S. Capponi, D. Poilblanc, and E. Arrigoni, Phys. Rev. B **57**, 6360 (1998).

[58] A. Vishwanath and D. Carpentier, Phys. Rev. Lett. **86**, 676 (2001).

[59] D. Boies, C. Bourbonnais, and A.-M. Tremblay, Phys. Rev. Lett. **74**, 968 (1995).

[60] E. Arrigoni, Phys. Rev. Lett. **80**, 790 (1998).

[61] L.G. Caron et al., Synthetic Metals **27B**, 123 (1988).

[62] S. Biermann, A. Georges, A. Lichtenstein, and T. Giamarchi, Phys. Rev. Lett. **87**, 276405 (2001), arXiv:cond-mat/0107633.

[63] Y.A. Firsov, Y.N. Prigodin, and C. Seidel, Rep. Prog. Phys. **126**, 245 (1985).

[64] S. Brazovskii and Y. Yakovenko, J. Phys. (Paris) Lett. **46**, L (1985).

[65] Y. Suzumura, J. Phys. Soc. Jpn. **54**, 2386 (1985).

[66] V.N. Prigodin and Y.A. Firsov, Sov. Phys. JETP **49**, 369 (1979).

[67] R. Duprat and C. Bourbonnais, Eur. Phys. J. B **21**, 219 (2001).

[68] H.J. Schulz, Phys. Rev. B **53**, 2959 (1996).

[69] H.-H. Lin, L. Balents, and M.P.A. Fisher, Phys. Rev. B **58**, 1794 (1998).

[70] K. Yamaji, J. Phys. Soc. of Japan **51**, 2787 (1982).

[71] J. Polchinski, Nucl. Phys. B **231**, 269 (1984).

[72] B. Dumoulin et al., Phys. Rev. Lett. **76**, 1360 (1996).

Appendix A One-Particle Self-Energy at the Two-Loop Level

A.1 Backward- and Forward-Scattering Contributions

Following the partial trace operation in the absence of Umklapp scattering, three-particle interactions are generated from the outer shell contraction $\delta S_\lambda^{(j)} \equiv \frac{1}{2}\langle S_{I,1}^2\rangle_{\bar{0},c}$. The corresponding expression reads

$$\begin{aligned}
\delta S_\lambda^{(j)} &= (\pi v_\mathrm{F})^2 \frac{T^2}{L^2} \sum_{\{\alpha,\alpha'\}} \sum_{\tilde{k}_{1,2},\tilde{k}'_2,\tilde{q}'} \sum_{\tilde{q}} \lambda^+_{\{\alpha,\alpha'\}}(\ell_{j-3}) G^0_+(\tilde{k}_1+\tilde{q}) \psi^*_{+,\alpha'_1}(\tilde{k}_1+\tilde{q}'+\tilde{q}) \\
&\quad \times \psi^*_{-,\alpha'_2}(\tilde{k}'_2-\tilde{q}') \psi^*_{-,\alpha_2}(\tilde{k}_2-\tilde{q}) \psi_{-,\alpha'_3}(\tilde{k}'_2) \psi_{-,\alpha_3}(\tilde{k}_2) \psi_{+,\alpha_4}(\tilde{k}_1) \\
&\quad + (\pi v_\mathrm{F})^2 \frac{T^2}{L^2} \sum_{\{\alpha,\alpha'\}} \sum_{\tilde{k}_1,\tilde{k}'_{1,2},\tilde{q}'} \sum_{\tilde{q}} \lambda^-_{\{\alpha,\alpha'\}}(\ell_{j-3}) G^0_-(\tilde{k}_2-\tilde{q}) \psi^*_{-,\alpha'_2}(\tilde{k}'_2-\tilde{q}') \\
&\quad \times \psi^*_{+,\alpha'_1}(\tilde{k}'_1+\tilde{q}') \psi^*_{+,\alpha_1}(\tilde{k}_1+\tilde{q}) \psi_{+,\alpha'_4}(\tilde{k}'_1) \psi_{+,\alpha_4}(\tilde{k}_1) \psi_{-,\alpha_3}(\tilde{k}_2)
\end{aligned} \tag{A.1}$$

where $g_{\{\alpha\}}$ is given by (4.2), $\lambda^+_{\{\alpha,\alpha'\}}(\ell_{j-3}) = g_{\{\alpha\}}(\ell_{j-3}) g_{\{\alpha'\}}(\ell_{j-3}) \delta_{\alpha_1\alpha'_4}$, $\lambda^-_{\{\alpha,\alpha'\}}(\ell_{j-3}) = g_{\{\alpha\}}(\ell_{j-3}) g_{\{\alpha'\}}(\ell_{j-3}) \delta_{\alpha_2\alpha'_3}$, and $\ell_{j-3} = \ln(E_0/E_0(\ell_{j-3}))$. Here the summation $\sum_{\tilde{q}}$ covers all ω_m but q is restricted such that k_1+q (k_2-q) in the propagator G^0_+ (G^0_-) is in the outer momentum shell at the step $(j-2)d\ell$ for $j \geq 3$. The two terms in (A.1), which depends on the step j, correspond to the diagrams of Fig. 3.8a. At the next partial integration $(j-1)d\ell$, the above contributions will in turn be contracted, leading to a second fermion line at the adjacent momentum shell (Fig. 3.8b). When this set of contractions is repeated as a function of j, contributions add to give an effective two-particle interaction of the form

$$\begin{aligned}
\delta S'_\lambda &\equiv \sum_{j=3}^{N-1} \langle \delta S_{\lambda,2}^{(j)} \rangle_{\bar{0},c} \\
&= (\pi v_\mathrm{F})^2 \frac{T^2}{L^2} \sum_{\tilde{k},\tilde{q}'} \sum_{j=3}^{N-1} \sum_{\{\alpha,\alpha'\}} \sum_{\tilde{k}'}{}^{*} \sum_{\tilde{q}} \lambda^+_{\{\alpha,\alpha'\}}(\ell_{j-3}) \\
&\quad \times \Big\{ \delta_{\alpha'_3\alpha_2} G^0_+(\tilde{k}+\tilde{q}) G^0_-(\tilde{k}'-\tilde{q}) \psi^*_{+,\alpha'_1}(\tilde{k}+\tilde{q}'+\tilde{q}) \\
&\qquad\qquad \times \psi^*_{-,\alpha'_2}(\tilde{k}'-\tilde{q}'-\tilde{q}) \psi_{-,\alpha_3}(\tilde{k}') \psi_{+,\alpha_4}(\tilde{k}) \\
&\qquad + \delta_{\alpha'_2\alpha_3} G^0_+(\tilde{k}+\tilde{q}) G^0_-(\tilde{k}'-\tilde{q}') \psi^*_{+,\alpha'_1}(\tilde{k}+\tilde{q}'+\tilde{q}) \\
&\qquad\qquad \times \psi^*_{-,\alpha_2}(\tilde{k}'-\tilde{q}'-\tilde{q}) \psi_{-,\alpha'_3}(\tilde{k}') \psi_{+,\alpha_4}(\tilde{k}) \Big\}
\end{aligned}$$

$$+ (\pi v_{\rm F})^2 \frac{T^2}{L^2} \sum_{\{\alpha,\alpha'\}} \sum_{\tilde{k}} \sum_{j=3}^{N-1} \sum_{\tilde{k}_1} \sum_{\tilde{q}} \lambda^+_{\{\alpha,\alpha'\}}(\ell_{j-3}) \delta_{\alpha'_1,\alpha_4} G^0_+(\tilde{k}_1 + \tilde{q})$$
$$\times G^0_+(\tilde{k}_1) \psi^*_{-,\alpha'_2}(\tilde{k}' + \tilde{q}) \psi^*_{-,\alpha_2}(\tilde{k} - \tilde{q}) \psi_{-,\alpha'_3}(\tilde{k}') \psi_{-,\alpha_3}(\tilde{k}) \quad \text{(A.2)}$$

for λ^+ and an analogous expression for λ^-. In the first two terms, the summation $\sum^*_{\tilde{k}'}$ covers all fermion frequencies $\omega_{n'}$ and its momentum interval is such that $k' - q$ is in the outer shell while k' is still in the inner shell. These expressions correspond to the two series of diagrams shown in Fig. 3.8c, while the last term (not shown in Fig. 3.8c) only involves outgoing fermions near $-k_{\rm F}$. The final contraction of the cascade is obtained by putting k' in the outer momentum shell at $Nd\ell$:

$$\langle \delta S'_{\lambda,2} \rangle_{\bar{0},c}$$
$$= 2(\pi v_{\rm F})^2 \frac{T^2}{L^2} \sum_{\alpha} \sum_{j=3}^{N-1} \sum^*_{\tilde{k}'} \sum_{\tilde{q}} (g_1^2(\ell_{j-3}) + g_2^2(\ell_{j-3}) - g_1(\ell_{j-3}) g_2(\ell_{j-3}))$$
$$\times \big\{ \big[G^0_+(\tilde{k}+\tilde{q}) G^0_-(\tilde{k}'-\tilde{q}) G^0_-(\tilde{k}') + G^0_+(\tilde{k}+\tilde{q}) G^0_-(\tilde{k}'+\tilde{q}) G^0_-(\tilde{k}') \big] \psi^*_{+,\alpha}(\tilde{k}) \psi_{+,\alpha}(\tilde{k})$$
$$+ \big[G^0_+(\tilde{k}' + \tilde{q}) G^0_-(\tilde{k} - \tilde{q}) G^0_-(\tilde{k}') + \cdots \big] \psi^*_{-,\alpha}(\tilde{k}) \psi_{-,\alpha}(\tilde{k}) \big\}. \quad \text{(A.3)}$$

The first two terms in the sum lead to a self-energy correction for fermions (on the $+k_{\rm F}$ branch) the momenta, k, of which are located at the top of the inner shell. The outer shell contractions leading to the last terms are self-energy corrections for fermions on the $-k_{\rm F}$ branch (only one such term is shown explicitly). Owing to the outer shell constraints put on the momentum, however, when the summations over fermion and boson frequencies are carried out in the low temperature limit, all these $-k_{\rm F}$ branch terms give a vanishingly small contribution in comparison with the first two terms and can be safely ignored.

$$\langle \delta S'_{\lambda,2} \rangle_{\bar{0},c} = - \sum_{\alpha} \Big\{ \int_{E_0(\ell_N)/2}^{E_0(\ell_{N-1})/2} \sum_{j=3}^{N-1} \int_{-E_0(\ell_N)/2 - E_0(\ell_{j-3})/2}^{-E_0(\ell_N)/2 - E_0(\ell_{j-2})/2}$$
$$+ \int_{-E_0(\ell_{N-1})/2}^{-E_0(\ell_N)/2} \sum_{j=3}^{N-1} \int_{E_0(\ell_N)/2 + E_0(\ell_{j-2})/2}^{E_0(\ell_N)/2 + E_0(\ell_{j-3})/2} \Big\} d\epsilon' d(v_{\rm F} q)$$
$$\times (g_1^2(\ell_{j-3}) + g_2^2(\ell_{j-3}) - g_1(\ell_{j-3}) g_2(\ell_{j-3}))$$
$$\times \frac{1}{[G^0_+(\tilde{k})]^{-1} - 2 v_{\rm F} q} \psi^*_{+\alpha}(\tilde{k}) \psi_{+\alpha}(\tilde{k}). \quad \text{(A.4)}$$

The integrand turns out to be independent of $\epsilon' \equiv \epsilon_-(k')$ and we will define $\delta E_0/2 = \int_{-\frac{1}{2}E_0(\ell_{N-1})}^{-\frac{1}{2}E_0(\ell_N)} d\epsilon' = \int_{\frac{1}{2}E_0(\ell_N)}^{\frac{1}{2}E_0(\ell_{N-1})} d\epsilon'$. Now, assuming that the coupling constants vary slowly as a function of j, this expression can be evaluated in the local approximation (Fig. 3.8d):

$$\begin{aligned}
&\langle \delta S'_{\lambda,2}\rangle_{\bar{0},c} \\
&= -\frac{\delta E_0}{2}\sum_\alpha \Big\{\int_{-(E_0(\ell)+E_0)/2}^{-E_0(\ell))} + \int_{E_0(\ell)}^{(E_0(\ell)+E_0)/2}\Big\} \\
&\qquad \times (g_1^2(\ell_q) + g_2^2(\ell_q) - g_1(\ell_q)g_2(\ell_q))\frac{d(v_{\rm F}q)}{[G_+^0(\tilde{k})]^{-1} - 2v_{\rm F}q}\psi^*_{+\alpha}(\tilde{k})\psi_{+\alpha}(\tilde{k}) \\
&\simeq -\frac{\delta E_0}{2}(g_1^2(\ell) + g_2^2(\ell) - g_1(\ell)g_2(\ell)) \\
&\times \sum_\alpha \Big\{\int_{-(E_0(\ell)+E_0)/2}^{-E_0(\ell)} + \int_{E_0(\ell)}^{(E_0(\ell)+E_0)/2}\Big\}\frac{d(v_{\rm F}q)}{[G_+^0(\tilde{k})]^{-1} - 2v_{\rm F}q}\psi^*_{+\alpha}(\tilde{k})\psi_{+\alpha}(\tilde{k}) \\
&\simeq \frac{1}{4}(g_1^2(\ell) + g_2^2(\ell) - g_1(\ell)g_2(\ell))t(\ell)d\ell \sum_\alpha [G_+^0(\tilde{k})]^{-1}\psi^*_{+\alpha}(\tilde{k})\psi_{+\alpha}(\tilde{k}) \quad \text{(A.5)}
\end{aligned}$$

to leading order in $[G_+^0(\tilde{k})]^{-1}$. Here $\ell_q = \ln E_0/|v_{\rm F}q|$, $t(\ell) = 1 - 2/(1 + e^\ell)$ and $E_0(\ell_{N-3}) \to E_0(\ell)$ in the limit $d\ell \to 0$. The final expression then becomes the outer shell contribution of the renormalization coefficient $z^{-1}(\ell)$ of $[G_+^0]^{-1}$ for the inner shell state k in the local approximation. The presence of the factor $t(\ell)$ indicates that the above result is not logarithmic over the entire range, especially at small ℓ where transients or scaling deviations exist [15]. These become vanishingly small at large ℓ and are neglected to logarithmic accuracy. A similar cascade of contractions applies to the second term of (A.1) with λ^- and this leads to the same renormalization for the one-particle propagator near $-k_{\rm F}$.

A.2 Umklapp Contribution

When Umklapp scattering is present, the contraction $\delta S_u^{(j)} \equiv \frac{1}{2}\langle (S_{I,1})^2\big|_{g_3}\rangle_{\bar{0},c}$, that involves g_3 alone, generates a three-particle interaction at the step $(j-2)d\ell$ with $j \geq 3$:

$$\begin{aligned}
&\delta S_u^{(j)} \\
&= (\pi v_{\rm F})^2\frac{T^2}{L^2}\sum_{\{\tilde{k},\widetilde{Q}'\}}\sum_{\widetilde{Q}'}\lambda_u^-(\ell_{j-3})\, G_-^0(\tilde{k}'_1 - \widetilde{Q}')\psi^*_{+,\alpha_1}(\tilde{k}_1 + \widetilde{Q}) \\
&\times \psi^*_{-,\alpha'_1}(\tilde{k}_1 - \widetilde{Q} - \widetilde{Q}' + \widetilde{G})\psi^*_{-,\alpha'_2}(\tilde{k}'_2 + \widetilde{Q}' - \widetilde{G})\psi_{-,\alpha_1}(\tilde{k}_1)\psi_{+,\alpha'_1}(\tilde{k}'_1)\psi_{+,\alpha'_2}(\tilde{k}'_2) \\
&+(\pi v_{\rm F})^2\frac{T^2}{L^2}\sum_{\{\tilde{k},\widetilde{Q}\}}\sum_{\widetilde{Q}}\lambda_u^+(\ell_{j-3})\, G_+^0(\tilde{k}_1+\widetilde{Q})\psi^*_{-,\alpha'_1}(\tilde{k}-\widetilde{Q}')\psi^*_{-,\alpha_1}(\tilde{k}_1+\widetilde{Q}+\widetilde{Q}'-\widetilde{G}) \\
&\qquad \times \psi^*_{+,\alpha_2}(\tilde{k}_2 - \widetilde{Q} + \widetilde{G})\psi_{+,\alpha'_1}(\tilde{k}'_1)\psi_{-,\alpha_1}(\tilde{k}_1)\psi_{-,\alpha_2}(\tilde{k}_2) \quad \text{(A.6)}
\end{aligned}$$

where $\lambda_u^\pm(\ell_{j-3}) = g_3^2(\ell_{j-3})$, $\widetilde{Q} = (2k_{\rm F} + q, \omega_m)$ and $\widetilde{G} = (4k_{\rm F}, 0)$. Here $\sum_{\widetilde{Q}}$ ($\sum_{\widetilde{Q}'}$) consists in summing all boson frequencies and those q (q') values of the momentum transfer for which $k_1 + q$ $(k_1' - q')$ is in the outershell at step $(j-2)d\ell$. Following the example of the incommensurate case, the next two contractions of the first term of (A.6) will lead to corrections to the one-particle term at $+k_{\mathrm{F}}$ with internal lines in three successive outer shells and the external k located at the top of the inner shell:

$$\sum_{j=3}^{N} \langle \delta S'^{(j)}_{u,2} \rangle_{\bar{0},c} = (\pi v_{\mathrm{F}})^2 \frac{T^2}{L^2} \sum_{\alpha} \sum_{j=3}^{N} \sum_{\tilde{k}'} \sum_{\tilde{q}} g_3^2(\ell_{j-3}) G_+^0(\tilde{k}' + \widetilde{Q}) G_+^0(\tilde{k}') \times G_-^0(\tilde{k} + \widetilde{Q} - \widetilde{G}) \psi^*_{+\alpha}(\tilde{k}) \psi_{+\alpha}(\tilde{k}) \quad (A.7)$$

The explicit evaluation runs parallel to the incommensurate case in (A.3)–(A.4). Thus in the local approximation, the variation of $g_3(\ell_{j-3})$ with j is neglected and the value of the coupling is taken at $(N-1)d\ell \to \ell$. The outer shell Umklapp contribution to the single-particle term at the inner shell state k of the action then becomes

$$\sum_{j=3}^{N} \langle \delta S'^{(j)}_{u,2} \rangle_{\bar{0},c} \simeq \frac{1}{8} g_3^2(\ell) t(\ell) d\ell \sum_{\alpha} [G_+^0(\tilde{k})]^{-1} \psi^*_{+\alpha}(\tilde{k}) \psi_{+\alpha}(\tilde{k}). \quad (A.8)$$

This result, which adds to $z^{-1}(d\ell)$, corresponds to the diagram of Fig. 3.8e. When we use the second term of (A.6), a similar expression holds for fermions at $-k_{\mathrm{F}}$.

4

An Introduction to Bosonization

D. Sénéchal

ABSTRACT After general comments on the relevance of field theory to condensed matter systems, the continuum description of interacting electrons in 1D is summarized. The bosonization procedure is then introduced heuristically, but the precise quantum equivalence between fermion and boson is also presented. Then the exact solution of the Tomonaga–Luttinger model is carried out. Two other applications of bosonization are then sketched. We end with a quick introduction to non-Abelian bosonization.

1 Quantum Field Theory in Condensed Matter

Before embarking on a technical description of what bosonization is and how it helps understanding the behavior of interacting electrons in one dimension, let us make some general comments on the usefulness of quantum field theory in the study of condensed matter systems. First of all, what is a quantum field theory? It has taken over half a century, from the late 1920s to the 1980s, for a satisfactory answer to this fundamental question to creep into the general folklore of theoretical physics, the most significant progress being the introduction of the renormalization group (RG). A field theory is by definition a physical model defined on the continuum as opposed to a lattice. However, quantum fluctuations being at work on all length scales, a purely continuum theory makes no sense, and a momentum cutoff Λ (or another physically equivalent procedure) has to be introduced in order to enable meaningful calculations. Such a *regularization* of the field theory invariably introduces a length scale Λ^{-1}, which is part of the definition of the theory as one of its parameters, along with various coupling constants, masses, and so on. A change in the cutoff Λ (through a trace over the high-momentum degrees of freedom) is accompanied by a modification of all other parameters of the theory. A field theory is then characterized not by a set of fixed parameter values, but by a RG trajectory in parameter space, which traces the changing parameters of the theory as the cutoff is lowered.

Most important is the concept of *fixed point*, i.e., of a theory whose parameters are the same whatever the value of the cutoff. Most notorious are free particle theories (bosons or fermions), in which degrees of freedom at different momentum scales are decoupled, so that a partial trace in a momentum shell has no impact on the remaining degrees of freedom. Theories close (in a perturbative sense) to fixed points see their parameters fall into three categories: relevant, irrelevant and marginal. Relevant parameters grow algebraically under renormalization, irrelevant parameters decrease algebraically, whereas marginal parameters undergo logarithmic variations. It used to be that only theories without irrelevant parameters were thought to make sense—and were termed *renormalizable*—for the following reason: If the momentum cutoff is taken to infinity, i.e., if the starting point of the renormalization procedure is at arbitrarily high energy, then an arbitrary number of irrelevant parameters can be added to the theory without measurable effect on the low-energy properties determined from experiments. Thus, the theory has no predictive power on its irrelevant parameters and the latter were excluded for this reason. In a more modern view, the cutoff Λ does not have to be taken to infinity, but has some natural value Λ_0, determined by a more microscopic theory (maybe even another field theory) which eventually supersedes the field theory considered at length scales smaller than Λ^{-1}. In the physics of fundamental interactions, the standard model may then be considered an effective field theory, superseded by something like string theory at the Planck scale ($\sim 10^{-33}$cm). Likewise, effective theories for the strong interactions are superseded by QCD at the QCD length scale ($\sim 10^{-13}$cm). In condensed matter physics, the natural cutoff is the lattice spacing ($\sim 10^{-8}$cm).

In practice, field theories should not be pushed too close to their natural cutoff Λ_0. It is expected that a large number (if not an infinity) of irrelevant couplings of order unity exist at that scale, and the theory then loses all predictive power. The general practice is to ignore irrelevant couplings altogether, and this is credible only well below the natural cutoff.[1] The price to pay for this reduction in parameters is that the finite number of marginal or relevant parameters remaining cannot be quantitatively determined from the underlying microscopic theory (i.e., the lattice model). However, the predictions of the field theory can (in principle) be compared with experiments and the parameters of the theory be inferred. After all, this is what the standard model of elementary particles is about. But even in cases where a quantitative determination of all the parameters of the theory is not possible, some universal (i.e., unaffected by the details of the microscopic theory) predictions are possible, and are the main target of fields theory of condensed matter systems.

The connection between condensed matter systems and relativistic field theories is particularly fruitful in one dimension, in good part because the

---

[1] In some applications, leading irrelevant parameters are physically important and must be kept.

finite electron density may be swept under the rug (or rather, the Fermi sea), leaving low-energy excitations enjoying a Lorentz-like invariance in the absence of interactions (relativistic field theories are generally applied at zero density). This is not true in higher dimensions, where the shape of the Fermi surface makes the connection with relativistic field theories more difficult. The development of bosonization was pursued in parallel by condensed matter and particle physicists. The former had the one-dimensional electron gas in mind, whereas the latter were first interested in a low-dimensional toy model for strong interactions, and later by the applications of bosonization to string theory. Reference [1] reproduces many important papers in the development of bosonization, along with a summary of the method (in a language more appropriate for particle theorists) and a few historical remarks.

2 A Word on Conformal Symmetry

2.1 Scale and Conformal Invariance

Let us briefly consider the action of the renormalization group (RG) on fermions. The low-energy limit is defined around the Fermi surface. The RG action consists in progressively tracing over degrees of freedom far from the Fermi surface, toward the Fermi surface. In one dimension, this process may be understood better by considering Fig. 4.1 below. After an RG step wherein the degrees of freedom located in the momentum shell $[\Lambda - \mathrm{d}\Lambda, \Lambda]$ have been traced over, one may choose to perform an infinitesimal momentum scale transformation $k \to (1 + \mathrm{d}\Lambda/\Lambda)k$ which brings the cutoff back to its initial value Λ before this RG step. This rescaling is necessary if one wants to compare the new coupling constants with the old ones (i.e., if one wants to define a RG trajectory) since we are then comparing theories with identical cutoffs Λ. In a (nearly) Lorentz invariant theory, it is natural to perform this scale transformation both in momentum and energy (or space and time).

In the context of one-dimensional electron systems, Lorentz invariance is an emerging symmetry, valid in the low-energy limit, when the linear approximation to the dispersion relation is acceptable. The Fermi velocity v then plays the role of the velocity of light: it is the only characteristic velocity of the system. Interactions may violate Lorentz invariance and cause the appearance of two characteristic velocities v_s and v_c (spin-charge separation). However, again in the low-energy limit, the system may then separate into disjoint sectors (spin and charge), each benefitting from Lorentz invariance, albeit with different "velocities of light." In the imaginary-time formalism, Lorentz invariance becomes ordinary spatial-rotation invariance.

A field theory with rotation, translation, and scale invariance is said to have *conformal invariance*. Fixed point theories, such as the free boson or the free fermion theories (and many interacting theories) are thus *conformal field theories*. A decent introduction to conformal field theory is beyond the scope of such a short tutorial. The subject is quite vast and has ramifications not only in the field of strongly correlated electrons, but also in the statistical physics of two-dimensional systems, in string theory, and in mathematical physics in general. We will simply mention minimal implications of conformal symmetry in bosonization. Of course, departures from fixed points break this symmetry partially, but their effect at weak coupling may be studied in the conformal symmetry framework with profit.

2.2 Conformal Transformations

By definition, a conformal transformation is a mapping of space-time unto itself that is locally equivalent to a rotation and a dilation. In two dimensions, and in terms of the complex coordinate z, it can be shown that the only such transformations have the form

$$z \to \frac{az+b}{cz+d}, \qquad ad-bc=1 \tag{2.1}$$

where the parameters a, b, c, d are complex numbers. In dimension d, the number of independent parameters of such transformations is $(d+1)(d+2)/2$, and $d=2$ is no exception. However, the case of two dimensions is very special, because of the possibility of *local* conformal transformations, which have all the characters of proper conformal transformations, except that they are not one-to-one (they do not map the whole complex plane onto itself). Any analytic mapping $z \to w$ on the complex plane is locally conformal, as we known from elementary complex analysis. Indeed, the local line element on the complex plane transforms as

$$\mathrm{d}w = \left(\frac{\mathrm{d}w}{\mathrm{d}z}\right)\mathrm{d}z. \tag{2.2}$$

The modulus of the derivative embodies a local dilation, and its phase a local rotation. This distinctive feature of conformal symmetry in two-dimensional space-time is what allows a complete solution of conformal field theories, even in circumstances that apparently break scale invariance. For instance, the entire complex plane (the space-time used with imaginary time at zero temperature) may be mapped onto a cylinder of circumference L via the complex mapping $z = \mathrm{e}^{2\pi w/L}$. This allows for the calculation of correlation functions in a system with a macroscopic length scale (a finite size L at zero temperature, or a finite-temperature $\beta = L/v$ in an infinite system) from the known solution in a scale-invariant situation. Mappings may also be performed from the upper half-plane unto finite, open regions; such mappings are particularly useful when studying boundary or impurity problems. On the formal side, this feature of conformal field theory makes the use of complex coordinates very convenient: In terms of x and imaginary time $\tau = it$, these coordinates may be defined as

$$\begin{aligned} z &= -i(x - vt) = v\tau - ix, \\ \bar{z} &= i(x + vt) = v\tau + ix \end{aligned} \tag{2.3}$$

with the following correspondence of derivatives:

$$\begin{aligned} \partial_z &= -\frac{i}{2}\left(\frac{1}{v}\partial_t - \partial_x\right), & \partial_x &= -i(\partial_z - \partial_{\bar{z}}), \\ \partial_{\bar{z}} &= -\frac{i}{2}\left(\frac{1}{v}\partial_t + \partial_x\right), & \partial_t &= iv(\partial_z + \partial_{\bar{z}}). \end{aligned} \tag{2.4}$$

A local operator $O(z, \bar{z})$ belonging to a conformal field theory is called *primary* (more precisely, quasi-primary) if it scales well under a conformal transformation, i.e., if the transformation

$$O'(\alpha z, \bar{\alpha}\bar{z}) = \alpha^{-h}\bar{\alpha}^{-\bar{h}}O(z, \bar{z}) \tag{2.5}$$

is a symmetry of correlation functions. The constants h and $\bar{h}$ are a fundamental property of the operator O and are called the right- and left-conformal dimensions, respectively. Under a plain dilation ($\alpha = \bar{\alpha}$), the operator scales as $O'(\alpha z, \alpha\bar{z}) = \alpha^{-\Delta}O(z, \bar{z})$, where $\Delta = h + \bar{h}$ is the ordinary scaling dimension of the operator. Under a rotation (or Lorentz transformation), for which $\alpha = \mathrm{e}^{i\theta}$ and $\bar{\alpha} = \mathrm{e}^{-i\theta}$, the operator transforms as $O'(\mathrm{e}^{i\theta}z, \mathrm{e}^{-i\theta}\bar{z}) = \mathrm{e}^{is\theta}O(z, \bar{z})$, where $s = h - \bar{h}$ is called the *conformal spin* of the operator.

The scaling dimension appears directly in the two-point correlation function of the operator, which is fixed by scaling arguments—up to multiplicative constant:

$$\langle O(z, \bar{z})O^\dagger(0, 0)\rangle = \frac{1}{z^{2h}}\frac{1}{\bar{z}^{2\bar{h}}}. \tag{2.6}$$

2.3 Effect of Perturbations

Fixed-point physics helps understanding what happens in the vicinity of the fixed point, in a perturbative sense. Consider for instance the perturbed action

$$S = S_0 + g\int \mathrm{d}t\,\mathrm{d}x\, O(x, t) \tag{2.7}$$

where S_0 is the fixed-point action and O an operator of scaling dimension Δ. Let us perform an infinitesimal RG step, as described in the first paragraph of this section, with a scale factor $\lambda = 1 + \mathrm{d}\Lambda/\Lambda$ acting on wavevectors; the inverse scaling factor acts on the coordinates, and if x and t now represent the new, rescaled coordinates, with the same short-distance cutoff as before the partial trace, then

$$\begin{aligned} S' &= S_0 + g \int \mathrm{d}(\lambda t)\, \mathrm{d}(\lambda x)\, O(\lambda x, \lambda t) \\ &= S_0 + g\lambda^{2-\Delta} \int \mathrm{d}t\, \mathrm{d}x\, O(x,t). \end{aligned} \tag{2.8}$$

We assumed that we are sufficiently close to the fixed point so that the effect of the RG trace in the shell $\mathrm{d}\Lambda$ is precisely the scaling of O with exponent Δ. By integrating such infinitesimal scale transformations, we find that the coupling constant g scales as

$$\frac{g}{g_0} = \left(\frac{\Lambda_0}{\Lambda}\right)^{\Delta-2} = \left(\frac{\xi_0}{\xi}\right)^{2-\Delta} \tag{2.9}$$

where ξ and ξ_0 are correlation lengths at the two cutoffs Λ and Λ_0. Thus, the perturbation is relevant if $\Delta < 2$, irrelevant if $\Delta > 2$ and marginal if $\Delta = 2$.

A relevant perturbation is typically the source of a gap in the low-energy spectrum. In a Lorentz-invariant system, a finite correlation length is associated with a mass gap $m \sim 1/\xi_0$. Let us suppose that the correlation length is of the order of the lattice spacing ($\xi \sim 1$) when the coupling constant g is of order unity. The above scaling relation then relates the mass m to the bare (i.e., "microscopic") coupling constant:

$$m \sim g_0^{1/(2-\Delta)}. \tag{2.10}$$

Given the warning issued in the introduction on the impossibility of knowing the bare parameters of the theory with accuracy, we should add that "bare" may mean here "at a not-so-high energy scale where the coupling g_0 is known by other means." In fact, this scaling formula works surprisingly well even if g_0 is taken as the bare coupling (at the natural cutoff Λ_0), provided g_0 is not too large.

Conformal field theory techniques may also be used to treat marginal perturbations, in performing the equivalent of one-loop calculations in conventional perturbation theory, with the so-called operator product expansion (OPE). An example calculation is given in the appendix (more can be found in [2–5]). However, perturbations with conformal spin $s \neq 0$ are more subtle, and typically lead to a change in k_F or incommensurabilities.

2.4 The Central Charge

A conformal field theory is characterized by a number c called the *central charge*. This number is roughly a measure of the number of degrees of freedom of the model considered. By convention, the free boson theory has central charge $c = 1$. So does the free complex fermion. A set of N

independent free bosons has central charge $c = N$. Conformal field theories with central charge $c < 1$ correspond to known critical statistical models, like the Ising model (known since Onsager's solution to be equivalent to a free Majorana fermion), the Potts model, etc. Models at $c = 1$ practically all correspond to a free boson or to slight modifications thereof called *orbifolds*, and have been completely classified [6]. A conformal field theory of integer central charge $c = N$ may have a representation in terms of N free bosons, although such representations are not known in general. Likewise, a conformal field theory of half-integer central charge may have a representation in terms of an integer number of real (i.e., Majorana) fermions.

3 Interacting Electrons in One Dimension

3.1 Continuum Fields and Densities

Let us consider noninteracting electrons on a lattice and define the corresponding (low-energy) field theory. This poses no problem, since the different momentum scales of this free theory are decoupled. The microscopic Hamiltonian takes the form

$$H_{\mathrm{F}} = \sum_k \varepsilon(k) c^\dagger(k) c(k) \tag{3.1}$$

where $c(k)$ is the electron annihilation operator at wavevector k (we ignore spin for the time being). The low-energy theory is defined in terms of creation and annihilation operators in the vicinity of the Fermi points (cf. Fig. 4.1) as follows:

$$\begin{aligned} \alpha(k) &= c(k_{\mathrm{F}} + k), \\ \alpha(-k) &= c(-k_{\mathrm{F}} - k), \\ \beta(k) &= c^\dagger(k_{\mathrm{F}} - k), \\ \beta(-k) &= c^\dagger(-k_{\mathrm{F}} + k) \end{aligned} \tag{3.2}$$

where k is positive. The noninteracting Hamiltonian then takes the form

$$H_{\mathrm{F}} = \int \frac{\mathrm{d}k}{2\pi} v|k| \{\alpha^\dagger(k)\alpha(k) + \beta^\dagger(k)\beta(k)\} \tag{3.3}$$

where v is the Fermi velocity, the only remaining parameter from the microscopic theory. The energy is now defined with respect to the ground state and the momentum integration is carried between $-\Lambda$ and Λ. The operators $\alpha(k)$ and $\beta(k)$, respectively, annihilate electrons and holes around the right Fermi point ($k > 0$) and the left Fermi point ($k < 0$). A momentum cutoff Λ is implied, which corresponds to an energy cutoff $v\Lambda$. Note that the continuum limit is in fact a low-energy limit: Energy considerations determine around which wavevector ($\pm k_{\mathrm{F}}$) one should expand. In position

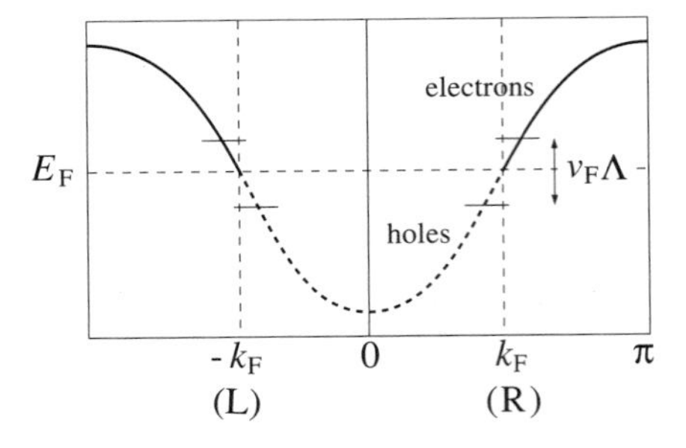

FIGURE 4.1. Typical tight-binding dispersion in 1D, illustrating left and right Fermi points and the linear dispersion in the vicinity of those points.

space, this procedure amounts to introducing slow fields ψ and $\bar{\psi}$ such that the annihilation operator at site n is[2]

$$\frac{c_x}{\sqrt{a}} = \psi(x)\mathrm{e}^{ik_\mathrm{F}x} + \bar{\psi}(x)\mathrm{e}^{-ik_\mathrm{F}x}. \tag{3.4}$$

The factor $\sqrt{a}$ is there to give the fields the proper delta-function anticommutator, and reflects their (engineering) dimension:

$$\begin{aligned}
\{\psi(x), \psi^\dagger(x')\} &= \delta(x - x'), \\
\{\bar{\psi}(x), \bar{\psi}^\dagger(x')\} &= \delta(x - x'), \\
\{\psi(x), \bar{\psi}^\dagger(x')\} &= 0.
\end{aligned} \tag{3.5}$$

3.1.1 Left-Right Separation

The mode expansions of the continuum fields are

$$\begin{aligned}
\psi(x) &= \int_{k>0} \frac{\mathrm{d}k}{2\pi} [\mathrm{e}^{ikx}\alpha(k) + \mathrm{e}^{-ikx}\beta^\dagger(k)], \\
\bar{\psi}(x) &= \int_{k<0} \frac{\mathrm{d}k}{2\pi} [\mathrm{e}^{ikx}\alpha(k) + \mathrm{e}^{-ikx}\beta^\dagger(k)].
\end{aligned} \tag{3.6}$$

The time dependence of $\alpha(k)$ and $\beta(k)$ is obtained through multiplying by the phase $\mathrm{e}^{-iv|k|t}$. In terms of the complex coordinates (2.3), the mode expansions for the time-dependent fields are then

$$\begin{aligned}
\psi(z) &= \int_{k>0} \frac{\mathrm{d}k}{2\pi} [\mathrm{e}^{-kz}\alpha(k) + \mathrm{e}^{kz}\beta^\dagger(k)], \\
\bar{\psi}(\bar{z}) &= \int_{k<0} \frac{\mathrm{d}k}{2\pi} [\mathrm{e}^{k\bar{z}}\alpha(k) + \mathrm{e}^{-k\bar{z}}\beta^\dagger(k)].
\end{aligned} \tag{3.7}$$

---

[2] We will generally use a bar ($^-$) to denote left-moving operators, and the same symbol without the bar for right-moving operators. A more common notation in condensed matter physics is the use of subscripts L and R. The (lighter) notation used here stresses the analogy with complex coordinates.

Thus the right-moving field ψ depends solely on the right-moving coordinate z, whereas the left-moving field $\bar{\psi}$ depends solely on $\bar{z}$.

The mode expansion (3.6) is misleading in one respect: it makes believe that the positive-wavevector modes are contiguous to the negative-wavevector modes in k-space, which is not the case (they are separated by roughly $2k_{\mathrm{F}}$). Thus right- and left-moving modes are clearly well-separated. Left and right electrons have density fluctuations

$$J(x) = \psi^{\dagger}(x)\psi(x), \qquad \bar{J}(x) = \bar{\psi}^{\dagger}(x)\bar{\psi}(x). \tag{3.8}$$

The total electronic density (with respect to the ground state) being $n_{\mathrm{tot}} = J + \bar{J}$.

3.1.2 Analogy with the Dirac Equation

It is a simple matter to express the continuum Hamiltonian (3.3) in terms of the fields ψ and $\bar{\psi}$:

$$H_{\mathrm{F}} = -iv \int \mathrm{d}x \; [\psi^{\dagger}\partial_x\psi - \bar{\psi}^{\dagger}\partial_x\bar{\psi}]. \tag{3.9}$$

This is exactly the Dirac Hamiltonian. Indeed, in $(1+1)$ dimensions, the Dirac matrices can be taken to be two-dimensional, for instance

$$\gamma^0 = \begin{pmatrix} 0 & 1 \\ -1 & 0 \end{pmatrix}, \qquad \gamma^1 = \begin{pmatrix} 0 & 1 \\ 1 & 0 \end{pmatrix} \tag{3.10}$$

and the two-component Dirac spinor can be taken to be

$$\Psi = \begin{pmatrix} \psi \\ \bar{\psi} \end{pmatrix}. \tag{3.11}$$

The Dirac Lagrangian density—with a mass term—being

$$\mathcal{L}_{\mathrm{D}} = i\bar{\Psi}(\gamma^0\partial_t - v\gamma^1\partial_x)\Psi - m\bar{\Psi}\Psi. \tag{3.12}$$

The canonical Hamiltonian density derived from this Lagrangian is therefore

$$H_{\mathrm{D}} = -iv[\psi^{\dagger}\partial_x\psi - \bar{\psi}^{\dagger}\partial_x\bar{\psi}] + m(\psi^{\dagger}\bar{\psi} - \bar{\psi}^{\dagger}\psi) \tag{3.13}$$

which coincides with (3.9) in the massless case.

Adding a mass term to the continuum Hamiltonian amounts to opening a gap $2m$ at the Fermi level. On the other hand, adding a term like

$$\mathcal{H}' = \psi^{\dagger}\psi + \bar{\psi}^{\dagger}\bar{\psi} = J + \bar{J} \tag{3.14}$$

to the Hamiltonian would simply shift the chemical potential, i.e., modify the value of k_{F}. This would of course require a redefinition of left and right modes, but the end result would again be a Hamiltonian of the form (3.9), with a slightly different velocity v.

In the language of Conformal Field Theory, the free fermion theory has central charge $c = 1$, and the conformal dimensions of the fermion fields are

$$\psi : (h, \bar{h}) = (\tfrac{1}{2}, 0), \qquad \bar{\psi} : (h, \bar{h}) = (0, \tfrac{1}{2}). \tag{3.15}$$

Thus, the mass term $\psi^\dagger\bar{\psi} - \bar{\psi}^\dagger\psi$ has scaling dimension $\Delta = 1$ and zero conformal spin; according to (2.10), this term gives rise to a gap proportional to m, as expected of course. On the other hand, the perturbation $\psi^\dagger\psi + \bar{\psi}^\dagger\bar{\psi}$, even if it also has scaling dimension 1, is a sum of terms with nonzero conformal spin. For such terms the analysis leading to (2.10) cannot be applied. Perturbations with conformal spin, such as this one, will typically change the slow wavevectors, cause incommensurability, etc.

3.2 Interactions

The general two-body interaction between 1D electrons (of one band) is

$$H_{\text{int.}} = \tfrac{1}{2}\sum_{\sigma,\sigma'}\sum_{k_1,k_2,k_3} V_{\sigma,\sigma'}(k_1,k_2,k_3)c^\dagger_\sigma(k_1)c^\dagger_{\sigma'}(k_2)c_{\sigma'}(k_3)c_\sigma(k_4) \tag{3.16}$$

where we have restored spin indices. In the low-energy limit, scattering processes are naturally restricted to the vicinity of the Fermi points, and fall into four kinematic types, illustrated in Fig. 4.2. In terms of continuum fields, the corresponding Hamiltonian densities are

$$\begin{aligned}
\mathcal{H}_1 &= vg_1\sum_\sigma \psi^\dagger_\sigma\bar{\psi}_\sigma\bar{\psi}^\dagger_{-\sigma}\psi_{-\sigma},\\
\mathcal{H}_2^c &= vg_{2,c}(J_\uparrow + J_\downarrow)(\bar{J}_\uparrow + \bar{J}_\downarrow),\\
\mathcal{H}_2^s &= vg_{2,s}(J_\uparrow - J_\downarrow)(\bar{J}_\uparrow - \bar{J}_\downarrow),\\
\mathcal{H}_3 &= \tfrac{1}{2}vg_3\sum_\sigma \psi^\dagger_\sigma\psi^\dagger_{-\sigma}\bar{\psi}_\sigma\bar{\psi}_{-\sigma} + \text{H.c.},\\
\mathcal{H}_4^c &= \tfrac{1}{2}vg_{4,c}[(J_\uparrow + J_\downarrow)^2 + (\bar{J}_\uparrow + \bar{J}_\downarrow)^2],\\
\mathcal{H}_4^s &= \tfrac{1}{2}vg_{4,s}[(J_\uparrow - J_\downarrow)^2 + (\bar{J}_\uparrow - \bar{J}_\downarrow)^2].
\end{aligned} \tag{3.17}$$

Remarks. 1. The incoming spins in $\mathcal{H}_1$ can be assumed to be antiparallel, since the case of parallel spins reduces to a density-density coupling of the form $\mathcal{H}_2^s$.

2. We have supposed that the couplings g_{1-4} are momentum independent, and this can be safely assumed in the low-energy limit. However, they are cutoff dependent, i.e., they are subject to a RG flow.

3. The interactions $\mathcal{H}_1$ and $\mathcal{H}_2^s$ are not separately spin-rotation invariant, but their combination is if $g_1 = -2g_{2,s}$. This will be shown in Section 7.2.

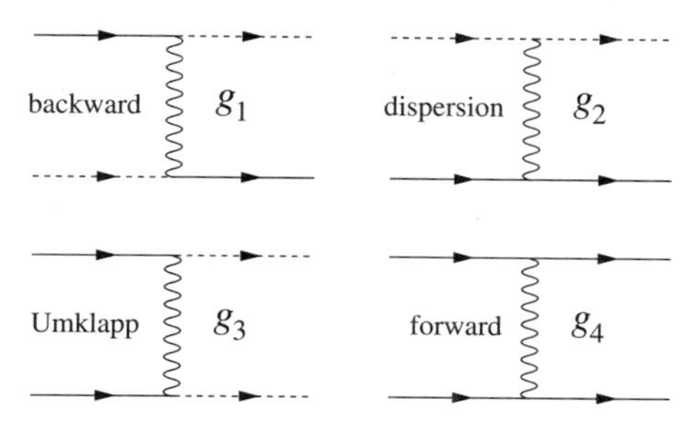

FIGURE 4.2. The four scattering processes for right-moving (continuous lines) and left-moving (dashed lines) electrons in one dimension: (i) g_4 is the amplitude for forward scattering; (ii) g_2 for dispersion (scattering of left onto right electrons); (iii) g_1 for backward scattering (each electron changes branch, but keeps its spin); (iv) g_3 for Umklapp scattering (two left electrons become right electrons or vice-versa; possible only at half-filling because of crystal momentum conservation). Spin indices are suppressed, but incoming electrons must have opposite spins for g_1 and g_3.

4. That $g_{2,\mu}$ and $g_{4,\mu}$ are density-density interactions is made more explicit in this real-space representation. If these couplings alone are nonzero (i.e., $g_1 = g_3 = 0$), the resulting model is called the *Tomonaga–Luttinger model* and can be solved exactly by bosonization.

The bosonized form of these interactions will be derived in Section 5.6.

4 Bosonization: A Heuristic View

4.1 Why is One-Dimension Special?

The basic idea behind bosonization is that particle-hole excitations are bosonic in character, and that somehow the greatest part (if not the totality) of the electron gas spectrum might be exhausted by these excitations. The question was raised by F. Bloch back in 1934, but Tomonaga realized in 1950 that it could only be true in one dimension. The reason is simple: Consider Fig. 4.3. On the left the creation of a particle-hole pair of momentum k is illustrated. On the right is the spectrum of particle-hole excitations created out of the Fermi sea: the energy of the pair is plotted versus the momentum of the pair (both quantities measured w.r.t. the ground state). Because of the linear one-particle dispersion near the Fermi level, the pairs have a narrow, quasiparticle-like dispersion near zero momentum: they can propagate coherently. In other words, the particle and hole have nearly the same group velocity at low energies and propagate together. Any weak particle-hole attraction is then bound to have dramatic effects, i.e., to bind the pair into a coherently propagating entity: a new particle.

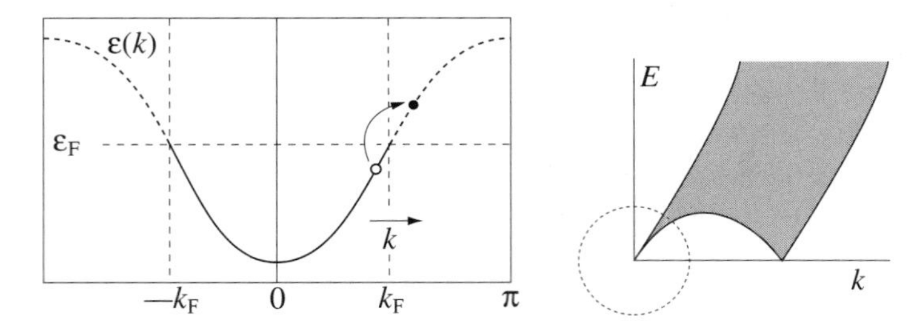

FIGURE 4.3. Particle-hole excitations in one dimension.

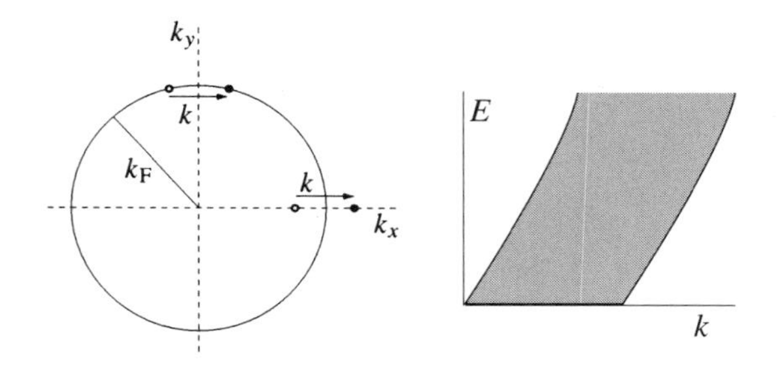

FIGURE 4.4. Particle-hole excitations in two dimensions.

In Fig. 4.4, the corresponding situation in two dimensions is illustrated, with a circular Fermi surface. There it is clear that a particle-hole pair with a given momentum k can have a continuous spectrum of energies, starting from zero. Thus, the particle-hole spectrum is a continuum throughout and interactions have a harder time forming coherently propagating particle-hole pairs.[3] In any case, defining the theory in terms of bosonic excitations is much less obvious in dimensions greater than one.

In the low-energy limit, particle-hole excitations in one dimension are exactly coherent and thus the density fluctuation fields J_σ and $\bar{J}_\sigma$ create propagating particles. The question is then whether any operator can be expressed in terms of J_σ and $\bar{J}_\sigma$, especially a single-particle creation or annihilation operator. The answer is yes: the right-moving fermion field may actually be expressed as

$$\psi_\sigma(x) = \frac{1}{\sqrt{2\pi}} \exp\left[2\pi i \int^x \mathrm{d}x'\, J_\sigma(x')\right]. \tag{4.1}$$

The rest of this section, and much of the following one, will be devoted to proving (5.28), equivalent to the above.

[3] This argument is partially weakened by the presence of nesting.

4.2 The Simple Boson

Before going any further we must define what a free boson is. This may be done fundamentally from a Lagrangian. For a massless boson field φ, this is

$$L_0 = \tfrac{1}{2}\int \mathrm{d}x \left[\frac{1}{v}(\partial_t\varphi)^2 - v(\partial_x\varphi)^2\right], \tag{4.2}$$

where v is the speed of the Bose particles. The corresponding Hamiltonian density is then

$$\mathcal{H}_0 = \tfrac{1}{2}v[\Pi^2 + (\partial_x\varphi)^2], \qquad \Pi = \frac{1}{v}\partial_t\varphi, \tag{4.3}$$

where Π is the field conjugate to φ, the two fields obeying the canonical commutation rules

$$\begin{gathered}[\varphi(x), \Pi(x')] = i\delta(x - x'),\\ [\varphi(x), \varphi(x')] = 0, \qquad [\Pi(x), \Pi(x')] = 0.\end{gathered} \tag{4.4}$$

The standard mode expansion for the fields φ and Π on the infinite line is

$$\begin{aligned}\varphi(x) &= \int \frac{\mathrm{d}k}{2\pi}\sqrt{\frac{v}{2\omega(k)}}[b(k)\mathrm{e}^{ikx} + b^\dagger(k)\mathrm{e}^{-ikx}],\\ \Pi(x) &= \int \frac{\mathrm{d}k}{2\pi}\sqrt{\frac{\omega(k)}{2v}}[-ib(k)\mathrm{e}^{ikx} + ib^\dagger(k)\mathrm{e}^{-ikx}],\end{aligned} \tag{4.5}$$

where $\omega(k) \equiv v|k|$. The creation and annihilation operators obey the commutation rules

$$[b(k), b^\dagger(k')] = 2\pi\delta(k - k'); \tag{4.6}$$

and the Hamiltonian may be expressed as

$$H = \int \frac{\mathrm{d}k}{2\pi}\omega(k)b^\dagger(k)b(k). \tag{4.7}$$

The ground state (or vacuum) $|0\rangle$ is annihilated by all the $b(k)$'s, and has zero energy by convention.

Like fermions, Bose particles in 1D either move to the left or to the right. Indeed, the time dependence of the annihilation operators is obtained through multiplying by the phase $\mathrm{e}^{-iv|k|t}$, if the boson is massless. A separation into left- and right-moving parts is then possible. We write

$$\varphi(x, t) = \phi(x - vt) + \bar{\phi}(x + vt) \tag{4.8}$$

where

$$\begin{aligned}\phi(z) &= \int_{k>0} \frac{\mathrm{d}k}{2\pi}\frac{1}{\sqrt{2k}}[b(k)\mathrm{e}^{-kz} + b^\dagger(k)\mathrm{e}^{kz}],\\ \bar{\phi}(\bar{z}) &= \int_{k>0} \frac{\mathrm{d}k}{2\pi}\frac{1}{\sqrt{2k}}[\bar{b}(k)\mathrm{e}^{-k\bar{z}} + \bar{b}^\dagger(k)\mathrm{e}^{k\bar{z}}].\end{aligned} \tag{4.9}$$

The variables z and $\bar{z}$ are defined as in (2.3). Note that wavevectors now take only positive values, since we have defined $\bar{b}(k) \equiv b(-k)$.

4.2.1 The Dual Field

It is customary to introduce the so-called *dual boson* $\vartheta(x,t)$, defined by the relation $\partial_x \vartheta = -\Pi = -(1/v)\partial_t \varphi$. In terms of right and left bosons ϕ and $\bar{\phi}$, this becomes (cf. (2.4))

$$\begin{aligned} \partial_x \vartheta &= -\frac{1}{v}\partial_t \varphi \\ &= -i(\partial_z + \partial_{\bar{z}})(\phi + \bar{\phi}) \\ &= -i(\partial_z - \partial_{\bar{z}})(\phi - \bar{\phi}) \\ &= \partial_x(\phi - \bar{\phi}) \end{aligned} \tag{4.10}$$

therefore $\theta = \phi - \bar{\phi}$, modulo an additive constant which we set to zero. One may then write

$$\phi = \tfrac{1}{2}(\varphi + \vartheta), \qquad \bar{\phi} = \tfrac{1}{2}(\varphi - \vartheta) \tag{4.11}$$

if one desires a definition of ϕ and $\bar{\phi}$ that stands independent from the mode expansion. Since ϑ is expressed nonlocally in terms of Π (i.e., through a spatial integral), the above expression makes it clear that the right and left parts ϕ and $\bar{\phi}$ are not true local fields relative to φ. The basic definition of ϑ in terms of Π implies the equal-time commutation rule

$$[\varphi(x), \vartheta(x')] = -i\theta(x - x'), \tag{4.12}$$

(please note that $\theta(x)$ is the step function $[\neq \vartheta]$) which further shows the nonlocal relation between φ and ϑ (the commutator is nonzero even at large distances).

4.3 Bose Representation of the Fermion Field

After this introduction of the boson field, let us proceed with a heuristic (and incomplete) derivation of the bosonization formula. The basic idea is the following (we ignore spin for the moment): the electron density n_{tot} being bilinear in the electron fields, it has Bose statistics. Let us suppose that it is the derivative of a boson field: $n_{\text{tot}} \propto \partial_x \varphi$. Then

$$\varphi(x) = \lambda \int_x^\infty dy\; n_{\text{tot}}(y), \qquad n_{\text{tot}}(x) = -\frac{1}{\lambda}\partial_x \varphi(x), \tag{4.13}$$

where λ is a constant to be determined. Creating a fermion at the position x' increases φ by λ if $x < x'$. This has the same effect as the operator

$$\exp\left[-i\lambda \int_{-\infty}^{x'} dy\; \Pi(y)\right], \tag{4.14}$$

which acts as a shift operator for $\varphi(x)$ if $x < x'$, because of the commutation relation

$$\left[\varphi(x), \int_{-\infty}^{x'} dy\, \Pi(y)\right] = \begin{cases} i & (x < x'), \\ 0 & (x > x'). \end{cases} \tag{4.15}$$

Thus, a creation operator $\psi^\dagger(x')$ or $\bar{\psi}^\dagger(x')$ may be represented by the above operator, times an operator that commutes with φ ($\mathrm{e}^{i\alpha\varphi}$, for instance). This extra factor must be chosen such that $\psi^\dagger$ depends on $x - vt$ only, and $\bar{\psi}^\dagger$ on $x + vt$ only. The left-right decomposition (4.8) and (4.11) are useful here. Since

$$\Pi = \frac{1}{v}\partial_t\varphi = -\partial_x\vartheta \tag{4.16}$$

we have

$$-i\lambda\int_{-\infty}^{x'} \mathrm{d}y\ \Pi(y) = i\lambda\vartheta(x') \tag{4.17}$$

from which purely right- and left-moving fields may be obtained by adding or subtracting $-\frac{1}{2}i\alpha\varphi(x')$. Thus, a sensible Ansatz for a boson representation of electron creation operators is

$$\psi^\dagger(x) = A\mathrm{e}^{2i\lambda\phi(x)}, \qquad \bar{\psi}^\dagger(x) = A\mathrm{e}^{-2i\lambda\bar{\phi}(x)} \tag{4.18}$$

where the constants A and λ are to be determined, for instance, by imposing the anticommutation relations (3.5). This will be done below.

The one-electron states $\psi^\dagger|0\rangle$, being generated by exponentials of boson creation operators, are *coherent states* of the boson field ϕ.[4] On the other hand, the elementary Bose excitations (the mesons, in field-theory jargon) are collective density (or spin) fluctuations, i.e., electron-hole excitations.

5 Details of the Bosonization Procedure

5.1 Left and Right Boson Modes

The mode expansions (4.5) may seem reasonable but are actually incomplete, for the following reasons:

1. The field being massless, the mode expansion is ill-defined at $k = 0$, because $\omega(k = 0) = 0$. The mode at $k = 0$, or *zero-mode*, has to be treated separately and turns out to be important. The existence of this mode apparently spoils the left-right separation of the field φ.

2. In order for bosonization to be rigorously defined, the field φ must have an angular character. In other words, the target space of the field must be a circle of radius R (to be kept general for the moment), so that φ and $\varphi + 2\pi R$ are identified. We say that the boson is *compactified* on a circle.

[4]The interactions g_1 and g_3 give the boson Hamiltonian the sine-Gordon form (cf. Section 5.6). In that case, one-electron states correspond to solitons (kinks) of the sine-Gordon theory.

3. A rigorous proof of bosonization procedure (at the spectrum level) is best obtained on a system of finite size L with periodic boundary conditions.

Taking the above remarks into account leads to the following improved mode expansion for the boson field $\varphi(x,t)$:

$$\varphi(x,t) = q + \frac{\pi_0 vt}{L} + \frac{\tilde{\pi}_0 x}{L} + \sum_{n>0} \frac{1}{\sqrt{4\pi n}} [b_n \mathrm{e}^{-kz} + b_n^\dagger \mathrm{e}^{kz} + \bar{b}_n \mathrm{e}^{-k\bar{z}} + \bar{b}_n^\dagger \mathrm{e}^{k\bar{z}}]. \quad (5.1)$$

Wavevectors are now quantized as $k = 2\pi n/L$ (n an integer). The creation and annihilation operators have a different normalization from (4.5), so as to obey the commutation rules

$$[b_n, b_m^\dagger] = \delta_{mn}, \qquad [\bar{b}_n, \bar{b}_m^\dagger] = \delta_{mn}. \quad (5.2)$$

The zero-mode is treated explicitly as a pair of canonical variables q and π_0 obeying the commutation rule

$$[q, \pi_0] = i. \quad (5.3)$$

Defining φ on a circle of radius R has two consequences:

1. The operator q is not single-valued, since it is an angular variable (q and $q + 2\pi R$ are identified). Only exponentials $\mathrm{e}^{inq/R}$, where n is an integer, are well defined. Because π_0 is conjugate to q, we have the relation

$$\mathrm{e}^{-inq/R} \pi_0 \mathrm{e}^{inq/R} = \pi_0 + \frac{n}{R} \quad (5.4)$$

(the exponential operators shifts the eigenvalues of π_0). Starting with the $\pi_0 = 0$ ground state, this means that the spectrum of π_0 is restricted to $\mathbf{Z}/R$.

2. Winding configurations are allowed; the constant $\tilde{\pi}_0$ is precisely defined as $\tilde{\pi}_0 = 2\pi Rm$, where m is the number of windings of φ as x goes from 0 to L. It turns out to be useful to introduce an *operator* $\tilde{\pi}_0$, defined by its eigenvalues $2\pi Rm$. This operator commutes with all other operators met so far, but one may define an operator $\tilde{q}$ conjugate to $\tilde{\pi}_0$, i.e., such that $[\tilde{q}, \tilde{\pi}_0] = i$.

The separation between left- and right-moving parts may then be done in spite of the existence of a zero mode:

$$\begin{aligned} \phi(z) &= Q + \frac{P}{2L}(vt - x) + \sum_{n>0} \frac{1}{\sqrt{4\pi n}} (b_n \mathrm{e}^{-kz} + b_n^\dagger \mathrm{e}^{kz}), \\ \bar{\phi}(\bar{z}) &= \bar{Q} + \frac{\overline{P}}{2L}(vt + x) + \sum_{n>0} \frac{1}{\sqrt{4\pi n}} (\bar{b}_n \mathrm{e}^{-k\bar{z}} + \bar{b}_n^\dagger \mathrm{e}^{k\bar{z}}) \end{aligned} \quad (5.5)$$

where we have defined the right and left zero-modes

$$\begin{aligned} Q &= \tfrac{1}{2}(q - \tilde{q}), & P &= \pi_0 - \tilde{\pi}_0, & [Q, P] &= i, \\ \bar{Q} &= \tfrac{1}{2}(q + \tilde{q}), & \overline{P} &= \pi_0 + \tilde{\pi}_0, & [\bar{Q}, \overline{P}] &= i. \end{aligned} \tag{5.6}$$

Note that the "artificial" operator $\tilde{q}$ drops from the combination $\varphi = \phi + \bar{\phi}$. The spectrum of P and $\overline{P}$ is

$$P = \left(\frac{n}{R} + 2\pi R m\right), \qquad \overline{P} = \left(\frac{n}{R} - 2\pi R m\right) \tag{5.7}$$

m and n being integers. Note that this spectrum is not a simple Cartesian product, which means that the left-right decomposition of the boson is not perfect, despite the expansion (5.5).

5.2 Proof of the Bosonization Formulas: Vertex Operators

Before investigating the constants A and λ of (4.18), serious questions must be asked on the precise definition of an exponential operator of the form $e^{i\alpha\phi}$ (such operators are called *vertex operators*, from their usage in string theory). Because ϕ is a fluctuating field, any power or more complicated function of ϕ requires a careful definition if it is not to have a divergent average value. The natural prescription is called *normal ordering* and consists in expressing ϕ as a mode expansion and putting all annihilation operators to the right of creation operators. Thus, the exponential operator has the following precise definition:

$$\begin{aligned} {:}e^{i\alpha\phi(x-vt)}{:} = e^{i\alpha Q} \exp\left[i\alpha \sum_{n>0} \frac{1}{\sqrt{4\pi n}} b_n^\dagger e^{kz}\right] \\ \times \exp\left[i\alpha \sum_{n>0} \frac{1}{\sqrt{4\pi n}} b_n e^{-kz}\right] e^{-i\alpha P(x-vt)/2L}, \end{aligned} \tag{5.8}$$

where P is considered an annihilation operator since the boson ground state $|0\rangle$ has zero momentum. The notation $:A:$ means that the operator A is normal ordered.

Because of normal ordering, vertex operators do not multiply like ordinary exponentials. Instead, one has the relation

$$e^{i\alpha\phi(z)} e^{i\beta\phi(z')} = e^{i\alpha\phi(z) + i\beta\phi(z')} e^{-\alpha\beta\langle\phi(z)\phi(z')\rangle}, \tag{5.9}$$

where $\langle\phi(z)\phi(z')\rangle$ is a simple ground-state expectation value if the left-hand side is an ordinary product, or a Green function if it is a time-ordered product. This is one of the most important formulas of this review, and it can be easily demonstrated using the Campbell–Baker–Hausdorff (CBH) formula:

$$e^A e^B = e^{A+B} e^{[A,B]/2} \qquad ([A, B] = \text{const.}). \tag{5.10}$$

Consider for instance a single harmonic oscillator a, and two operators

$$A = \alpha a + \alpha' a^\dagger , \qquad B = \beta a + \beta' a^\dagger . \tag{5.11}$$

The CBH formula allows for a combination of the normal-ordered exponentials:

$$\begin{aligned} :\mathrm{e}^A::\mathrm{e}^B: &= \mathrm{e}^{\alpha' a^\dagger}\mathrm{e}^{\alpha a}\mathrm{e}^{\beta' a^\dagger}\mathrm{e}^{\beta a} \\ &= \mathrm{e}^{\alpha' a^\dagger}\mathrm{e}^{\beta' a^\dagger}\mathrm{e}^{\alpha a}\mathrm{e}^{\beta a}\mathrm{e}^{\alpha\beta'} \\ &= :\mathrm{e}^{A+B}:\mathrm{e}^{\langle 0|AB|0\rangle} . \end{aligned} \tag{5.12}$$

The last equality also applies if $(a, a^\dagger)$ is replaced, in the same order, by a pair (p, q) of canonically conjugate operators ($[q, p] = i$). It also applies to a combination of independent oscillators, such as the boson fields ϕ or $\bar\phi$. Finally, it applies to a time-ordered product as well as to an ordinary product, as long as $\langle 0|AB|0\rangle$ is also replaced by a time-ordered product. Thus, the identity (5.9) is proven.

It remains to calculate the Green function $\langle\phi(z)\phi(z')\rangle$. This can be done in several ways. For instance, by using the mode expansion (5.5) and taking the limit $L \to \infty$:

$$\begin{aligned} \langle\phi(z)\phi(0)\rangle = -\frac{i}{2L}(vt - x) + \frac{1}{4\pi}\sum_{n>0}\frac{1}{n}\mathrm{e}^{-2\pi n z/L} &= -\frac{z}{2L} - \frac{1}{4\pi}\ln\frac{2\pi z}{L} \\ &\to -\frac{1}{4\pi}\ln z + \text{const.} \qquad (L \to \infty) . \end{aligned} \tag{5.13}$$

We will adopt the normalization

$$\langle\phi(z)\phi(z')\rangle = -\frac{1}{4\pi}\ln(z - z') \tag{5.14}$$

and drop the constant term, which basically defines an overall length scale. Likewise, one finds

$$\langle\bar\phi(\bar z)\bar\phi(\bar z')\rangle = -\frac{1}{4\pi}\ln(\bar z - \bar z') . \tag{5.15}$$

Another way of computing the boson Green function is to notice that the nonchiral Green function $\mathcal{G}(x, \tau) = \langle\varphi(x, \tau)\varphi(0, 0)\rangle$ must obey the two-dimensional Poisson equation

$$\nabla^2\mathcal{G} = \delta(x)\delta(v\tau) \tag{5.16}$$

in Euclidian space-time. The solution to that equation is readily obtained in polar coordinates, up to an additive constant (the correct normalization is obtained by applying Gauss' theorem around the origin):

$$\mathcal{G}(x, v\tau) = -\frac{1}{4\pi}\ln(x^2 + v^2\tau^2) = -\frac{1}{4\pi}\ln(z\bar z) . \tag{5.17}$$

The result is thus equivalent to what can be inferred from the mode expansion, since

$$\langle \varphi(x,\tau)\varphi(0,0)\rangle = \langle \phi(z)\phi(0)\rangle + \langle \bar{\phi}(\bar{z})\bar{\phi}(0)\rangle. \tag{5.18}$$

Formula (5.9) can thus be rewritten as

$$\mathrm{e}^{i\alpha\phi(z)}\mathrm{e}^{i\beta\phi(z')} = \mathrm{e}^{i\alpha\phi(z)+i\beta\phi(z')}(z-z')^{\alpha\beta/4\pi}. \tag{5.19}$$

We are now in a position to demonstrate the boson-fermion equivalence at the level of the (anti)commutation relations. In fact, it is simpler to determine the constants A and λ of (4.18) by comparing Green functions rather than simply looking at equal time (anti)commutators (this avoids the equal time singularity). The electron propagator is readily calculated from the mode expansion (3.6)[5]:

$$\begin{aligned}
\langle \psi(z)\psi^\dagger(z')\rangle &= \int_{k>0}\frac{\mathrm{d}k}{2\pi}\int_{q>0}\frac{\mathrm{d}q}{2\pi}\langle 0|\alpha(k)\beta^\dagger(q)|0\rangle \mathrm{e}^{-kz+qz'}\\
&= \int_{k>0}\frac{\mathrm{d}k}{2\pi}\mathrm{e}^{-k(z-z')}\\
&= \frac{1}{2\pi}\frac{1}{z-z'},
\end{aligned} \tag{5.20}$$

where we have supposed that $\tau > \tau'$, which garantees the convergence of the integral and follows from the prescribed time-ordered product. Likewise,

$$\langle \psi^\dagger(z)\psi(z')\rangle = \frac{1}{2\pi}\frac{1}{z-z'}. \tag{5.21}$$

The correct anticommutator may be obtained from the equal time limit:

$$\begin{aligned}
\{\psi(x,0),\psi^\dagger(x',0)\} &= \lim_{\varepsilon\to 0}(\langle \psi(\varepsilon - ix)\psi^\dagger(-ix')\rangle + \langle \psi^\dagger(\varepsilon - ix')\psi(-ix)\rangle)\\
&= \frac{1}{2\pi}\lim_{\varepsilon\to 0}\left(\frac{1}{\varepsilon - i(x-x')} + \frac{1}{\varepsilon - i(x'-x)}\right)\\
&= \frac{1}{2\pi}\lim_{\varepsilon\to 0}\frac{2\varepsilon}{(x-x')^2+\varepsilon^2} = \delta(x-x')
\end{aligned} \tag{5.22}$$

(ψ or $\bar{\psi}$ appearing with a single argument are considered functions of z or $\bar{z}$). One easily checks that the space-time Green function (5.20) has the expected expression in momentum-frequency space:

$$\frac{1}{2\pi}\frac{1}{z-z'} \to \frac{1}{\omega - v|k|}. \tag{5.23}$$

---

[5]For a calculation in the path-integral formalism, see [7].

To demonstrate (4.18) and fix the values of A and λ, one needs to calculate

$$\langle\psi(z)\psi^\dagger(z')\rangle = A^2\langle \mathrm{e}^{-2i\lambda\phi(z)}\mathrm{e}^{2i\lambda\phi(z')}\rangle. \tag{5.24}$$

From the relation (5.19), this is

$$\begin{aligned}\langle\psi(z)\psi^\dagger(z')\rangle &= A^2\langle \mathrm{e}^{-2i\lambda(\phi(z)-\phi(z'))}\rangle(z-z')^{-\lambda^2/\pi} \\ &= A^2(z-z')^{-\lambda^2/\pi}.\end{aligned} \tag{5.25}$$

Note that the normal-ordered expectation value

$$\langle \mathrm{e}^{-i\alpha\phi(z)-i\beta\phi(z')}\rangle \tag{5.26}$$

is nonzero only if $\alpha+\beta=0$. This can be inferred from the action of the zero-mode operator Q:[6]

$$\langle 0|\mathrm{e}^{i(\alpha+\beta)Q}\mathrm{e}^{-i(\alpha+\beta)P/2L}|0\rangle \tag{5.27}$$

vanishes if $\alpha+\beta\neq 0$, since the exponential of Q acting on $|0\rangle$ yields another eigenstate of P, with eigenvalue $\alpha+\beta$, orthogonal to $|0\rangle$. The condition $\alpha+\beta=0$ is called the neutrality condition.[7] By comparing (5.25) with (5.20), we conclude that the constant λ must be $\sqrt{\pi}$ and that A must be $1/\sqrt{2\pi}$. The correct bosonization formula is then

$$\begin{aligned}\psi(x) &= \frac{1}{\sqrt{2\pi}}\mathrm{e}^{-i\sqrt{4\pi}\phi(x)}, & \psi^\dagger(x) &= \frac{1}{\sqrt{2\pi}}\mathrm{e}^{i\sqrt{4\pi}\phi(x)}, \\ \bar{\psi}(x) &= \frac{1}{\sqrt{2\pi}}\mathrm{e}^{i\sqrt{4\pi}\bar{\phi}(x)}, & \bar{\psi}^\dagger(x) &= \frac{1}{\sqrt{2\pi}}\mathrm{e}^{-i\sqrt{4\pi}\bar{\phi}(x)}.\end{aligned} \tag{5.28}$$

Notice the change of sign in the exponent between ψ and $\bar{\psi}$. Notice also that the choice of additive constant leading to (5.15) influences the value of A only, the end result being the same irrespective of that choice.

Remarks. 1. It is also instructive to check the anticommutation relation

$$\{\psi(x),\psi(x')\} = 0. \tag{5.29}$$

This is done simply by noting that the Green function

$$\langle\psi(z)\psi(z')\rangle = \frac{1}{2\pi}\langle \mathrm{e}^{-i\sqrt{4\pi}\phi(z)}\mathrm{e}^{-i\sqrt{4\pi}\phi(z')}\rangle \tag{5.30}$$

vanishes, by the neutrality condition. The anticommutator, obtained in the equal-time limit, also vanishes.

[6] It is also a consequence of the U(1) symmetry $\varphi\to\varphi+a$, where a is a constant.

[7] This terminology comes from an analogy with the Coulomb gas system of two-dimensional statistical mechanics.

2. The vertex operator $\mathrm{e}^{i\alpha\phi}$ is a scaling field. Its conformal dimension h_α may be retrieved from the exponent of the Green function

$$\langle \mathrm{e}^{i\alpha\phi(z)}(\mathrm{e}^{i\alpha\phi(0)})^\dagger\rangle = \frac{1}{z^{\alpha^2/4\pi}} \tag{5.31}$$

and from the relation (2.6). The same analysis may be applied to a left-moving vertex operator $\mathrm{e}^{i\bar{\alpha}\bar{\phi}}$. Thus,

$$\begin{aligned} h(\alpha) &= \frac{\alpha^2}{8\pi}, \qquad & \bar{h}(\alpha) &= 0, \\ h(\bar{\alpha}) &= 0, \qquad & \bar{h}(\bar{\alpha}) &= \frac{\bar{\alpha}^2}{8\pi}. \end{aligned} \tag{5.32}$$

Roughly speaking, the boson field has scaling dimension zero and its powers series may produce operators with nonzero scaling dimension because of the necessary normal ordering.

5.3 Bosonization of the Free-Electron Hamiltonian

The bosonization formulas (5.28) are the basis of a fermion-to-boson translation of various operators. However, this translation process is often subtle, since normal ordering of these operators is required. Take for instance the chiral fermion densities J and $\bar{J}$. Normal ordering can be constructed from the mode expansion. But a shorter and more elegant way to achieve it is by *point splitting*, i.e., by defining

$$J(z) = \lim_{\varepsilon\to 0}[\psi^\dagger(z+\varepsilon)\psi(z) - \langle\psi^\dagger(z+\varepsilon)\psi(z)\rangle], \tag{5.33}$$

where ε has positive time (so as to appear as shown after time ordering). This limit must be taken after applying the bosonization formulas and the relation (5.9):

$$\begin{aligned} J &= \frac{1}{2\pi}\lim_{\varepsilon\to 0}\left[\mathrm{e}^{i\sqrt{4\pi}\phi(z+\varepsilon)}\mathrm{e}^{-i\sqrt{4\pi}\phi(z)} - \frac{1}{\varepsilon}\right] \\ &= \frac{1}{2\pi}\lim_{\varepsilon\to 0}\left[\mathrm{e}^{i\sqrt{4\pi}[\phi(z+\varepsilon)-\phi(z)]}\frac{1}{\varepsilon} - \frac{1}{\varepsilon}\right] \\ &= \frac{1}{2\pi}\lim_{\varepsilon\to 0}\left[\mathrm{e}^{i\varepsilon\sqrt{4\pi}\partial_z\phi(z)}\frac{1}{\varepsilon} - \frac{1}{\varepsilon}\right] \\ &= \frac{1}{2\pi}i\sqrt{4\pi}\partial_z\phi = \frac{i}{\sqrt{\pi}}\partial_z\varphi \end{aligned} \tag{5.34}$$

where we have Taylor-expanded the exponential on the last line. We proceed likewise for the left-moving density, except that the sign is opposite. To summarize:

$$J = \frac{i}{\sqrt{\pi}}\partial_z\varphi, \qquad \bar{J} = -\frac{i}{\sqrt{\pi}}\partial_{\bar{z}}\varphi \tag{5.35}$$

(note that $\partial_z \phi = \partial_z \varphi$ and $\partial_{\bar{z}} \bar{\phi} = \partial_{\bar{z}} \varphi$).

The density fluctuations J and $\bar{J}$ are also components of a conserved current from the boson point of view. Indeed, the compactified boson has a U(1) symmetry: the Lagrangian is invariant under the shift $\varphi \to \varphi + a$. By Noether's theorem, this symmetry implies the existence of a conserved quantity, with a density proportional to $\partial_t \varphi$ and current proportional to $v^2 \partial_x \varphi$. It is customary to define right and left "currents" exactly as in (5.35), so that the continuity equation reduces to $\partial_z \bar{J} + \partial_{\bar{z}} J = 0$. In fact, the critical (i.e., massless) and one-dimensional nature of the theory enhance this U(1) symmetry to a *chiral* U(1) symmetry, by which the left and right currents are separately conserved:

$$\partial_{\bar{z}} J = 0, \qquad \partial_z \bar{J} = 0. \tag{5.36}$$

The same short-distance expansion as in (5.34) may be used to demonstrate explicitly the equivalence of the electron Hamiltonian (3.9) with the boson Hamiltonian (4.3), except that a higher-order expansion is necessary. More explicitly, an expression like $\psi^\dagger \partial_x \psi$ must be evaluated as

$$\psi^\dagger \partial_x \psi = -i \lim_{\varepsilon \to 0} \{ \psi^\dagger(z+\varepsilon) \partial_z \psi(z) - \langle \psi^\dagger(z+\varepsilon) \partial_z \psi(z) \rangle \}. \tag{5.37}$$

The limit is evaluated by using the following short-distance products (to order ε^2), derived from the bosonization formulas and the relation (5.9):

$$\begin{aligned} \psi(z')\psi^\dagger(z) &= \frac{1}{2\pi\varepsilon} + i\frac{1}{\sqrt{\pi}}\partial_z\varphi + i\frac{\varepsilon}{\sqrt{4\pi}}\partial_z^2\varphi - \varepsilon(\partial_z\varphi)^2, \\ \psi^\dagger(z')\psi(z) &= \frac{1}{2\pi\varepsilon} - i\frac{1}{\sqrt{\pi}}\partial_z\varphi - i\frac{\varepsilon}{\sqrt{4\pi}}\partial_z^2\varphi - \varepsilon(\partial_z\varphi)^2, \\ \bar{\psi}(\bar{z}')\bar{\psi}^\dagger(\bar{z}) &= \frac{1}{2\pi\bar{\varepsilon}} - i\frac{1}{\sqrt{\pi}}\partial_{\bar{z}}\varphi - i\frac{\bar{\varepsilon}}{\sqrt{4\pi}}\partial_{\bar{z}}^2\varphi - \bar{\varepsilon}(\partial_{\bar{z}}\varphi)^2, \\ \bar{\psi}^\dagger(\bar{z}')\bar{\psi}(\bar{z}) &= \frac{1}{2\pi\bar{\varepsilon}} + i\frac{1}{\sqrt{\pi}}\partial_{\bar{z}}\varphi + i\frac{\bar{\varepsilon}}{\sqrt{4\pi}}\partial_{\bar{z}}^2\varphi - \bar{\varepsilon}(\partial_{\bar{z}}\varphi)^2, \end{aligned} \tag{5.38}$$

where $\varepsilon = z' - z$ and $\bar{\varepsilon} = \bar{z}' - \bar{z}$, and the fields on the r.h.s. are considered functions of $(z, \bar{z})$ only, the $(z', \bar{z}')$ dependence residing in the powers of ε and $\bar{\varepsilon}$ only. The limit (5.37) may be obtained by differentiating with respect to z or $\bar{z}$ the above expansions, and proceeding to a straightforward substitution. The Hamiltonian (3.9) then becomes

$$\begin{aligned} H_{\mathrm{F}} &= -v \int \mathrm{d}x \; [(\partial_z\varphi)^2 + (\partial_{\bar{z}}\varphi)^2] \\ &= \pi v \int \mathrm{d}x \; (J^2 + \bar{J}^2) \end{aligned} \tag{5.39}$$

which is precisely the boson Hamiltonian (4.3), after using the relations (2.4). Thus the equivalence between a free fermion and a free boson is demonstrated at the level of the Hamiltonian. Note that we have not demonstrated this equivalence at the Lagrangian level; this cannot be done with the techniques used here.

5.4 Spectral Equivalence of Boson and Fermion

That the free boson (at a certain radius R) is equivalent to a free fermion requires that the spectra of the two theories be identical. This can be tested by comparing the grand partition functions of the two models. Let us try to do it for left and right excitations separately. Consider the right-moving fermion $\psi(x)$, and the corresponding Hamiltonian and fermion number:

$$H_{\rm F} = \sum_{k>0} vk(\alpha_k^\dagger \alpha_k + \beta_k^\dagger \beta_k), \qquad N_{\rm F} = \sum_{k>0} (\alpha_k^\dagger \alpha_k - \beta_k^\dagger \beta_k) \tag{5.40}$$

where the operators α_k and β_k have different normalizations from those of (3.3) and obey the anticommutation rules

$$\{\alpha_k, \alpha_q^\dagger\} = \delta_{kq}, \qquad \{\alpha_k, \alpha_q\} = 0 \tag{5.41}$$

(and likewise for β_k). The spectrum of the Hamiltonian is encoded in the grand partition function

$$\mathcal{Z}_{\rm F} = \sum_{\rm states} {\rm e}^{-\beta(H-\mu N)}. \tag{5.42}$$

We have defined the theory on a cylinder, so as to have discrete wavenumbers. We shall use antiperiodic boundary conditions, known in string theory as *Neveu–Schwarz boundary conditions:*

$$\psi(x+L) = -\psi(x). \tag{5.43}$$

The wavenumbers are then half-integer moded: $k = 2\pi(n+\frac{1}{2})/L$, where n is a positive integer (by contrast, periodic boundary conditions, also known as *Ramond boundary conditions*, lead to integer-moded wavenumbers and to the presence of a fermion zero-mode). Each mode being independent, the fermion grand partition function factorizes as

$$\mathcal{Z}_{\rm F} = \prod_{n=1}^{\infty} (1+q^{n+1/2}t)(1+q^{n+1/2}t^{-1}), \quad q \equiv {\rm e}^{-2\pi v\beta/L}, \quad t \equiv {\rm e}^{\beta\mu}. \tag{5.44}$$

Let us now turn to the boson. We first have to write down an expression for the Hamiltonian in terms of the modes appearing in (5.5). A straightforward calculation yields

$$H_{\rm B} = \frac{P^2}{4L} + \frac{2\pi}{L}\sum_{n>0} n b_n^\dagger b_n \tag{5.45}$$

for the right-moving part. Let us then notice that the fermion number N_F is proportional to the zero-mode P. More precisely,

$$N_F = \frac{P}{\sqrt{4\pi}}. \tag{5.46}$$

This can be shown from the mode expansion (5.5) and by the equivalence $J = -\partial_x \phi / \sqrt{\pi}$, which implies

$$N_F = \int dx \; J = -\frac{1}{\sqrt{\pi}}(\phi(L) - \phi(0)) = \frac{P}{\sqrt{4\pi}}. \tag{5.47}$$

Each mode b_n being independent, the boson grand partition function is then

$$\begin{aligned} \mathcal{Z}_B &= \left\{ \prod_{n=1}^{\infty} (1 + q^n + q^{2n} + q^{3n} + \cdots) \right\} \sum_P q^{P^2/8\pi} t^{P/\sqrt{4\pi}} \\ &= \left\{ \prod_{n=1}^{\infty} \frac{1}{1-q^n} \right\} \sum_P q^{P^2/8\pi} t^{P/\sqrt{4\pi}}. \end{aligned} \tag{5.48}$$

So far we have not determined the boson radius R, nor the spectrum of P. We now invoke Jacobi's triple product formula:

$$\prod_{n=1}^{\infty} (1 - q^n)(1 + q^{n-1/2} t)(1 + q^{n-1/2} t^{-1}) = \sum_{n \in \mathbf{Z}} q^{n^2/2} t^n. \tag{5.49}$$

Applied to the fermion partition function, this remarkable identity shows that

$$\mathcal{Z}_F = \left\{ \prod_{n=1}^{\infty} \frac{1}{1-q^n} \right\} \sum_n q^{n^2/2} t^n. \tag{5.50}$$

In order to have $\mathcal{Z}_F = \mathcal{Z}_B$, we therefore need P to take the values $P = n\sqrt{4\pi}$. This is not quite the spectrum (5.7). The closest it comes to is obtained at radius $R = 1/\sqrt{4\pi}$, where (5.7) becomes

$$P = \sqrt{4\pi}(n + \tfrac{1}{2}m), \qquad \overline{P} = \sqrt{4\pi}(n - \tfrac{1}{2}m). \tag{5.51}$$

In fact, because of the imperfect left-right separation of the boson spectrum (5.7), the equivalence with a perfectly left-right separated fermion theory is impossible. However, a state-by-state correspondence between the two theories exists if some constraints are imposed on the fermion spectrum. Explicitly:

1. One needs to include both periodic and antiperiodic boundary conditions in the fermion theory: the Hilbert space is then the direct sum of two sectors.

2. The periodic sector must contain a single zero-mode ψ_0 (the same for left- and right-moving fermions), such that $\{\psi_0, \psi_0^\dagger\} = 1$. The operator $\psi_0^\dagger$ does not change the energy of a state, but increases the fermion number by one. This mixes (in a weak way) the left and right-moving spectra.

3. The number of fermions in the antiperiodic sector must be even (odd-number states are thrown out).

These constraints will not be demonstrated here; a detailed proof can be found in [7,8].

That the simple-minded correspondence between left-moving fermions and left-moving bosons is impossible should not bother us too much. The complications mentioned above basically involve boson or fermion zero-modes and boundary conditions. The nonzero frequency modes are not affected and the fermion-boson correspondence works well for those modes, as reflected in the Jacobi triple product formula (5.49). First and foremost, we should retain from this exercise that the boson radius must be $R = 1/\sqrt{4\pi}$ in order for the fermion-boson equivalence to hold.

5.5 *Case of Many Fermion Species: Klein Factors*

Suppose now that we have more than one fermion species, labeled by a greek index: ψ_μ. Bosonization would then require an equal number of boson species ϕ_μ and the formulas (5.28) would still be applicable, except for the fact that they do not provide for the anticommutation of different fermion species. In fact, this problem arises for a single species as well, since the left and right fermions ψ and $\bar\psi$ must anticommute, whereas the fields ϕ and $\bar\phi$ commute.[8] A possible solution lies in the introduction of additional anticommuting factors, the so-called *Klein factors*, in the bosonization formulas:

$$\begin{aligned}
\psi_\mu(x) &= \frac{1}{\sqrt{2\pi}}\eta_\mu \mathrm{e}^{-i\sqrt{4\pi}\phi_\mu(x)}, \qquad & \psi_\mu^\dagger(x) &= \frac{1}{\sqrt{2\pi}}\eta_\mu \mathrm{e}^{i\sqrt{4\pi}\phi_\mu(x)}, \\
\bar\psi_\mu(x) &= \frac{1}{\sqrt{2\pi}}\bar\eta_\mu \mathrm{e}^{i\sqrt{4\pi}\bar\phi_\mu(x)}, \qquad & \bar\psi_\mu^\dagger(x) &= \frac{1}{\sqrt{2\pi}}\bar\eta_\mu \mathrm{e}^{-i\sqrt{4\pi}\bar\phi_\mu(x)}.
\end{aligned} \tag{5.52}$$

The Klein factors η_μ and $\bar\eta_\mu$ are Hermitian and obey the Clifford algebra:

$$\{\eta_\mu, \eta_\nu\} = \{\bar\eta_\mu, \bar\eta_\nu\} = 2\delta_{\mu\nu}, \qquad \{\eta_\mu, \bar\eta_\nu\} = 0. \tag{5.53}$$

Klein factors act on a Hilbert space distinct from the boson Hilbert space generated by the modes of (5.5). This Hilbert space expansion must be

[8] In our periodic quantization scheme. On the infinite line, with vanishing boundary conditions at $x = \pm\infty$, this is not true and $[\phi(x), \bar\phi(x')] = \frac{1}{4}i$. This guarantees the anticommutation $\{\psi(x), \bar\psi(x')\} = 0$.

compensated by some sort of "gauge fixing": the Hamiltonian must be diagonal in this Klein factor space, as well as required physical observables. Then, one Klein eigenstate is chosen and the rest of the Klein Hilbert space decouples.

An alternative to Klein factors, which requires no extra Hilbert space, is to include in the bosonization formula for ψ_μ the factor

$$\exp\left[i\pi\sum_{\nu<\mu} N_{\mathrm{F},\nu}\right] = \exp\left[i\frac{\sqrt{\pi}}{2}\sum_{\nu<\mu} P_\nu\right] \tag{5.54}$$

(P_ν is the conjugate to the zero mode Q_μ, cf. (5.5)). Because of the commutation rule $[Q_\mu, P_\nu] = i\delta_{\mu\nu}$, the CBH formula (5.10) makes the different fermions species anticommute. This must be properly extended to left-moving fermions, for instance by enumerating them consecutively after the right fermions (unless one uses the infinite-line quantization scheme).

5.6 Bosonization of Interactions

Let us finally turn our attention to the basic interactions (4.2). This requires introducing spin: we need two bosons $\varphi_\uparrow$ and $\varphi_\downarrow$. The boson fields may be naturally combined into spin and charge components:

$$\varphi_c = \frac{1}{\sqrt{2}}(\varphi_\uparrow + \varphi_\downarrow), \qquad \varphi_s = \frac{1}{\sqrt{2}}(\varphi_\uparrow - \varphi_\downarrow) \tag{5.55}$$

and likewise for the chiral components $\phi_{c,s}$ and $\bar{\phi}_{c,s}$.

5.6.1 Backward Scattering

The backward scattering term (g_1) does not need normal ordering, since all four Fermi fields anticommute. A straightforward application of the bosonization formulas is sufficient:

$$\begin{aligned}\mathcal{H}_1 &= \frac{vg_1}{(2\pi)^2}\sum_\sigma \eta_\sigma\bar{\eta}_\sigma\bar{\eta}_{-\sigma}\eta_{-\sigma}\exp[i\sqrt{4\pi}(\phi_\sigma + \bar{\phi}_\sigma - \bar{\phi}_{-\sigma} - \phi_{-\sigma})] \\ &= \frac{vg_1}{(2\pi)^2}\left[\eta_\uparrow\bar{\eta}_\uparrow\bar{\eta}_\downarrow\eta_\downarrow \mathrm{e}^{i\sqrt{8\pi}\varphi_s} + \eta_\downarrow\bar{\eta}_\downarrow\bar{\eta}_\uparrow\eta_\uparrow \mathrm{e}^{-i\sqrt{8\pi}\varphi_s}\right].\end{aligned} \tag{5.56}$$

Since all the Klein factors present anticommute, $\eta_\uparrow\bar{\eta}_\uparrow\bar{\eta}_\downarrow\eta_\downarrow = \eta_\downarrow\bar{\eta}_\downarrow\bar{\eta}_\uparrow\eta_\uparrow$ and it remains

$$\mathcal{H}_1 = \frac{vg_1}{2\pi^2}\eta_\uparrow\bar{\eta}_\uparrow\bar{\eta}_\downarrow\eta_\downarrow\cos(\sqrt{8\pi}\varphi_s). \tag{5.57}$$

The product of Klein factors poses a slight conceptual problem. The Klein Hilbert space (on which the η_σ and $\bar{\eta}_\sigma$ act) is the representation space of a Clifford algebra, and has a minimal dimension which depends on the number of fermion species. With four species (two spins, times left and right), the minimal dimension is four, exactly like Dirac matrices in four-dimensional space-time. For instance, the following representation is possible, in terms of tensor products of Pauli matrices:

$$\begin{aligned} \eta_\uparrow &= \sigma_1 \otimes \sigma_1, & \eta_\downarrow &= \sigma_3 \otimes \sigma_1, \\ \bar\eta_\uparrow &= \sigma_2 \otimes \sigma_1, & \bar\eta_\downarrow &= 1 \otimes \sigma_2. \end{aligned} \tag{5.58}$$

One readily checks that this is a faithful representation of the Clifford algebra (5.53). Then, the above Klein product is

$$\eta_\uparrow \bar\eta_\uparrow \bar\eta_\downarrow \eta_\downarrow = 1 \otimes \sigma_3. \tag{5.59}$$

This matrix is diagonal with eigenvalues ± 1. If no other term in the Hamiltonian contains a nondiagonal Klein product, one may safely pick a Klein eigenstate and stick with it, forgetting the rest of the Klein Hilbert space. This is some kind of gauge choice.[9] In the case at hand, we may chose the eigenvalue $+1$ and write

$$\mathcal{H}_1 = \frac{vg_1}{2\pi^2} \cos\big(\sqrt{8\pi}\varphi_s\big). \tag{5.60}$$

5.6.2 Forward Scattering

The next interaction term (g_2) is quite simple to bosonize, since it involves only density operators,

$$\mathcal{H}_2^c = \frac{vg_{2,c}}{\pi} \sum_{\sigma,\sigma'} \partial_z \varphi_\sigma \partial_{\bar z} \varphi_{\sigma'} = \frac{2vg_{2,c}}{\pi} \partial_z \varphi_c \partial_{\bar z} \varphi_c, \tag{5.61}$$

and likewise for the spin interaction,

$$\mathcal{H}_2^s = \frac{2vg_{2,s}}{\pi} \partial_z \varphi_s \partial_{\bar z} \varphi_s. \tag{5.62}$$

Likewise, the forward terms (g_4) are

$$\begin{aligned} \mathcal{H}_4^c &= -\frac{vg_{4,c}}{2\pi} \left[(\partial_z \varphi_\uparrow + \partial_z \varphi_\downarrow)^2 + (\partial_{\bar z} \varphi_\uparrow + \partial_{\bar z} \varphi_\downarrow)^2 \right] \\ &= -\frac{vg_{4,c}}{\pi} \left[(\partial_z \varphi_c)^2 + (\partial_{\bar z} \varphi_c)^2 \right] \end{aligned} \tag{5.63}$$

and

$$\mathcal{H}_4^s = -\frac{vg_{4,s}}{\pi} \left[(\partial_z \varphi_s)^2 + (\partial_{\bar z} \varphi_s)^2 \right]. \tag{5.64}$$

[9] One has to be careful when calculating various correlation functions, since the operators involved may also contain Klein factors, and simultaneous diagonality of all relevant operators with interactions is important if one wants to forget about the Klein Hilbert space.

5.6.3 Umklapp Scattering

Finally, the Umklapp term (g_3) is treated in the same way as g_1, this time with the charge boson involved:

$$\begin{aligned}\mathcal{H}_3 &= \frac{vg_3}{2(2\pi)^2}[(\eta_\uparrow\eta_\downarrow\bar{\eta}_\uparrow\bar{\eta}_\downarrow + \eta_\downarrow\eta_\uparrow\bar{\eta}_\downarrow\bar{\eta}_\uparrow)e^{i\sqrt{4\pi}(\phi_\uparrow+\phi_\downarrow+\bar{\phi}_\uparrow+\bar{\phi}_\downarrow)} + \text{H.c.}] \\ &= \frac{vg_3}{(2\pi)^2}\left[\eta_\uparrow\eta_\downarrow\bar{\eta}_\uparrow\bar{\eta}_\downarrow e^{i\sqrt{8\pi}\varphi_c} + \text{H.c.}\right] \\ &= \frac{vg_3}{2\pi^2}\eta_\uparrow\eta_\downarrow\bar{\eta}_\uparrow\bar{\eta}_\downarrow\cos(\sqrt{8\pi}\varphi_c).\end{aligned} \tag{5.65}$$

After a mere transposition, the Klein prefactor is the same as (5.59). Therefore,

$$\mathcal{H}_3 = \frac{vg_3}{2\pi^2}\cos(\sqrt{8\pi}\varphi_c). \tag{5.66}$$

6 Exact Solution of the Tomonaga–Luttinger Model

6.1 *Field and Velocity Renormalization*

The power of bosonization lies mainly in its capacity of representing fermion interactions, namely, $\mathcal{H}_2$ and $\mathcal{H}_4$, as parts of a noninteracting boson Hamiltonian. The continuum electron model defined by the Hamiltonian density

$$\mathcal{H}_{\text{TL}} = \mathcal{H}_{\text{F}} + \mathcal{H}_{2,c} + \mathcal{H}_{2,s} + \mathcal{H}_{4,c} + \mathcal{H}_{4,s} \tag{6.1}$$

is called the *Tomonaga–Luttinger model* (TL model). The free electron part $\mathcal{H}_{\text{F}}$ may be expressed as the sum of free boson theories for the spin and charge bosons, with the same velocity v. As seen in Section 5.6, each interaction term is expressed either in terms of the spin boson or of the charge boson. Thus, the total Hamiltonian is the sum of charge and a spin Hamiltonians, i.e., the two sectors (charge and spin) are completely decoupled[10]:

$$\mathcal{H}_{\text{TL}} = \mathcal{H}_c + \mathcal{H}_s. \tag{6.2}$$

When expressed in terms of the spin and charge bosons—more precisely, in terms of the conjugate field Π_μ and of $\partial_x\varphi_\mu$—the Hamiltonian of each sector becomes

$$\mathcal{H}_\mu = \mathcal{H}_0^\mu + \mathcal{H}_2^\mu + \mathcal{H}_4^\mu, \tag{6.3}$$

where

$$\mathcal{H}_{0,\mu} = \tfrac{1}{2}v[\Pi_\mu^2 + (\partial_x\varphi_\mu)^2],$$

[10] This is true at the field level, but not completely true at the spectrum level, where things are more subtle, as shown above. Obviously the number of spin and charge excitations are not independent, since the total number of electrons is conserved.

$$\begin{aligned}\mathcal{H}_{2,\mu} &= -\frac{vg_{2,\mu}}{2\pi}[\Pi_\mu^2 - (\partial_x\varphi_\mu)^2], \\ \mathcal{H}_{4,\mu} &= \frac{vg_{4,\mu}}{2\pi}\left[\Pi_\mu^2 + (\partial_x\varphi_\mu)^2\right],\end{aligned}$$

where $\mu = c, s$. Combining these expressions, one finds

$$\mathcal{H}_\mu = \tfrac{1}{2}v_\mu\left[K_\mu\Pi_\mu^2 + \frac{1}{K_\mu}(\partial_x\varphi_\mu)^2\right], \tag{6.4}$$

where

$$K_\mu = \sqrt{\frac{\pi - g_{2,\mu} + g_{4,\mu}}{\pi + g_{2,\mu} + g_{4,\mu}}}, \qquad v_\mu = v\sqrt{\left(1 + \frac{g_{4,\mu}}{\pi}\right)^2 - \left(\frac{g_{2,\mu}}{\pi}\right)^2}. \tag{6.5}$$

The constant K_μ is simply a renormalization of the field φ_μ. This is most obvious in the associated Lagrangian:

$$L_\mu = \frac{1}{2K_\mu}\int \mathrm{d}x\left[\frac{1}{v_\mu}(\partial_t\varphi_\mu)^2 - v_\mu(\partial_x\varphi_\mu)^2\right]. \tag{6.6}$$

The Hamiltonian (6.4) may be brought back in the canonical form (4.3) simply by introducing rescaled operators

$$\varphi'_\mu = \frac{1}{\sqrt{K_\mu}}\varphi_\mu, \qquad \Pi'_\mu = \sqrt{K_\mu}\Pi_\mu, \tag{6.7}$$

which still obey the canonical commutation relations. Clearly, the field φ'_μ has radius

$$R_\mu = \frac{R}{\sqrt{K_\mu}} = \frac{1}{\sqrt{4\pi K_\mu}}. \tag{6.8}$$

Remarks. 1. Let us point out that we have supposed the couplings $g_{2,\mu}$ and $g_{4,\mu}$ to be momentum independent. However, this is not necessary: the above solution of the Tomonaga–Luttinger model may be repeated with momentum-dependent $g_{2,\mu}(q)$ and $g_{4,\mu}(q)$, for the simple reason that the boson Hamiltonian is quadratic and that different momenta are therefore decoupled. In particular, the above formula for the velocity renormalization may be used to write down the exact dispersion relation of charge excitations:

$$\omega(k) = v_c|k| = v|k|\sqrt{\left(1 + \frac{g_{4,\mu}(k)}{\pi}\right)^2 - \left(\frac{g_{2,\mu}(k)}{\pi}\right)^2}. \tag{6.9}$$

For simplicity, we shall ignore from now on such a momentum dependence of the couplings, mainly because it is generally irrelevant (in the RG sense).

2. It is important that the perturbations $\mathcal{H}_2$ and $\mathcal{H}_4$ of (6.3) be considered as perturbations of the Hamiltonian, not of the Lagrangian. If the conjugate momentum Π were replaced by $\partial_t\varphi/v$ and the resulting expressions considered as perturbations of the Lagrangians, the field and velocity renormalization would be quite different. The reason is that, as they are expressed here, these perturbations depend on $\partial_t\varphi_\mu$ and would change the definition of the conjugate momentum, were they to be applied to the Langrangian.

6.2 Left-Right Mixing

The perturbations $\mathcal{H}_{2,\mu}$ of (6.3) mix right and left bosons. One then expects that the new eigenstates will be linear combinations of left and right excitations of the original boson. Indeed, rescaling the field φ ($\varphi' = \varphi/\sqrt{K}$) implies the opposite rescaling of its conjugate momentum Π—and therefore of the dual field ϑ—in order to preserve the canonical commutation rules: $\vartheta' = \vartheta\sqrt{K}$ (we drop the spin/charge index for the time being). Thus, the left and right bosons ϕ and $\bar{\phi}$ are not simply rescaled by K, they are mixed:

$$\begin{aligned} \phi = \tfrac{1}{2}(\varphi + \vartheta) &\to \phi' = \tfrac{1}{2}\left(\frac{1}{\sqrt{K}}\varphi + \sqrt{K}\vartheta\right), \\ \bar{\phi} = \tfrac{1}{2}(\varphi - \vartheta) &\to \bar{\phi}' = \tfrac{1}{2}\left(\frac{1}{\sqrt{K}}\varphi - \sqrt{K}\vartheta\right). \end{aligned} \tag{6.10}$$

Expressing the old right and left bosons (ϕ and $\bar{\phi}$) in terms of the new ones (ϕ' and $\bar{\phi}'$) and defining ξ as $K = \mathrm{e}^{2\xi}$, one finds

$$\begin{aligned} \phi &= \phi' \cosh\xi + \bar{\phi}' \sinh\xi, \\ \bar{\phi} &= \phi' \sinh\xi + \bar{\phi}' \cosh\xi. \end{aligned} \tag{6.11}$$

In terms of the mode expansion (5.5), this left-right mixing is a Bogoliubov transformation of the creation and annihilation operators, if one collects like powers of e^{ikx} at equal times:

$$\begin{aligned} b_n &= b'_n \cosh\xi + \bar{b}'^{\dagger}_n \sinh\xi, \\ \bar{b}_n &= b'^{\dagger}_n \sinh\xi + \bar{b}'_n \cosh\xi. \end{aligned} \tag{6.12}$$

Thus, the fermion field $\psi_\uparrow$, a pure right-moving field in the unperturbed system, becomes a mixture of right- and left-moving fields in the Tomonaga–Luttinger model, if $K_\mu \neq 1$ (i.e., if $g_{2,\mu} \neq 0$). Consider for instance the case $K_s = 1$, $K_c \neq 1$:

$$\begin{aligned} \psi_\uparrow(x,\tau) &= \frac{1}{\sqrt{2\pi}}\eta_\uparrow \mathrm{e}^{-i\sqrt{4\pi}\phi_\uparrow} \\ &= \frac{1}{\sqrt{2\pi}}\eta_\uparrow \mathrm{e}^{-i\sqrt{2\pi}\phi_c}\mathrm{e}^{-i\sqrt{2\pi}\phi_s} \\ &= \frac{1}{\sqrt{2\pi}}\eta_\uparrow \mathrm{e}^{-i\sqrt{2\pi}\cosh\xi\phi'_c}\mathrm{e}^{-i\sqrt{2\pi}\sinh\xi\bar{\phi}'_c}\mathrm{e}^{-i\sqrt{2\pi}\phi_s}. \end{aligned} \tag{6.13}$$

This is expected, since the interaction $\mathcal{H}_2$ surrounds the free electron with a cloud of right *and* left electrons.

6.3 Correlation Functions

6.3.1 Green Function

Once the renormalizations (6.5) are obtained, the calculation of physical quantities (correlation functions) is a straighforward task in principle, at least in real space. Let us start with the one-particle Green function

$$G_\uparrow(x,\tau) = \langle c_\uparrow(x,\tau)c_\uparrow^\dagger(0,0)\rangle. \tag{6.14}$$

Near zero wavevector, in terms of continuum fields, this becomes[11]

$$G_\uparrow(x,\tau) = \langle \psi_\uparrow(x,\tau)\psi_\uparrow^\dagger(0,0)\rangle + \langle \bar{\psi}_\uparrow(x,\tau)\bar{\psi}_\uparrow^\dagger(0,0)\rangle. \tag{6.15}$$

Let us concentrate on the right-moving part, in the case $K_s = 1$ and $K_c \neq 1$ for simplicity. Using (6.13), it becomes

$$\begin{aligned}
\langle \psi_\uparrow(x,\tau)\psi_\uparrow^\dagger(0,0)\rangle &= \frac{1}{2\pi}\langle \mathrm{e}^{-i\sqrt{2\pi}\cosh\xi\phi_c'(z_c)}\mathrm{e}^{i\sqrt{2\pi}\cosh\xi\phi_c'(0)}\rangle \\
&\quad\times \langle \mathrm{e}^{i\sqrt{2\pi}\sinh\xi\bar{\phi}_c'(\bar{z}_c)}\mathrm{e}^{-i\sqrt{2\pi}\sinh\xi\bar{\phi}_c'(0)}\rangle \\
&\quad\times \langle \mathrm{e}^{-i\sqrt{2\pi}\phi_s'(z_s)}\mathrm{e}^{i\sqrt{2\pi}\phi_s'(0)}\rangle \\
&= \frac{1}{2\pi}\frac{1}{(v_c\tau - ix)^{\frac{1}{2}\cosh^2\xi}}\frac{1}{(v_c\tau + ix)^{\frac{1}{2}\sinh^2\xi}}\frac{1}{(v_s\tau - ix)^{1/2}} \\
&= \frac{1}{2\pi}\frac{1}{(v_c\tau - ix)^{1/2}}\frac{1}{|v_c\tau - ix|^{\theta_c}}\frac{1}{(v_s\tau - ix)^{1/2}}
\end{aligned} \tag{6.16}$$

where the velocity difference between spin and charge demands a distinction between spin and charge complex coordinates (z_s and z_c). We have introduced the exponent[12]

$$\theta_c = \tfrac{1}{4}\left(K_c + \frac{1}{K_c} - 2\right). \tag{6.17}$$

Once the Fourier transform is taken in time and space, this Green function leads to extended spectral weight,[13] as illustrated in Fig. 4.5. This spectral weight has power-law singularities near $\omega = \pm v_c q$ and $\omega = v_s q$, with exponents related to θ_c. From this spectral function, it can be shown that the

---

[11] There is also a singular component near $k = 2k_\mathrm{F}$ is $K < 1$.

[12] The notation $\gamma_c = \frac{1}{2}\theta_c$ is also used, as is $\alpha_c = \theta_c$.

[13] For a detailed calculation of the spectral weight from the above space-time Green function, see [4]

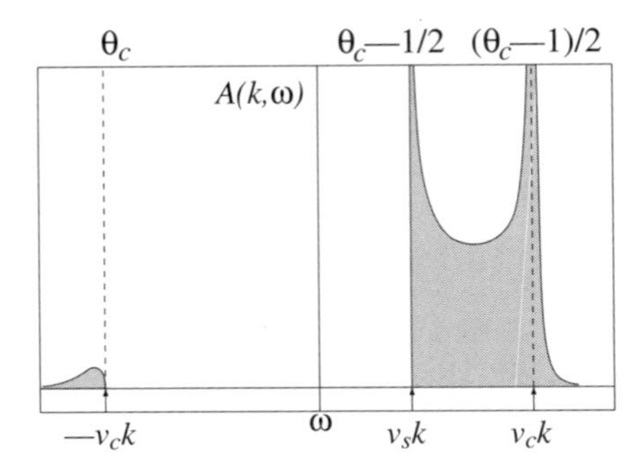

FIGURE 4.5. Typical spectral function $A(k,\omega)$ of the Tomonaga–Luttinger model, with $K_s = 1$. The exponents appearing on top refer to the three singularities: $(\omega + v_c q)^{\theta_c}$, $(\omega - v_s q)^{\theta_c - 1/2}$, and $(\omega - v_c q)^{(\theta_c - 1)/2}$.

momentum distribution function $n(k)$ has an algebraic singularity at the Fermi level:

$$n(k) = n(k_{\rm F}) - \text{const.} \times \text{sgn}(k - k_{\rm F})|k - k_{\rm F}|^{\theta_c} \tag{6.18}$$

and so does the one-particle density of states: $N(\omega) \sim |\omega|^{\theta_c}$.

6.3.2 Compressibility

Next, let us consider the compressibility near $q = 0$:

$$\chi(x,\tau) = \langle n_{\rm tot}(x,\tau) n_{\rm tot}(0,0)\rangle. \tag{6.19}$$

Since $n_{\uparrow,\rm tot} = \partial_x \varphi_\uparrow/\sqrt{\pi}$, we have

$$n_{\rm tot} = \sqrt{\frac{2}{\pi}}\partial_x\varphi_c \quad \text{whereas} \quad S_z = \sqrt{\frac{1}{2\pi}}\partial_x\varphi_s \tag{6.20}$$

is the z component of the uniform spin density. The compressibility is then

$$\begin{aligned}\chi(x,\tau) &= \frac{2}{\pi}\langle\partial_x\varphi_c(x,\tau)\partial_x\varphi_c(0,0)\rangle \\ &= \frac{2K_c}{\pi}\langle\partial_x\varphi_c'(x,\tau)\partial_x\varphi_c'(0,0)\rangle \\ &= -\frac{2K_c}{\pi}[\langle\partial_z\varphi_c'(x,\tau)\partial_z\varphi_c'(0,0)\rangle + \langle\partial_{\bar z}\varphi_c'(x,\tau)\partial_{\bar z}\varphi_c'(0,0)\rangle] \\ &= \frac{K_c}{2\pi^2}\left[\frac{1}{(v_c\tau - ix)^2} + \frac{1}{(v_c\tau + ix)^2}\right].\end{aligned} \tag{6.21}$$

This, of course, is the quasi-uniform component of the compressibility. The corresponding result for the magnetic susceptibility (still at $K_s = 1$) is

$$\langle S_z(x,\tau)S_z(0,0)\rangle = \frac{1}{8\pi^2}\left[\frac{1}{(v_s\tau - ix)^2} + \frac{1}{(v_s\tau + ix)^2}\right], \tag{6.22}$$

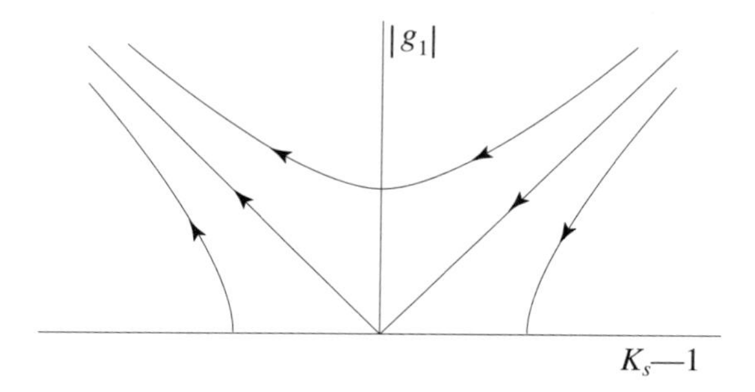

FIGURE 4.6. RG flow with backward scattering g_1 near the point $K_s = 1$.

where v_s is the spin velocity.

The compressibility also has a charge-density-wave component at $\pm 2k_{\rm F}$, determined by the correlations of the operator

$$\begin{aligned}\mathcal{O}_{\rm CDW} &= {\rm e}^{-2ik_{\rm F}x}\sum_\sigma \psi^\dagger_\sigma \bar{\psi}_\sigma \\ &= \frac{1}{2\pi}{\rm e}^{-2ik_{\rm F}x}\sum_\sigma \eta_\sigma \bar{\eta}_\sigma {\rm e}^{-i\sqrt{4\pi}\varphi_\sigma} \\ &= \frac{1}{2\pi}{\rm e}^{-2ik_{\rm F}x}{\rm e}^{i\sqrt{2\pi}\varphi_c}\left[\eta_\uparrow\bar{\eta}_\uparrow {\rm e}^{i\sqrt{2\pi}\varphi_s} + \eta_\downarrow\bar{\eta}_\downarrow {\rm e}^{-i\sqrt{2\pi}\varphi_s}\right]. \end{aligned} \tag{6.23}$$

The CDW compressibility is then

$$\begin{aligned}\langle \mathcal{O}_{\rm CDW}(x,\tau)\mathcal{O}^\dagger_{\rm CDW}(0,0)\rangle &= \frac{1}{2\pi^2}{\rm e}^{-2ik_{\rm F}x}\langle {\rm e}^{i\sqrt{2\pi}\varphi_c(z)}{\rm e}^{-i\sqrt{2\pi}\varphi_c(0)}\rangle \\ &= \frac{1}{2\pi^2}{\rm e}^{-2ik_{\rm F}x}\frac{1}{-ix|^{K_c}}\frac{1}{|v_s\tau - ix|}. \end{aligned} \tag{6.24}$$

Notice that the Klein factors disappear from the correlation function, as they do in general from a product of the type $\mathcal{O}\mathcal{O}^\dagger$.

6.4 Spin or Charge Gap

The interactions $\mathcal{H}_1$ and $\mathcal{H}_3$ do not lend themselves to an exact solution like $\mathcal{H}_2$ and $\mathcal{H}_4$ do. However, they are subject to RG arguments, at least at weak coupling.

Let us first consider the backward scattering term (5.60). Combined with the spin boson Hamiltonian (for $K_s \neq 1$), the perturbation defines the well-known sine-Gordon model, one of the most studied one-dimensional field theories[14]:

$$\mathcal{H}_{\rm SG} = \tfrac{1}{2}v\big[\Pi_s'^2 + (\partial_x\varphi_s')^2\big] + \frac{vg_1}{2\pi^2}\cos\big(\sqrt{8\pi K_s}\varphi_s'\big) \tag{6.25}$$

(we used the rescaled boson φ_s' in order to give the unperturbed part its standard normalization). If $K_s = 1$, the cosine term is marginal, as can be checked from (5.32): $\Delta = h + \bar{h} = 1 + 1 = 2$. One would naively conclude that it is irrelevant if $K_s > 1$, and relevant if $K_s < 1$, thus leading to a gap in the spin excitations. The story is a little more complicated by the existence at $g_1 = 0$ not of a fixed *point*, but of a fixed *line*, parametrized by K_s. It turns out that K_s is subjected to RG flow away from that line and one must instead consider the set of coupled RG equations

$$\frac{\mathrm{d}K_s}{\mathrm{d}\ell} = -\frac{1}{2\pi^2}K_s^2 g_1^2 \qquad \frac{\mathrm{d}g_1}{\mathrm{d}\ell} = -2g_1(K_s - 1) \tag{6.26}$$

first obtained by Kosterlitz and Thouless [9]. These equations are demonstrated in Appendix A, from the short distance product, or OPE, of the perturbations involved. The RG flow is illustrated in Fig. 4.6. Note that, at weak coupling, the system flows toward strong coupling if $|g_1] > 2\pi(K_s - 1)$ (g_1 is then said to be *marginally relevant*) while it flows to $g_1 = 0$ if $|g_1] < 2\pi(K_s - 1)$ (g_1 is then said to be *marginally irrelevant*). Thus, we expect a spin gap for g_1 infinitesimal and $K_s < 1$.[15]

This expectation is fullfilled exactly at the special value $K_s = \frac{1}{2}$. There, the cosine term has scaling dimension 1 and is precisely what one would obtain from bosonizing the Dirac mass term of (3.13), except that the corresponding fermion represents not electrons, but massive spinons. This $K_s = 1$ theory is called the Luther–Emery model [10] and is a free theory for spinons, even though it is a complicated interacting theory for electrons (and who knows what happens in the charge sector at the same time...).

The above analysis can be repeated for the Umklapp term (5.66), which occurs only at half filling. One simply has to replace g_1 by g_3 and K_s by K_c. Thus, a charge gap develops if $K_c < 1$ (i.e., if $g_{2,c}$ is repulsive) or if $|g_3] > 2\pi(K_c - 1)$. Again, free massive holons occur at $K_c = \frac{1}{2}$.

---

[14] For the sake of argument, I have neglected the velocity renormalization here ($v_s = v$). It would lead to an additional renormalization of g_1, defined by $vg_1 = v_s g_1'$, which is neglected at low coupling.

[15] Note that the sign of g_1 is of no consequence: it is the result of a gauge choice through Klein factors. The arbitrariness in the sign of g_1 may be traced back to its expression in terms of left and right fermions: changing the sign of either $\psi_\uparrow$, $\bar{\psi}_\uparrow$, $\psi_\downarrow$, or $\bar{\psi}_\downarrow$ is of no consequence in the noninteracting case, but changes the sign of g_1 and g_3, without affecting those of g_2 or g_4.

7 Non-Abelian Bosonization

7.1 Symmetry Currents

The bosonization procedure described in Section 5 is often called *Abelian bosonization*, because a single compactified boson φ has a $U(1)$ symmetry $\varphi \to \varphi + a$ and the group $U(1)$ is Abelian (i.e., commutative). Abelian bosonization may be used to describe the spin sector of a 1D electron gas, but this description is not manifestly spin-rotation invariant. Indeed, the chiral currents associated to the three spin components are easily shown to be

$$\begin{aligned} J^z &= \tfrac{1}{2}\psi^\dagger_\alpha(\sigma_3)_{\alpha\beta}\psi_\beta = \frac{i}{\sqrt{2\pi}}\partial_z\varphi_s, \\ J^x &= \tfrac{1}{2}\psi^\dagger_\alpha(\sigma_1)_{\alpha\beta}\psi_\beta = \frac{i}{2\pi}\eta_\uparrow\eta_\downarrow \sin\left(\sqrt{8\pi}\phi_s\right), \\ J^y &= \tfrac{1}{2}\psi^\dagger_\alpha(\sigma_2)_{\alpha\beta}\psi_\beta = -\frac{i}{2\pi}\eta_\uparrow\eta_\downarrow \cos\left(\sqrt{8\pi}\phi_s\right). \end{aligned} \tag{7.1}$$

The z component manifestly has a special role in this representation. While the correlations $\langle J^a(x)J^a(0)\rangle$ ($a = x, y, z$) all decay with the same power law, they do not have the same normalization, a somewhat unsatisfactory feature.

Fortunately, there exists a bosonization procedure that is manifestly symmetric under the spin rotation group SU(2), or under any Lie group for that matter. This *non-Abelian bosonization* expresses a set of fermion fields in terms of a matrix field belonging to a representation of a Lie algebra, instead of one or more simple boson fields. The corresponding boson theory, the Wess–Zumino–Witten (WZW) model, is far less familiar than the ordinary boson. But in all situations where symmetry considerations (including symmetry breaking) are important, non-Abelian bosonization is the method of choice. It is of course unrealistic to give a rigorous and complete introduction to the WZW model and non-Abelian bosonization within this brief tutorial. Reference [7] may be consulted for an in-depth discussion.

Let us first return to Abelian bosonization and insist on the role of the U(1) current, before generalizing to larger symmetry groups. The currents associated with the U(1) symmetry are given in (5.35) for a single species of fermions and coincide with the chiral fermion density fluctuations. For the charge U(1) symmetry, it is the total ($\uparrow$+$\downarrow$) density fluctuation that matters, and the charge currents are therefore

$$J_c = \frac{i}{\sqrt{2\pi}}\partial_z\varphi_c, \qquad \bar{J}_c = -\frac{i}{\sqrt{2\pi}}\partial_{\bar{z}}\varphi_c. \tag{7.2}$$

In terms of the U(1) group element $g = \mathrm{e}^{i\varphi_c}$, these currents are proportional to $g^{-1}\partial_z g$ and $g^{-1}\partial_{\bar{z}} g$. The charge Hamiltonian can then be expressed in terms of these currents, as in (5.39):

$$H = \frac{\pi v_c}{2} \int \mathrm{d}x \; (J_c^2 + \bar{J}_c^2). \tag{7.3}$$

This is enough to define the theory, provided that (i) the commutation relations of the currents are specified,

$$[J_c(x), J_c(x')] = -\frac{i}{\pi}\delta'(x - x'), \tag{7.4}$$

and (ii) the vertex operators are introduced, with their proper commutations with the currents. Note: The above current commutation relation can be demonstrated in the same way as we have recovered the fermion anticommutation relations (5.22), from the current-current correlation function

$$\langle J_c(z) J_c(z') \rangle = \frac{1}{8\pi^2} \frac{1}{(z - z')^2} \tag{7.5}$$

obtained from (5.14) by differentiating w.r.t. z and z'.

Likewise, non-Abelian bosonization (for SU(2)) may be introduced starting with the various components of the spin currents J_a defined above in terms of two electron fields ψ_σ, and expressed in terms of a group element g as

$$J(z) = (\partial_z g) g^{-1} = \tfrac{1}{2} \sum_a J_a \sigma_a, \qquad \bar{J}(\bar{z}) = g^{-1}(\partial_{\bar{z}} g) = \tfrac{1}{2} \sum_a \bar{J}_a \sigma_a. \tag{7.6}$$

The Hamiltonian of the SU(2) WZW model can be expressed in terms of these SU(2) currents:

$$H = \frac{2\pi v}{k+2} \sum_{a=1}^{3} \int \mathrm{d}x \; [{:}J^a(x) J^a(x){:} + {:}\bar{J}^a(x) \bar{J}^a(x){:}] \tag{7.7}$$

where the integer k is called the *level* of the WZW model, and where the currents now obey the following commutation rules:

$$\begin{aligned}
[J^a(x), J^b(x')] &= -\frac{ik}{2\pi}\delta_{ab}\delta'(x - x') + i\varepsilon_{abe} J^e(x)\delta(x - x'), \\
[\bar{J}^a(x), \bar{J}^b(x')] &= -\frac{ik}{2\pi}\delta_{ab}\delta'(x - x') + i\varepsilon_{abe} \bar{J}^e(x)\delta(x - x'), \\
[J^a(x), \bar{J}^b(x')] &= 0.
\end{aligned} \tag{7.8}$$

Once the currents are Fourier expanded as

$$J^a(z) = \frac{1}{L} \sum_n J_n^a \mathrm{e}^{i2\pi n z/L}. \tag{7.9}$$

The above commutators translate into the following set of commutation rules, known as a *Kac–Moody algebra:*

$$\begin{aligned} [J^a_m, J^b_n] &= i\varepsilon_{abe} J^e_{m+n} + \tfrac{1}{2} k n \delta_{ab} \delta_{m+n,0}, \\ [\bar{J}^a_m, \bar{J}^b_n] &= i\varepsilon_{abe} \bar{J}^e_{m+n} + \tfrac{1}{2} k n \delta_{ab} \delta_{m+n,0}. \end{aligned} \tag{7.10}$$

States of the WZW model are created by applying the operators J^a_n $(n < 0)$ on the vacuum $|0\rangle$ and on a finite number of spin states.

The formulation of the WZW in terms of currents is not as fundamental as its definition in the Lagrangian formalism, in terms of a field $g(x,t)$ taking its values in SU(2) (or in another Lie group). The model must be such that the currents are chiral, i.e., $\partial_z \bar{J} = \partial_{\bar{z}} J = 0$, and must have conformal invariance. The needed action is

$$\begin{aligned} S[g] = \frac{k}{8\pi} \int \mathrm{d}^2x \ \mathrm{tr}(\partial^\mu g^{-1} \partial_\mu g) \\ - \frac{ik}{12\pi} \int_B \mathrm{d}^3x \, \varepsilon_{\mu\nu\lambda} \, \mathrm{tr}(g^{-1}\partial^\mu g \, g^{-1}\partial^\nu g \, g^{-1}\partial^\lambda g) \end{aligned} \tag{7.11}$$

where the second term is topological: it is integrated in a three-dimensional manifold B whose boundary is the two-dimensional space-time, but its value depends only on the field configuration on the boundary of B. The *level* k must be an integer in order for the WZW model to have conformal invariance and chiral currents. Working with the action (7.11) is somewhat unwieldy and almost never done in practice, once key identities (Ward identities) have been derived, leading, among others, to the commutators (7.8).

Bosonization cannot be done in terms of the currents alone. The group elements g occur in the representation of left-right products of fermions, as specified by the *Witten formula* [11]:

$$\begin{aligned} \psi_\alpha \bar{\psi}^\dagger_\beta &= \frac{1}{2\pi} g_{\alpha\beta} \mathrm{e}^{-i\sqrt{2\pi}\varphi_c}, \\ \bar{\psi}_\alpha \psi^\dagger_\beta &= \frac{1}{2\pi} (g^\dagger)_{\alpha\beta} \mathrm{e}^{i\sqrt{2\pi}\varphi_c} \end{aligned} \tag{7.12}$$

where the presence of the charge boson φ_c is still necessary in order to represent the charge component of the products. This formula leads to the following representation of the charge and spin densities:

$$\begin{aligned} 2\pi n_{\mathrm{tot}}(x) &= J + \bar{J} + \left[\mathrm{e}^{-2ik_{\mathrm{F}}x} \mathrm{e}^{i\sqrt{2\pi}\varphi_c} \, \mathrm{tr}\, g + \mathrm{H.c.}\right], \\ 2\pi S^a_{\mathrm{tot}}(x) &= J^a + \bar{J}^a + \left[\mathrm{e}^{-2ik_{\mathrm{F}}x} \mathrm{e}^{i\sqrt{2\pi}\varphi_c} \, \mathrm{tr}(g\sigma_a) + \mathrm{H.c.}\right]. \end{aligned} \tag{7.13}$$

The WZW model is a conformal field theory, with a set of well-identified primary (or scaling) fields. It is completely soluble, in the sense that the structure of its Hilbert space is completely known and correlation functions obey linear differential equations. The central charge of the level-k SU(2) WZW model is

$$c = \frac{3k}{k+2} \tag{7.14}$$

and it contains scaling fields of all spins from $s = 0$ to $s = k/2$, with conformal dimensions

$$h = \bar{h} = \frac{s(s+1)}{k+2}. \tag{7.15}$$

The spin part of the 1D electron gas with one band is described by the simplest of all WZW models: SU(2) at level 1. The matrix field g corresponds to $s = \frac{1}{2}$ and has conformal dimensions $(\frac{1}{4}, \frac{1}{4})$. Its components may be expressed in terms of the spin boson φ_s, if we follow Witten's formula and the Abelian bosonization formulas:

$$g = \begin{pmatrix} \mathrm{e}^{-i\sqrt{2\pi}\varphi_s} & \mathrm{e}^{-i\sqrt{2\pi}\vartheta_s} \\ \mathrm{e}^{i\sqrt{2\pi}\varphi_s} & \mathrm{e}^{i\sqrt{2\pi}\vartheta_s} \end{pmatrix}. \tag{7.16}$$

The equivalence of the $k = 1$ SU(2) WZW model with the spin boson theory is possible because the $k = 1$ model has central charge unity. Such a correspondence between a WZW model and simple bosons is possible only if c is an integer.

Other WZW models describe more exotic critical systems. The biquadratic spin-1 chain, with Hamiltonian

$$H = \sum_n \mathbf{S}_n \cdot \mathbf{S}_{n+1} + \beta(\mathbf{S}_n \cdot \mathbf{S}_{n+1})^2 \tag{7.17}$$

is critical at $\beta = 1$ and $\beta = -1$. At $\beta = -1$, the low-energy limit is described by the $k = 2$ SU(2) WZW model, which has central charge $c = \frac{3}{2}$. This model may be expressed in terms of three Majorana fermions [12, 13]. At $\beta = +1$, the low-energy limit is described by the $k = 1$ SU(3) WZW model. A recent attempt at extending Zhang's SO(5) theory of antiferromagnetism and d-wave superconductivity to one-dimensional systems is formulated in terms of WZW models [14].

7.2 Application to the Perturbed Tomonaga–Luttinger Model

The noninteracting part of the Tomonaga–Luttinger Hamiltonian may be expressed as follows in terms of charge and spin currents:

$$\mathcal{H}_0 = \frac{\pi v}{2}[J_c^2 + \bar{J}_c^2] + \frac{2\pi v}{3}[\mathbf{J}^2 + \bar{\mathbf{J}}^2]. \tag{7.18}$$

It is the sum of a free charge boson and of a $k = 1$ SU(2) WZW model. The various interactions bosonized in Section 5.6 have the following expressions in terms of currents:

$$\begin{aligned}
\mathcal{H}_1 &= -2vg_1(J_x\bar{J}_x + J_y\bar{J}_y),\\
\mathcal{H}_2^c &= vg_{2,c}J_c\bar{J}_c,\\
\mathcal{H}_4^c &= \tfrac{1}{2}vg_{4,c}[J_c^2 + \bar{J}_c^2],\\
\mathcal{H}_2^s &= 4vg_{2,s}J_z\bar{J}_z,\\
\mathcal{H}_4^s &= \frac{2v}{3}g_{4,s}[\mathbf{J}^2 + \bar{\mathbf{J}}^2],\\
\mathcal{H}_3 &= \frac{vg_3}{2\pi^2}\cos(\sqrt{8\pi}\varphi_c).
\end{aligned} \tag{7.19}$$

This correspondence may be established with the help of (7.1). It is then manifest that $g_{4,c}$ and $g_{4,s}$ only renormalize the velocity. Moreover, if $g_1 = -2g_{2,s}$, $\mathcal{H}_1$ and $\mathcal{H}_2^s$ combine into a single, rotation-invariant interaction:

$$\mathcal{H}_1 + \mathcal{H}_2^s = 4vg_{2,s}\mathbf{J}\cdot\bar{\mathbf{J}} \qquad (g_1 = -2g_{2,s}). \tag{7.20}$$

7.2.1 The Spin Sector

Let us restrict ourselves to this isotropic case, at half-filling, so that $g_3 \neq 0$. As we have seen in Fig. 4.6, the RG flow will bring g_3 to strong coupling if $|g_3| > -2g_{2,c}$ and the charge sector will have a gap. At the same time, the coupling g_1 flows along the separatrix of Fig. 4.6, toward the fixed point $g_1 = 0, K_s = 1$ (a detailed RG analysis, following the method described in Appendix A, shows that the separatrix is precisely determined by the condition of rotation invariance $g_1 = -2g_{2,s}$). Thus, the low-energy limit of the half-filled perturbed TL model is the $k = 1$ WZW model, provided $|g_3| > -2g_{2,c}$ and $g_1 = -2g_{2,s}$.

The appearance of a gap in the charge sector has another consequence, that of changing the scaling dimensions of some operators. Specifically, let us consider the $2k_{\rm F}$ component of the spin density, which becomes the staggered magnetization at half-filling:

$$\begin{aligned}
S_a^{2k_{\rm F}} &= \tfrac{1}{2}\bar{\psi}_i^\dagger(\sigma_a)_{ij}\psi_j + \text{H.c.}\\
&= -\frac{1}{4\pi}g_{ji}(\sigma_a)_{ij}e^{i\sqrt{2\pi}\varphi_c} + \text{H.c.}\\
&= -\frac{1}{4\pi}\operatorname{tr}(\sigma_a g)e^{i\sqrt{2\pi}\varphi_c} + \text{H.c.}
\end{aligned} \tag{7.21}$$

where we have applied Witten's formula (7.12). The cosine Umklapp term gives φ_c a nonzero expectation value, such that $\langle\cos\sqrt{4\pi}\varphi_c\rangle \neq 0$ and $\langle\sin\sqrt{4\pi}\varphi_c\rangle = 0$. Neglecting fluctuations around that expectation value, one may write the staggered magnetization density as

$$S_a^{2k_{\rm F}} = \text{const.}\big(\operatorname{tr}(\sigma_a g) + \text{H.c.}\big). \tag{7.22}$$

According to (7.15) with $s = \frac{1}{2}$, this operator has scaling dimension $\Delta = h + \bar{h} = \frac{1}{2}$, although it had $\Delta = 1$ initially (before the onset of the charge gap). Essentially, the charge contribution to the fluctuations, i.e., the operator $e^{\pm i\sqrt{2\pi}\varphi_c}$, also of scaling dimension $\Delta = \frac{1}{2}$, is frozen; this makes the operator $S_a^{2k_F}$ more relevant. The staggered magnetization correlation is therefore decreasing like $1/r$, instead of the $1/r^2$ decay of uniform correlations:

$$\langle S_a^{2k_F}(x,t) S_a^{2k_F}(0,0)\rangle = \frac{\text{const.}}{|x^2 - v^2t^2|^{1/2}}. \tag{7.23}$$

7.2.2 Other Perturbations

One may add various relevant or marginally relevant perturbations to the model (7.18):

1. An Ising-like anisotropy would contribute a term of the form $\mathcal{H}_{\text{Ising}} = \lambda J_z \bar{J}_z$. This term, a correction to $g_{2,s}$ is marginally relevant if $\lambda > 0$: it brings the flow above the separatrix of Fig. (4.6). On the other hand, it is marginally irrelevant if $\lambda < 0$, and brings the system towards an XY fixed point.

2. An explicit dimerization, maybe in the form of a staggered hopping term in the underlying Hubbard model. This perturbation would be proportional to the CDW (i.e., $2k_F$) component of the electron density. In terms of the WZW field g, this may be obtained by replacing the Pauli matrices of (7.21) by the unit matrix. At half-filling, the freezing of $\cos\sqrt{4\pi}\varphi_c$ would leave the single scaling field $\operatorname{tr} g$ in the perturbation:

$$\mathcal{H}_{\text{dim.}} = \lambda \operatorname{tr} g. \tag{7.24}$$

This term has scaling dimension $\Delta = \frac{1}{2}$ and would lead, according to (2.10), to a spin gap of order $m \sim \lambda^{2/3}$.

3. A uniform external magnetic field B. This couples to the uniform magnetization density $J_z + \bar{J}_z$, producing a perturbation with nonzero conformal spin. It breaks rotation invariance and is best treated within Abelian bosonization. In terms of the spin boson φ_s, the perturbation is $\lambda\partial_x\varphi_s$, where $\lambda \propto B$. The spin Hamiltonian is then

$$\mathcal{H} = \tfrac{1}{2}v\left[\Pi_s^2 + (\partial_x\varphi_s)^2\right] + \lambda\partial_x\varphi_s \tag{7.25}$$

(we have supposed $K_s = 1$, i.e., rotation invariance in the absence of external field). This perturbation can be eliminated from the Hamiltonian by redefining

$$\varphi_s \to \varphi_s - \frac{\lambda}{v}x. \tag{7.26}$$

In terms of the original electrons, this is almost equivalent to changing the value of k_F (almost, because it should have no effect on the charge sector). Basically, the uniform magnetic field introduces some incommensurability in the spin part of the system.

8 Other Applications of Bosonization

8.1 The Spin-$\frac{1}{2}$ Heisenberg Chain

The spin-$\frac{1}{2}$ Heisenberg model in one dimension may also be treated by bosonization in the low-energy limit. The microscopic Hamiltonian is

$$H = \sum_n \{ J(S_n^x S_{n+1}^x + S_n^y S_{n+1}^y) + J_z S_n^z S_{n+1}^z \}. \tag{8.1}$$

This Hamiltonian may be mapped to a spinless fermion problem via the Jordan–Wigner transformation:

$$S_n^+ = c_n^\dagger \exp\left(i\pi \sum_m^{n-1} c_m^\dagger c_m \right), \qquad S_n^z = c_n^\dagger c_n - \tfrac{1}{2} \tag{8.2}$$

where the operators c_n and $c_m^\dagger$ obey standard anticommutation relations. In terms of these operators, the Heisenberg Hamiltonian is

$$H = \sum_n \left\{ -\frac{J}{2}(c_n^\dagger c_{n+1} + \text{H.c.}) + J_z(c_n^\dagger c_n - \tfrac{1}{2})(c_{n+1}^\dagger c_{n+1} - \tfrac{1}{2}) \right\}. \tag{8.3}$$

The fermion number $\sum_n c_n^\dagger c_n$ is related to the total magnetization M, in such a way that half-filling corresponds to $M = 0$, the appropriate condition in the absence of external magnetic field. The continuum limit may then be taken and the bosonization procedure applied. The spinless fermion is then endowed with interactions proportional to J_z. Two of these interaction terms are of the g_2 and g_4 type and renormalize the boson field φ and the velocity v. From comparing with known exact exponents, one derives the following relation between the radius R of the boson and the bare interaction strength J_z:

$$R = \frac{1}{\sqrt{2\pi}} \left\{ 1 - \frac{1}{\pi} \cos^{-1}\left(\frac{J_z}{J} \right) \right\}^{1/2}. \tag{8.4}$$

Various correlation functions may be calculated, once the correspondence between spin components and the boson field is known [2]:

$$\begin{aligned} S^z(x) &= \frac{1}{2\pi R} \partial_x \varphi + (-1)^x \text{cst.} \mathrm{e}^{i\varphi/R}, \\ S^-(x) &= \mathrm{e}^{-2\pi i R\vartheta} + (-1)^x \text{cst.} \{ \mathrm{e}^{i[\varphi/R - 2\pi R\vartheta]} + \mathrm{e}^{i[-\varphi/R - 2\pi R\vartheta]} \}. \end{aligned} \tag{8.5}$$

For instance,

$$\begin{aligned}\langle S^z(x,t)S^z(0,0)\rangle &= \frac{1}{16\pi^3 R}(\langle \partial_z\varphi(x,t)\partial_z\varphi(0,0)\rangle + \langle \partial_{\bar z}\varphi(x,t)\partial_{\bar z}\varphi(0,0)\rangle) \\ &\qquad + \mathrm{cst.}(-1)^x\langle \mathrm{e}^{i\varphi(x,t)/R}\mathrm{e}^{i\varphi(0,0)/R}\rangle \\ &= \frac{1}{16\pi^3 R}\left[\frac{1}{(x-vt)^2}+\frac{1}{(x+vt)^2}\right] \\ &\qquad + \mathrm{cst.}(-1)^x\frac{1}{|x^2+v^2t^2|^{1/4\pi R^2}} . \end{aligned} \tag{8.6}$$

Note that the staggered part has a nonuniversal exponent, which depends on J_z/J. At the isotropic point $J_z = J$, $R = 1/\sqrt{2\pi}$ and this exponent is $\frac{1}{2}$, exactly like the non-Abelian bosonization prediction (7.23).

8.2 Edge States in Quantum Hall Systems

The two-dimensional electron gas (2DEG) in a magnetic field has been the object of intense theoretical investigation since the discovery of the quantum Hall effect (integral and fractional) in the early 1980's. While the integer Hall effect may be understood (in the bulk) on the basis of weakly correlated electrons, the fractional quantum Hall effect is a strongly correlated problem. Quantum Hall plateaus correspond to gapped states, but electric conduction nevertheless occurs through electrons at the edge of the 2D gas, the so-called *edge states*. One must imagine a 2DEG immersed in a magnetic field B and further confined to a finite area by an in-plane electic field E (in, say, a radial configuration). The field causes a persistent current along the edge of the gas:

$$\mathbf{j} = \sigma_{xy}\hat{\mathbf{z}}\wedge\mathbf{E}, \tag{8.7}$$

where σ_{xy} is the Hall conductance, equal to νe^2, where ν is the filling fraction. The edge electrons drift in one direction at the velocity $v = cE/B$. We will adopt a hydrodynamic description of the edge, following [15]. Let x is the coordinate along the edge and $h(x)$ the transverse displacement of the edge with respect to the ground state shape of the gas (cf. Fig. 4.7). If $n = \nu/2\pi\ell_B^2$ is the electron density (ℓ_B is the magnetic length), then the linear electron density in an edge excitation is $J(x) = nh(x)$. The energy of the excitation is the electrostatic energy associated with the edge deformation:

$$H = \int \mathrm{d}x\, \tfrac{1}{2}ehJE = \frac{\pi v}{\nu}\int \mathrm{d}x\, J^2 . \tag{8.8}$$

The field J being from the start treated as a collective mode, bosonization is natural. The main difference with the 1D electron gas is that only one half (left of right) of the boson is necessary, since the edge excitations only

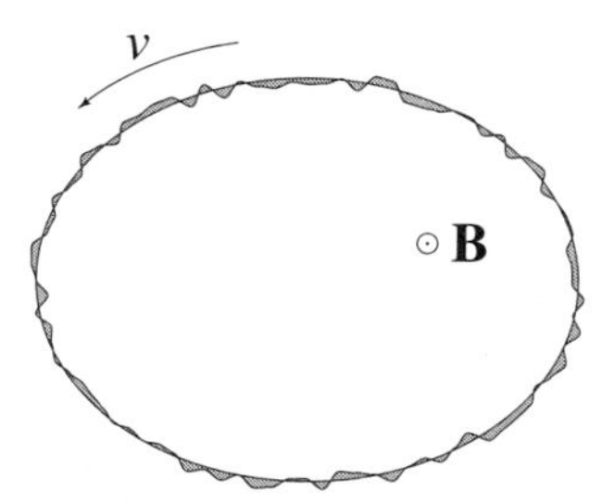

FIGURE 4.7. Fluctuation of the edge of a quantum Hall system.

propagate clockwise or counterclockwise, depending on the field direction. We therefore assert

$$J = \frac{1}{\sqrt{\pi}}\partial_x\phi = -\frac{i}{\sqrt{\pi}}\partial_z\phi, \tag{8.9}$$

and the Hamiltonian density for edge excitations is

$$\mathcal{H} = -\frac{v}{\nu}(\partial_z\phi)^2. \tag{8.10}$$

This Hamiltonian is chiral, i.e., involves right-moving modes only. The edge electron must be represented by a field $\psi(x)$ such that

$$[J(x), \psi(x')] = -\delta(x - x')\psi(x). \tag{8.11}$$

From the short-distance product of J with $\mathrm{e}^{-i\alpha\phi}$, one shows that the correct representation for ψ is

$$\psi(x) = \frac{1}{\sqrt{2\pi}}\mathrm{e}^{i\sqrt{4\pi}\phi/\nu}. \tag{8.12}$$

From this we derive the one-electron Green function

$$G(x,t) = \frac{1}{2\pi}\frac{1}{(x - vt)^{1/\nu}}. \tag{8.13}$$

Note that this coincides with a free-electron propagator in the case $\nu = 1$ only, i.e., when the first Landau level is exactly filled. Even in the integer effect, the edge system is a Luttinger liquid if $m = 1/\nu > 1$, albeit a *chiral* Luttinger liquid. In momentum space, the Green function (8.13) is

$$G(k,\omega) \propto \frac{(\omega + vk)^{m-1}}{\omega - vk} \tag{8.14}$$

and the momentum-integrated density of states near the Fermi level is

$$N(\omega) \propto |\omega|^{m-1}. \tag{8.15}$$

8.3 And More...

This short tutorial has concentrated on basic Luttinger liquid physics. Time and space are lacking to introduce the many physical systems that have been treated with bosonization. Here we will just indicate a few useful references.

The bosonization formalism may also be used to treat disorder. The problem of an isolated impurity in a Luttinger liquid is treated in [4] and [16], following Kane and Fisher's original work [17]. The more complicated problem of Anderson localization, resulting from the coherent scattering off many impurities, is also described summarily in [16]. Formally related to impurity scattering is the question of electron-phonon interaction in a Luttinger liquid, a review of which may be found in [4].

The Kondo problem, or the strong interaction of magnetic impurities with conduction electrons, has also been treated using bosonization, and a strong dose of conformal field theory, by Affleck and Ludwig. Reference [18] is a comprehensive review of the progress accomplished with the help of conformal methods. Most interesting in this application is the three-dimensional nature of the initial problem, before it is mapped to a one dimension.

Spin chains of spin $s > \frac{1}{2}$ can be constructed from $2s$ fermion species with a Hund-type coupling. This is the approach followed in [19], also reviewed in [2]. The spin-1 chain has been treated as a perturbed $k = 2$ SU(2) WZW model in [13].

Coupled spin chains, or spin ladders, have been studied in Abelian bosonization and are reviewed in [16]. A non-Abelian bosonization study of two coupled spin chains is performed in [5]. Doped spin ladders are coupled Luttinger liquids. Bosonization studies may be found in [20] and [21]. Reference [14] looks at this problem from the perspective of SO(5) symmetry. The latter two references treat the question of d-wave fluctuations in a doped ladder.

The reader is referred to [4, 16] for additional references (we apologize for our poor bibliography). More applications of bosonization to condensed matter can be found in [4]. Other important review articles on the subject include [16] and [2]. A historical perspective is given in the collection of preprints [1].

9 Conclusion

Let us conclude by highlighting the principal merits and limitations of bosonization:

1. Bosonization is a *nonperturbative* method. The Tomonaga–Luttinger (TL) model, a continuum theory of *interacting* fermions, can be translated into a theory of noninteracting bosons and solved exactly.

2. Bosonization is a method for translating a fermionic theory into a bosonic theory, and eventually retranslating part of the latter into a new fermionic language (cf. the Luther–Emery model or the use of Majorana fermions). This translation process is exact in the continuum limit, but does not warrant an exact solution of the model, except in a few exceptional cases (e.g., the TL model). For the rest, one must rely on renormalization-group analyses, which generally complement bosonization.

3. The bosonized theory may have decoupled components, like spin and charge, which are not manifest in fermionic language. Thus, spin-charge separation is an exact prediction of bosonization (even beyond the TL model). When one of the components (spin or charge) becomes massive, this favors the emergence of physical operators with new scaling dimensions, like the staggered magnetization of (7.22).

4. The contact with microscopic model is not obvious, since an infinite number of irrelevant parameters stand in the way. But bosonization leads to universal or quasi-universal predictions, and one can argue that such predictions are preferable to the exact solution of a particular microscopic Hamiltonian of uncertain relevance.

5. In the rare cases where a microscopic Hamiltonian has an exact solution (e.g., the 1D Hubbard model), the solution is so complex that dynamical quantities cannot be explicitly calculated. Bosonization can then be used as a complementary approach. Exact values of thermodynamic quantities of the microscopic theory may be used to fix the parameters (e.g., v_c, v_s, K_c, K_s) of the boson field theory, and the latter may be used to calculate dynamic quantities (e.g., the spectral weight) or asymptotics of correlation functions [22, 23].

6. The low-energy limit corresponds to a few regions in momentum-space where bosonization has something to say: the vicinity of wavevectors that are small multiples of $k_{\rm F}$.

7. Finally, bosonization is limited to one-dimensional systems, despite attempts at generalizing it to higher dimensions [24, 25].

Acknowledgments: The author expresses his gratitude to the Centre de recherches mathématiques for making this workshop possible, and to the organizers for inviting him. Support from NSERC (Canada) and FCAR (Québec) is gratefully acknowledged.

10 References

[1] M. Stone, ed., *Bosonization*, World Scientific, Singapore, 1994, a collection of reprints.

[2] I. Affleck, in *Fields, Strings and Critical Phenomena*, eds. E. Brézin and J. Zinn-Justin (Elsevier, Amsterdam, 1989), p. 564.

[3] J. Cardy, *Scaling and Renormalization in Statistical Physics*, Cambridge Lecture Notes Phys. (Cambridge University Press, Cambridge, 1996). (See particularly Section 5.2 for a discussion of the perturbative Renormalization Group.)

[4] A. Gogolin, A. Nersesyan, and A. Tsvelik, *Bosonization and Strongly Correlated Systems* (Cambridge University Press, Cambridge, 1998).

[5] D. Allen and D. Sénéchal, Phys. Rev. B **55** (1997), 299.

[6] P. Ginsparg, Nuclear Phys. B **295** (1988), 153–170.

[7] P.D. Francesco, P. Mathieu, and D. Sénéchal, *Conformal Field Theory* (Springer Verlag, New York, 1997). (See also the errata page at `www.physique.usherb.ca/~dsenech/cft.htm`.)

[8] P. Ginsparg, in *Fields, Strings and Critical Phenomena*, eds. E. Brézin and J. Zinn-Justin (Elsevier, Amsterdam, 1989).

[9] J. Kosterlitz and D. Thouless, J. Phys. C: Solid State Phys. **6** (1973), 1181.

[10] A. Luther and V. Emery, Phys. Rev. Lett. **33** (1974), 589.

[11] E. Witten, Comm. Math. Phys. **92** (1984), 455–472.

[12] A. Zamolodchikov and V. Fateev, Sov. J. Nuclear Phys. **43** (1986), 657.

[13] A. Tsvelik, Phys. Rev. B **42** (1990), 10499.

[14] D. Shelton and D. Sénéchal, Phys. Rev. B **58** (1998), 6818.

[15] X.-G. Wen, Internat. J. Modern Phys. B **6** (1992), 1711–1762.

[16] H. Schulz, G. Cuniberti, and P. Pieri, *Fermi Liquids and Luttinger Liquids*, Lecture Notes of the Chia Laguna (Italy), Summer School, September 1997, `cond-mat/9807366`.

[17] C. Kane and M. Fisher, Phys. Rev. B **46** (1992), 1220.

[18] I. Affleck, Acta Phys. Polon. B **26** (1995), 1869, `cond-mat/9512099`.

[19] I. Affleck and F. Haldane, Phys. Rev. B **36** (1987), 5291.

[20] M. Fabrizio, Phys. Rev. B **48** (1993), 15838.

[21] H. Schulz, Phys. Rev. B **53** (1996), R2959.

[22] H. Schulz, Phys. Rev. Lett. **64** (1990), 2831.

[23] N. Kawakami and S.-K. Yang, Prog. Theoret. Phys. **107** (1992), 59.

[24] F. Haldane, in *Perspectives in Many-Particle Physics*, eds. R. Broglia and J.R. Schrieffer, (North Holland, Amsterdam 1994).

[25] H.-J. Kwon, A. Houghton, and B. Marston, Phys. Rev. B **52** (1995), 8002.

Appendix A RG Flow and Operator Product Expansion

In this appendix we will illustrate how to calculate the RG flow of coupling constants using the so-called operator product expansion (OPE) of local scaling operators. We consider a fixed-point action S_0 describing a conformal field theory, and a set of perturbations with coupling constants g_i:

$$S[\Phi] = S_0[\Phi] + \sum_i g_i \int \mathrm{d}x\,\mathrm{d}\tau\, O_i(x,\tau). \tag{A.1}$$

Here Φ denotes a field or a collection of fields that defines the theory, in the sense of the action and of the path integral measure.

The local operators O_i are in principle functions of Φ. They are scaling operators (primary fields and their "conformal descendants"); i.e., they have definite conformal dimensions. In a conformal field theory, products of these operators at two positions may always be expressed as a power series of inverse distance:

$$O_i(\mathbf{x})O_j(\mathbf{x}') = \sum_k C_{ij}^k O_k(\mathbf{x}') \frac{1}{|\mathbf{x}-\mathbf{x}'|^{\Delta_i+\Delta_j-\Delta_k}}, \tag{A.2}$$

where the "coefficients" $O_k(\mathbf{x}')$ are themselves scaling operators. It is assumed here that the set of operators O_i forms a "closed algebra" under this short-distance expansion (this is a fundamental structure in any conformal field theory).

It can be shown that the the coefficients C_{ij}^k fix the quadratic terms in the RG flow of coupling constants. Specifically,

$$\frac{\mathrm{d}g_k}{\mathrm{d}\ell} = (2-\Delta_k)g_k - \sum_{i,j} C_{ij}^k g_i g_j. \tag{A.3}$$

Rather than giving a proof of this relation, we refer the reader to the excellent monograph by Cardy [3]. The quadratic terms, of course, matter most in the case of marginal perturbations ($\Delta_k = 2$).

Let us apply (A.3) to the spin sector of the 1D electron gas subject to the interactions g_1 and g_2, in order to recover the flow equations (6.26).

Let us first note that, when expanded to first order in $g_{2,s}$ from (6.5), those flow equations become

$$\frac{\mathrm{d}g_1}{\mathrm{d}\ell} = \frac{2}{\pi} g_1 g_2 \qquad \frac{\mathrm{d}g_2}{\mathrm{d}\ell} = \frac{1}{2\pi} g_1^2 \tag{A.4}$$

(we dropped the "s" index from now on: it is understood that we deal with the spin sector). The corresponding operators are

$$O_1 = \frac{1}{2\pi^2}\cos(\sqrt{8\pi}\varphi) \qquad O_2 = \frac{2}{\pi}\partial\varphi\bar{\partial}\varphi. \tag{A.5}$$

The corresponding OPEs are obtained by applying Wick's theorem. Their most singular terms are

$$\begin{aligned}
O_1(z,\bar{z})O_1(w,\overline{w}) \sim{}& \frac{1}{16\pi^4}\frac{1}{(z-w)^2(\bar{z}-\overline{w})^2} \\
&- \frac{1}{2\pi^3}\left[\frac{1}{(z-w)^2}(\bar{\partial}\varphi)^2 + \frac{1}{(\bar{z}-\overline{w})^2}(\partial\varphi)^2\right] \\
&- \frac{1}{\pi^3}\frac{1}{(z-w)(\bar{z}-\overline{w})}\partial\varphi\bar{\partial}\varphi, \\
O_2(z,\bar{z})O_2(w,\overline{w}) \sim{}& \frac{1}{4\pi^4}\frac{1}{(z-w)^2(\bar{z}-\overline{w})^2} \\
&- \frac{1}{\pi^3}\left[\frac{1}{(z-w)^2}(\bar{\partial}\varphi)^2 + \frac{1}{(\bar{z}-\overline{w})^2}(\partial\varphi)^2\right], \\
O_2(z,\bar{z})O_1(w,\overline{w}) \sim{}& -\frac{1}{2\pi^4}\frac{1}{(z-w)(\bar{z}-\overline{w})}\cos(\sqrt{8\pi}\varphi).
\end{aligned} \tag{A.6}$$

This last OPE is derived from the relation

$$\partial\varphi(z)\mathrm{e}^{i\alpha\varphi(w,\overline{w})} \sim -\frac{i\alpha}{4\pi}\frac{1}{z-w}\mathrm{e}^{i\alpha\varphi(w,\overline{w})}, \tag{A.7}$$

which is obtained by applying Wick's theorem to the series expansion of $\mathrm{e}^{i\alpha\varphi}$ (see, e.g., [7]). We may rewrite these OPEs as

$$\begin{aligned}
O_1(z,\bar{z})O_1(w,\overline{w}) &\sim \text{const.} - \frac{1}{2\pi^2}\frac{1}{(z-w)(\bar{z}-\overline{w})}O_2(w,\overline{w}) + \text{conf. spin}, \\
O_2(z,\bar{z})O_2(w,\overline{w}) &\sim \text{const.} + \text{conf. spin}, \\
O_2(z,\bar{z})O_1(w,\overline{w}) &\sim -\frac{2}{\pi^2}\frac{1}{(z-w)(\bar{z}-\overline{w})}O_1(w,\overline{w})
\end{aligned} \tag{A.8}$$

where "conf. spin" denotes terms with conformal spin, which do not affect the quadratic terms of the RG flow. From (A.3), we may then write the flow equations:

$$\frac{\mathrm{d}g_1}{\mathrm{d}\ell} = \frac{2}{\pi^2} g_1 g_2 \qquad \frac{\mathrm{d}g_2}{\mathrm{d}\ell} = \frac{1}{2\pi^2} g_1^2. \tag{A.9}$$

These differ from (A.4) only by a rescaling of the flow variable ℓ by π (different renormalization-group techniques may lead to such variations).

5

Disordered Quantum Solids

T. Giamarchi and E. Orignac

ABSTRACT Due to the peculiar non-Fermi liquid of one-dimensional systems, disorder has particularly strong effects. We show that such systems belong to the more general class of disordered quantum solids. We discuss the physics of such disordered interacting systems and the methods that allow us to treat them. In addition to, by now standard, renormalization group methods, we explain how a simple variational approach allows us to treat these problems even when the RG fails. We discuss various physical realizations of such disordered quantum solids both in one and higher dimensions (Wigner crystal, Bose glass). We investigate in detail the interesting example of a disordered Mott insulator and argue that intermediate disorder can lead to a novel phase, the Mott glass, intermediate between a Mott and and Anderson insulator.

1 Introduction

Disorder effects are omnipresent in condensed matter physics, where one has to struggle very hard to deal with clean systems. Quite remarkably fermionic system exhibit marked differences with disorder in classical systems. Indeed the very existence of a Fermi energy E_{F} "reduces" the effects of disorder since the relevant parameter now become the relative strength of the disorder compared to the Fermi energy D/E_{F} or the mean free path compared to the Fermi length $k_{\text{F}}l$. Of course nature is not as simple as that, and quantum effects lead in fact also to reinforcement of the disorder effects and in low dimensions turn a free electron system into an insulator, as pointed out by Mott and Twose [1].

We have now gained a very good understanding of the properties of such disordered free-electron systems. To tackle them an arsenal of methods ranging from diagrams [2, 3], scaling theory [4], replicas [5, 6], and supersymmetry [7] have been developed. But life becomes much less simple when interactions among fermions are taken into account. Although it is very intuitive to think that when interactions are small the noninteracting problem is a good starting point, such an intuition turns out to be wrong for a number of reasons: (i) even if in the pure system interactions can be "removed" from the system by resorting to Fermi liquid theory [8] this is not the case when disorder is present. Because disorder renders electrons slowly diffusive rather than ballistic, they feel the effects of the interactions much more strongly with explosive results [9]. Effective interactions increase when we are looking at low-energy properties and Fermi liquid theory breaks down. The noninteracting physics is thus only relevant for very-low-disorder systems. (ii) When the dimension is small or the interactions strong to start with (as in systems undergoing Mott transitions) it is of course impossible to start from the noninteracting limit and one has to solve the full problem. What puts us at a disadvantage here is that most of the techniques useful for the noninteracting case also fail as soon as interactions are included: (i) The supersymmetric method, which rests by construction on the quadratic nature of the Hamiltonians, is useless. (ii) It is now difficult really to determine classes of diagram to sum.

Disorder can still be averaged over by using replicas, but then one is left with a complicated (and untractable) theory. Renormalization group attempts have been made with some success [10, 11] but also with the problem that the coupling constants diverge, so that the low-energy fixed point remains elusive. Many other attempts to treat this very complicated problem exist in the literature and it is impossible to list them all [12]. Note that numerical studies are also hampered with regards to this problem both because of the difficulty in taking the interactions into account (which more or less imposes either Monte Carlo or exact diagonalization) and then performing the complicated disorder averaging with enough statistics or large system sizes to get reliable results.

A very peculiar situation occurs when one considers one dimension. On one hand, one expects the difficulty to be maximum here. The interactions lead to very strong effects and destroy any trace of Fermi liquid, giving rise to what is known as a Luttinger liquid. The disorder is also extremely strong, giving rise in the noninteracting case to a system so localized that the localization length is simply the mean free path and a diffusive regime is absent. On the other hand, one is in $d = 1$ in a much better situation to tackle the problem since the interactions can be treated essentially exactly, using for example techniques such as bosonization [13–17] so one only needs good techniques to tackle the disorder. Quite interestingly, the physics of interacting and disordered one-dimensional systems is one of disordered quantum solids. Higher-dimensional examples of such systems are the Wigner crystal, Charge Density Waves, and Bose glass. In these notes, we will explain the techniques allowing us to treat such systems, ranging from simple physical arguments to a quite sophisticated variational approach. In order to remain pedagogical and keep the algebra simple, we mostly discuss the technicalities on the simplest example of spinless fermions. We briefly discuss the specific physical realizations and give references so that the reader can look in more detail at the physical properties of these specific systems.

Before we embark with the physics, let us point out that these notes result from the synthesis of various lectures. Some arbitrary choice of material had to be made in order to keep some level of clarity. Even if we have made some effort to cover various interesting topics, these notes cannot pretend to be as exhaustive as a full review. We thus apologize in advance to anybody whose pet problem (or paper) is not covered in these few pages.

2 Disordered Interacting Fermions

2.1 Model

Let us consider spinless fermions hopping on a lattice with a kinetic energy t and an interaction V:

$$H = -t \sum_i (c_i^\dagger c_{i+1} + H.c.) + \sum_{i>j} V_{i-j}(n_i - \bar{n})(n_j - \bar{n}), \qquad (2.1)$$

where n_i ($\bar{n}$) is the local (average) electron density, and the rest of the notation is standard ($t = 1$ is the unit of energy). For nearest-neighbor interactions this is the well-known $t - V$ model. When using a Jordan–Wigner transformation [18] to express the fermions in term of spins this latter model maps to an XXZ spin chain. In addition to (2.1) we want to submit the fermions to a disorder. We concentrate here on site disorder. Again most results/methods will carry over for randomness in the hopping. The randomness is simply

$$H_{\text{int}} = \sum_i \mu_i (n_i - \bar{n}), \qquad (2.2)$$

where μ_i is a random variable of zero mean.

2.2 Pure System

The low-energy properties of the pure system (2.1) are by now well understood. We will thus review the bosonization of (2.1) only very briefly to fix the notations and refer the reader to the various reviews on the subject [13–17]. To get the low-energy physics, it is enough to focus on the excitations around the Fermi points. In the continuum limit $i \to x$ this amounts to expressing the fermion field in terms of a slowly varying (with respect to the lattice spacing) field of right (with momentum close to $+k_{\text{F}}$) and left (with momentum close to $-k_{\text{F}}$) movers:

$$c_i^\dagger \to \psi^\dagger(x) = e^{-ik_{\text{F}}x}\psi_+^\dagger(x). + e^{ik_{\text{F}}x}\psi_-^\dagger(x) \qquad (2.3)$$

In term of these fields (2.1) becomes

$$H = -i\hbar v_{\mathrm{F}} \int dx\, (\psi_+^\dagger \partial_x \psi_+(x) - \psi_-^\dagger \partial_x \psi_-(x)) + \int dx_1\, dx_2\, V(x_1 - x_2)\rho(x_1)\rho(x_2), \quad (2.4)$$

where the density reads

$$\rho(x) = \psi_+^\dagger \psi_+ + \psi_-^\dagger \psi_- + (e^{-i2k_{\mathrm{F}}x} \psi_+^\dagger \psi_- + \mathrm{H.c.}). \quad (2.5)$$

The remarkable feature in $d = 1$ is that all the excitations of the system can be reexpressed in terms of the fluctuations of density. If one introduces a field $\phi(x)$ describing the long-wavelength part of the density, (2.5) reads

$$\rho(x) = \rho_0 - \frac{1}{\pi}\nabla\phi(x) + \frac{1}{(2\pi\alpha)}(e^{-i2k_{\mathrm{F}}x + 2\phi(x)} + \mathrm{H.c.}), \quad (2.6)$$

where α is a lattice spacing, $\rho_0 = \bar{n}/\alpha$ and

$$\nabla\phi(r) = \sum_{\pm} \psi_\pm^\dagger(x)\psi_\pm(x) \quad (2.7)$$

$$\nabla\theta(r) = \sum_{\pm} \pm\psi_\pm^\dagger(x)\psi_\pm(x). \quad (2.8)$$

θ is a similar field but is associated with the long-wavelength part of the current. ϕ and $\Pi = (1/\pi)\nabla\theta$ are canonically conjugate. Both the Hamiltonian and the fermion operator can be expressed in terms of these two fields:

$$H = \frac{1}{2\pi}\int dx \left[uK(\nabla\theta(x))^2 + \frac{u}{K}(\nabla\phi(x))^2 \right] \quad (2.9)$$

$$\psi_\pm(r) = \frac{1}{\sqrt{2\pi\alpha}} e^{-i(\pm\phi(r) - \theta(r))}. \quad (2.10)$$

Using (2.10) in (2.5), one easily recovers (2.6). For the free Hamiltonian (2.1) (with $V = 0$), one has in (2.9) $u = v_{\mathrm{F}}$ and $K = 1$. What makes the boson representation so useful is the fact that even in the presence of interactions the bosonized form of (2.1) remains (2.9) but with renormalized (Luttinger Liquid) parameters u and K.

2.3 Disorder

Similarly the disorder can be rewritten in the boson representation. It is natural to separate the Fourier components with wavevectors close to $q \sim 0$ (η) and $q \sim \pm 2k_{\mathrm{F}}$ (ξ, ξ^*). The disorder becomes thus

$$H_{\mathrm{dis}} = \eta(x)[\psi_+^\dagger\psi_+ + \psi_-^\dagger\psi_-] + (\xi(x)\psi_+^\dagger\psi_- + \mathrm{H.c.}) \quad (2.11)$$

where η and ξ are now two independent random potentials. η is real but when the system is incommensurate ($2k_{\mathrm{F}} \neq \pi$) ξ is complex. The potentials η and ξ correspond respectively to the forward scattering and the backward scattering of the fermions on impurities. Equation (2.11) reads in the boson representation

$$H_{\mathrm{dis}} = -\eta(x)\frac{1}{\pi}\nabla\phi(x) + \left[\frac{\xi(x)}{(2\pi\alpha)}e^{i2\phi(x)} + \mathrm{H.c.}\right]. \tag{2.12}$$

For incommensurate systems, the forward scattering can easily be absorbed by a gauge transformation on the fermions. In the boson language this amounts to shift $\phi \to \phi - 2K\left(\int^x dy\, \eta(y)\right)/u$. Since ξ is complex and hence has a random phase this shift does not affect the backscattering term (this will be even more explicit on the replicated Hamiltonian (6.3) below). Thus the forward scattering can be accounted for completely. Its main effect is to lead to an exponential decay of the disorder-averaged density correlation functions. Since the current and the superconducting correlation functions depends only on θ (or equivalently in an action representation on $\partial_t\phi$) they are not affected by the shift. Transport properties thus depend on the backscattering alone, as is obvious on a physical basis since forward scattering cannot change the current. For commensurate potentials the situation is more complicated since ξ is real ($e^{+i2k_{\mathrm{F}}} = e^{-i2k_{\mathrm{F}}}$) and the forward scattering now *does* affect the backscattering term (in other words, by combining a forward scattering and a scattering on the lattice one can generate a back-scattering term). This has consequences that we will examine in more detail in Section 7. For the moment, let us focus on the incommensurate case and get rid of the forward scattering.

3 Tackling the Disorder

Disordered one-dimensional particles are thus described by

$$S/\hbar = \int dx\, d\tau \left[\frac{1}{2\pi K}\left[\frac{1}{v}(\partial_\tau\phi)^2 + v(\partial_x\phi)^2\right] + \frac{\xi(x)}{2\pi\alpha\hbar}e^{i2\phi(x)} + \mathrm{H.c.}\right], \tag{3.1}$$

where we have rewritten (2.9) and (2.11) as an action. We have reintroduced $\hbar$ and other pesky constants to show explicitly the various physical limits. Note that although we are mainly concerned here with fermions, (2.9) describes in fact nearly every one-dimensional disordered problem ranging from dirty bosons [19–21] to disordered spin chains [22], since all these problems have essentially the same boson representation. We will examine these different systems in more detail in Section 4. Equation (3.1) emphasizes the physics of the problem. The electron system can be viewed as a charge density wave [23, 24] since the density varies as (2.6):

$$\rho(x,\tau) = \rho_0 \cos\bigl(2k_{\mathrm{F}}x - 2\phi(x,\tau)\bigr). \tag{3.2}$$

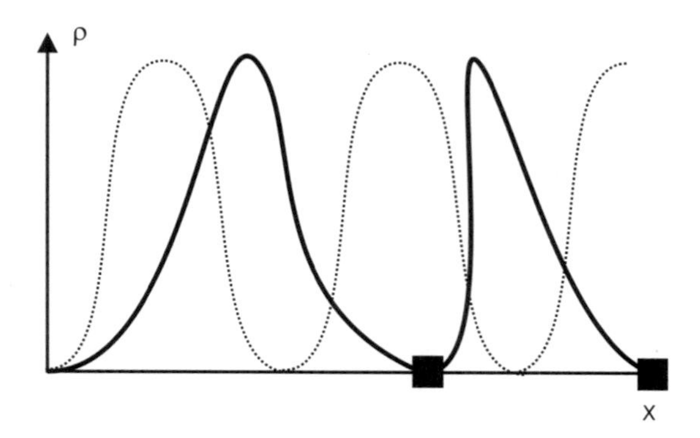

FIGURE 5.1. Impurities (black square) will pin the density wave. The phase ϕ will have to change compared to the undistorted density (dotted line), which costs kinetic energy.

The elastic term in (3.1) wants the phase of the density wave to be as constant as possible, and have a nice sinusoidally modulated density. The disorder term, on the contrary, wants to pin this charge density on the impurities by distorting the phase, as shown in Fig. 5.1.

The problem of localization of interacting fermions is thus very similar to the one of the pinning of classical charge density waves [25]. The charge density wave is here intrinsic to the one-dimensional interacting electron gas and not due to a coupling to phonons. The main features are nevertheless similar, the main difference being the fact that the effective mass of the "CDW" is much smaller in the absence of the electron-phonon coupling and hence the importance of quantum fluctuations is much higher. Various quantities of special interest have a quite simple expression in terms of the bosons. The conductivity is simply given by

$$\sigma(\omega) = \frac{i}{\omega}\left[\langle \partial_\tau \phi \partial_\tau \phi \rangle_{i\omega_n, q=0}\right]_{i\omega_n \to \omega + i\delta}, \tag{3.3}$$

whereas the compressibility is given by a similar correlation function but with different limits

$$\kappa = \frac{1}{\pi^2} \langle \partial_x \phi \partial_x \phi \rangle_{i\omega_n=0, q}. \tag{3.4}$$

The correlations of the $2k_\mathrm{F}$ part of the density and of the superconducting order parameter $\psi_+^\dagger \psi_-^\dagger$ are, respectively, given by

$$\chi_\rho(r) = \langle e^{i2\phi(r)} e^{-i2\phi(0)} \rangle \tag{3.5}$$

$$\chi_s(r) = \langle e^{i2\theta(r)} e^{-i2\theta(0)} \rangle, \tag{3.6}$$

where $r = (x, \tau)$ and $\pi\Pi(x) = \nabla\theta(x)$.

Given the highly nonlinear nature of the coupling to disorder, (3.1) is quite tough to solve. On the other hand, if $\xi(x)$ was just a constant we would have a simple sine-Gordon theory for which a great deal is known. We will thus tackle the problem by increasingly sophisticated methods.

3.1 Chisel and Hammer

In the absence of quantum fluctuations, ϕ would be a classical field and we would have a good idea of what happens. This is the way Fukuyama and Lee [25] looked at this problem. Such an approximation is of course very good for "phononic" charge density waves [23] since the quantum term is $\Pi^2/(2M)$ and thus very small. For fermions this corresponds to the "classical" limit $\hbar \to 0$, $K \to 0$ keeping $\overline{K} = K/\hbar$ fixed, and thus to very repulsive interactions. In that case, we can ignore all quantum fluctuations, and look for a *static* solution for ϕ. It is of course crucial for the existence of such a solution that the disorder do not depend on time. This solution $\phi_0(x)$ describes the static distortion of the phase imposed by the random potential. In the absence of kinetic energy $(\nabla\phi)^2$, it would be easy to "determine" ϕ_0. If we write the random field ξ as an amplitude $|\xi(x)|$ and a random phase 2ζ, the disorder term becomes

$$\int dx\, |\xi(x)| e^{i2(\phi(x)-\zeta(x))} + \text{c.c.} \tag{3.7}$$

The optimum is thus for $\phi_0(x)$ to follow the random phase on each point. For point-like impurities located on random positions R_i, $|\xi|$ would just be the strength of each impurity potential and $\zeta = k_{\rm F} R_i$. Thus $\phi_0(x) = \zeta(x)$ is the generalization to any type of disorder (and in particular to the Gaussian disorder so dear to the theorist) of the physics expressed in Fig. 5.1: get the density minimum at each impurity. In presence of kinetic energy, following the random phase would cost too much kinetic energy. We do not know exactly now how to determine the optimal $\phi_0(x)$, but we can do some scaling arguments. Let us assume that ϕ remains constant for a lengthscale $L_{\rm loc}$. On this lengthscale ϕ takes the value that optimizes the disorder term, which now reads

$$E_{\rm dis} = \left[\int_0^{L_{\rm loc}} \xi(x)\right] e^{i2\phi} + \text{c.c.}. \tag{3.8}$$

If we take for example a Gaussian distribution for ξ,

$$\overline{\xi(x)\xi^*(x')} = \mathcal{D}\delta(x-x'), \tag{3.9}$$

one gets because of the average of a complex random variable on a box of size $L_{\rm loc}$ that the disorder contributes as

$$E_{\rm dis} = -\sqrt{\mathcal{D}L_{\rm loc}} e^{i(2\phi_0 - 2\Xi)}, \tag{3.10}$$

where Ξ is some phase. It clear that the optimum energy is reached if ϕ_0 adjusts to this (now unknown) phase. The global energy gain now scales as $\sqrt{L_{\rm loc}}$. Between two segments of size $L_{\rm loc}$ the phase has to distort to reach the next optimal value. The distortion being of the order of 2π the cost in kinetic energy reads

$$E_{\rm kin} \propto \frac{1}{L_{\rm loc}}. \tag{3.11}$$

Minimizing the total cost shows that the length over which ϕ_0 remains constant is given by

$$L_{\text{loc}} \propto \left(\frac{1}{\mathcal{D}}\right)^{1/3} . \tag{3.12}$$

This tells us that the system *does* pin on the impurities and that below L_{loc} the system looks very much like an undistorted system. Since at the scale L_{loc}, ϕ_0 varies randomly the $2k_{\text{F}}$ density, density correlations will decay exponentially with a characteristic size L_{loc}. It is thus very tempting to associate L_{loc} with the Anderson localization length. Note that for the free fermion point, $L_{\text{loc}} \propto 1/\mathcal{D}$ instead of (3.12), so the above formula is clearly missing a piece of physics when K is not zero. Nevertheless from this simple scaling argument we have obtained—(i) the fact that classical CDW or very repulsive fermions are pinned (localized) by disorder; (ii) the localization length; and (iii) the fact that the ground state should contain a static distortion of the phase due to the disorder. Unfortunately, we have no other information on ϕ_0, which is certainly a drawback.

Even with our limited knowledge of the statics we can nevertheless try to extract the dynamics. Let us assume that all deformations of the phase which are not contained in the static distortion are small and thus that we can write

$$\phi(x,\tau) = \phi_0(x) + \delta\phi(x,\tau) \tag{3.13}$$

with $\delta\phi(x,\tau) \ll \phi_0(x)$ in a very vague sense since we deal with random variables. One can try to expand the random term in power of $\delta\phi$

$$\begin{aligned} S_{\text{dis}} &= \int d\tau\, dx\, |\xi(x)| \cos\Big(2\big(\phi(x,\tau) - \zeta(x)\big)\Big) \\ &\simeq -2 \int d\tau\, dx\, |\xi(x)| \cos\Big(2\big(\phi_0(x) - \zeta(x)\big)\Big) \big(\delta\phi(x,\tau)\big)^2 . \end{aligned} \tag{3.14}$$

One can thus use in principle (3.14) to compute the various physical quantities. Note that the conductivity (3.3) will *not* depend *directly* on the statics solution ϕ_0 since $\partial_t \phi_0 = 0$, so we can hope to compute it. Of course the dependence of the fluctuations $\delta\phi$ in ϕ_0 is hidden in (3.14). If ϕ_0 was following the random phase at every point, then the disorder term would just lead to a mass term for the fluctuations and the optical conductivity would show a gap. In fact this is not true at every point so (3.14) leads to a distribution of masses for the fluctuations. Unfortunately the knowledge of ϕ_0 is too crude to compute the conductivity accurately and depending on what exactly is L_{loc} one can find either a gap, a nonanalytic behavior, or a $\sigma(\omega) \sim \omega^2$ behavior at small frequencies [25]. Based on physical intuition, Fukuyama and Lee opted for the latter [25], but the method shows its limitations here and does not allow a reliable calculation of the physical quantities. More precise calculations of ϕ_0 and conductivity can be performed in the classical limit $K \to 0$ using a transfer matrix formalism [26].

One thing that can be obtained from the partial knowledge of the fluctuations is the effect of quantum fluctuations [27]. Indeed, when quantum fluctuations are present the expansion (3.14) is not valid any more since the cosine should be normal ordered before it can be expanded. This leads to

$$\cos\bigl(2(\phi_0 - \zeta + \delta\phi)\bigr) \simeq -2\cos\bigl(2(\phi_0 - \zeta)\bigr)e^{-2\langle\delta\phi^2\rangle}(\delta\phi)^2, \tag{3.15}$$

where the average $\langle\rangle$ has now to be computed self-consistently using (3.15). This leads to a modified localization length [27] of the form

$$L_{\text{loc}} \propto \left(\frac{1}{\mathcal{D}}\right)^{1/(3-2K)} \tag{3.16}$$

This expression for the localization length suggests that a delocalization transition is induced by the quantum fluctuations and occurs at $K = \frac{3}{2}$. In the fermion language, this corresponds to extremely attractive interactions.

3.2 Starting From the Metal: RG

The previous method starts directly from the localized phase. It provides some limited information about this phase, but suffers from serious limitations. An alternative approach is to start from the pure Luttinger liquid and investigate the effects of disorder perturbatively, and build a renormalization group analysis. The RG provides us with the best possible description of the delocalized phase and the critical properties of the transition. It also gives a very accurate description of the localized phase *up to* length-scales of the order of the localization length L_{loc}. Here again we describe the method for simplicity on spinless fermions and discuss more complex systems in Section 4.

To get a hint of the RG equations, let us expand the disorder term (3.1) to second order. This leads to

$$\int d\tau\, dx \int d\tau'\, dx'\, \xi(x)\xi^*(x')e^{2i(\phi(r)-\phi(r'))}. \tag{3.17}$$

It is easy to see that at the tree level, (3.17) scales as $\mathcal{D}L^{3-2K}$. This leads to the scaling of the disorder

$$\frac{\partial\mathcal{D}}{dl} = (3-2K)\mathcal{D}. \tag{3.18}$$

This expresses in fact the dressing of the scattering on the disorder by the interactions and has been derived using either diagrams or RG [19, 20, 28–33]. In itself it seems to confirm the result of (3.16), i.e., the existence of a transition. Note that the advantage of the bosonization derivation is to allow us to reach the nonperturbative point in interactions where such a metal-insulator transition would take place.

In fact (3.18) would not allow us in itself really to determine the metal-insulator transition point. This can be seen by using the RG to compute the finite temperature (or finite frequency) conductivity of the system [19, 20]. The idea is simply to renormalize until the cutoff is of the order of the thermal length $l_T \sim u/T$ corresponding to $e^{l^*} \sim l_T/\alpha$. At this lengthscale the disorder can be treated in the Born approximation. As the conductivity is a physical quantity it is not changed under renormalization and we have

$$\sigma(n(0), D(0), 0) = \sigma(n(l), D(l), l) = \sigma_0 \frac{n(l)D(0)}{n(0)D(l)} = \sigma_0 \frac{e^l D(0)}{D(l)}, \qquad (3.19)$$

where $\sigma(n(l), D(l), l) = \sigma(l)$, and $n(l)$ are, respectively, the conductivity and the electronic density at the scale l. $\sigma_0 = e^2 v_{\mathrm{F}}^2 / 2\pi\hbar\mathcal{D}$ is the conductivity in the Born approximation, expressed with the initial parameters. Using (3.18), one gets from (3.19),

$$\sigma(T) \sim \frac{1}{\mathcal{D}} T^{2-2K}. \qquad (3.20)$$

This result is the direct consequence of the renormalization of the scattering on impurities by interactions (3.18). One immediately sees that (3.20) *alone* would lead to a paradox since (3.18) gives a localized-delocalized boundary at $K = \frac{3}{2}$ whereas (3.20) gives perfect conductivity above $K = 1$ (i.e., the noninteracting point). One could also immediately see that if one introduces a new variable such as

$$\widetilde{\mathcal{D}} = e^{-al}\mathcal{D}, \qquad (3.21)$$

the dimension of such a variable would be $(3 - a - 2K)$, leaving the location of the transition point as determined from (3.18) quite arbitrary. Although such a transformation seems arbitrary, if one considers that the disorder stems from impurities with a concentration n_i and a strength V, the limit of Gaussian disorder corresponds simply in taking $n_i \to \infty$ with $V \to 0$ keeping $\mathcal{D} = n_i V^2$ fixed. Thus the choice $a = 1$ in (3.21) simply corresponds to $\tilde{\mathcal{D}} = V^2$, i.e., writing an RG equation for the impurity strength.

The answer to this simple paradox is of course that (3.18) should be complemented by another RG equation. In addition to renormalizing $\mathcal{D}$, (3.17) generates as well quadratic terms that renormalize the free part of the Hamiltonian, i.e., the velocity v and the Luttinger parameter K. Details can be found in [19, 20]. The main equation is the renormalization of the Luttinger parameter K and reads

$$\frac{\partial K}{\partial l} = -K^2 \mathcal{D}/2. \qquad (3.22)$$

This equation describes the renormalization of the interactions by the disorder. Both RG equations (3.18) and (3.22) have a diagrammatic representation shown in Fig. 5.2.

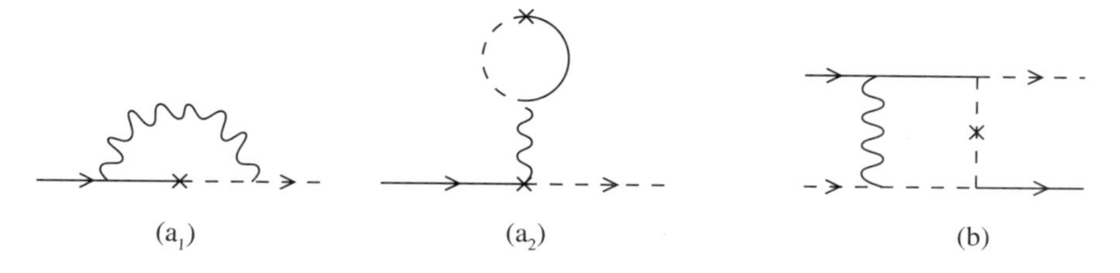

FIGURE 5.2. Diagrams describing the renormalization of the disorder by the interactions (a) and the renormalization of the interactions by the disorder (b). Solid and dotted lines are fermions with $\pm k_{\mathrm{F}}$, the wiggly line is the interaction and the cross is the impurity scattering.

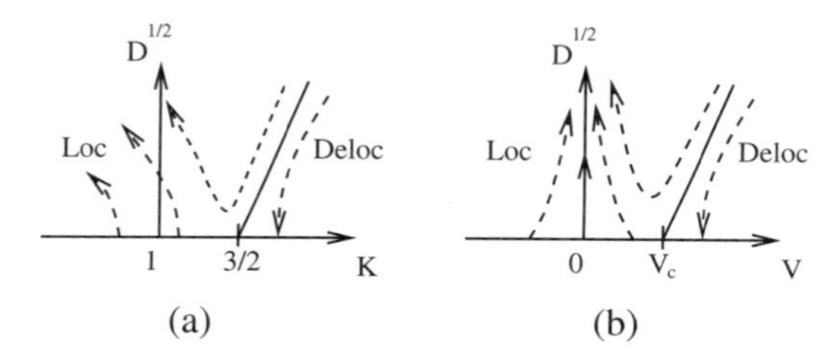

FIGURE 5.3. Phase diagram and flow for spinless fermions in presence of disorder. (a) is the flow in $\mathcal{D}$ and K variables. (b) the flow in the $\mathcal{D}$ and interactions g. Disorder kills inelastic interactions.

Using the flow (3.18) and (3.22), one can easily check that two phases exist as shown on Fig. 5.3.

Large K corresponds to the delocalized phase where the disorder is irrelevant and the system is a Luttinger liquid with renormalized coefficients u^* and K^*. All correlation functions decay as power laws, and because $K^* > \frac{3}{2}$ the system is dominated by superconducting fluctuations. On the transition line the exponent flows to the universal value $K^* = \frac{3}{2}$. Below this line, $\mathcal{D}$ flows to large values, indicating that the disorder is relevant. This phase is the localized phase. This is obvious for physical reasons but can also be guessed from the exact solution known for the noninteracting line $K = 1$ (and any $\mathcal{D}$) which belongs to this phase. As can be seen from (3.18) and (3.22) the transition is Berezinskii–Kosterlitz–Thouless (BKT)-like [34, 35] in the K, $\sqrt{\mathcal{D}}$ variables.

In addition to the phase diagram itself, a host of physical properties can be extracted from the RG. The simplest one is the localization length. One can use that for $\mathcal{D}(l) \sim 1$, the localization length is of the order of the (renormalized) lattice spacing αe^l. The full determination needs an integration of both (3.18) and (3.22). Close to the transition the divergence of the localization length is BKT-like (setting $K = \frac{3}{2} + \eta$):

$$L_{\mathrm{loc}} \sim e^{2\pi/\sqrt{9\mathcal{D}-\eta^2}} \tag{3.23}$$

. Deep in the localized regime, and for weak disorder, a good approximation is to neglect the renormalization of K in (3.18). A trivial integration of (3.18) then gives back (3.16). This we see that the SCHA calculation corresponds in fact, both for the phase diagram and for the localization length, to the limit of infinitesimal disorder.

Out of the RG, one can also extract, using (3.19), the behavior of the temperature or frequency dependence of the conductivity. In the localized phase, this can only be used up to the energy scale corresponding to the localization length, i.e., $E_{\text{pin}} = k_B T_{\text{pin}} = \hbar\omega_{\text{pin}}, \omega_{\text{pin}} = v/L_{\text{loc}}$. Below this lengthscale another method than the weak coupling RG should be used. We will come back to that point in Section 6. Here again, although the full flow should be taken into account one can get an approximate formula by ignoring the renormalization of K, which leads to (3.20). For $K < \frac{3}{2}$ (including the noninteracting point) any small but *finite* disorder *grows*, renormalizing the exponents and ultimately leading to a decrease of the conductivity, even if one started initially from $K > 1$. A very crude way of taking into account both equations (3.18) and (3.22) would be to say that one can still use (3.20) but with scale-dependent exponents (see [19, 20]):

$$\sigma(T) \sim T^{2-2K(T)}. \tag{3.24}$$

This renormalization of exponents and the faster decay of conductivity is in fact the signature of Anderson localization. The equivalent frequency dependence is

$$\sigma(\omega) \propto \omega^{2K-4}. \tag{3.25}$$

Here again this formula breaks down below the scale $\omega_{\text{pin}} \sim v/L_{\text{loc}}$, which is the pinning frequency. Similarly, below L_{loc}, correlation functions can again be computed using the RG, but of course the asymptotic behavior cannot be obtained.

4 Other Systems and RG

Despite its limitations to physics above E_{pin}, the RG is an extremely efficient method given its simplicity. It allows in addition a perfect description of the delocalized phase and of the critical behavior, something unattainable through the methods of Section 3.1 and allows for interesting extensions.

First let us note that (3.18) and (3.22) also describe the case of a single impurity [36, 37]. Indeed, in that case one can go back to the definition $\mathcal{D} = n_i V^2$ and take the limit $n_i \to 0$, i.e., $\mathcal{D} \to 0$. Equation (3.22) shows that in that case K cannot be renormalized since a single impurity cannot change the thermodynamic behavior. Only (3.18) remains, leading directly to a temperature dependence of the form (3.20), and a localized-delocalized transition at $K = 1$. More details on such a relation between the two problems and on the remaining open question can be found in [38].

Quite remarkably, the set (3.18)–(3.22) seems wrong. Indeed K naively depends on the (inelastic) interactions. Perturbatively, for the pure systems $K = 1 - V/(2\pi v_{\mathrm{F}})$. If one starts with $K = 1$, i.e., for the noninteracting system, it would thus seem from (3.22) that *elastic* scattering on the impurities can generate *inelastic* fermion-fermion interactions. The solution of this paradox is hidden in the precise way the RG procedure is built. In order to have the elastic nature of the scattering on impurities, the time integrations in (3.17) should be done independently for τ and τ'. When one performs the RG, one introduces a cutoff and imposes $|\tau - \tau'| > \alpha$. Thus a part is left out of (3.17): viz.,

$$\mathcal{D} \int dx \int_{|\tau-\tau'|<\alpha} d\tau\, d\tau'\, \rho(x,\tau)\rho(x,\tau') \simeq 2\mathcal{D}\alpha \int dx \int d\tau\, \rho(x,\tau)\rho(x,\tau), \quad (4.1)$$

which is exactly an inelastic interaction term. Thus, in fact, K contains not only the original inelastic interactions V but also a small correction coming from the disorder itself. In order to determine the flow for V it is thus necessary to take this small correction into account [19, 20] which gives the flow of Fig. 5.3b. One thus sees that the elastic case $V = 0$ indeed remains elastic and also that for spinless fermions, the perturbative flow seems to indicate that the inelastic interactions are reduced by the disorder. This is compatible with the physical image that one would get at strong disorder: fermions localize individually and since the overlap of wavefunctions is exponentially small, so is the effect of interactions. One could thus naively expect that below L_{loc} the effect of interactions are strong but disappear above L_{loc}. As we will see in Section 6, the variational approach confirms this image. For fermions with spins (to be discussed below), we expect only the interaction in the charge sector to vanish. Interactions in the spin sector do remain and lead to random exchange.

Of course many more physical systems can be studied by this method. This is the case for spin chains, which can directly be mapped onto spinless fermions [18]. A spin chain under a random magnetic field along z [22] is exactly the problem that we solved in the previous section, with K being the anisotropy ($K = 1$ for an XY chain and $K = \frac{1}{2}$ for an Heisenberg one). However, although for fermion problems it is unlikely that one is at a commensurate filling this is the natural situation for a spin chain (since the magnetization is zero in the absence of external field, the filling of the equivalent spinless fermion system is $n = \frac{1}{2}$). A spin chain with random exchange is thus like a *commensurate* fermionic system with random hopping. We will come back to this peculiar case in Section 7. Quite remarkably, in one dimension, bosons [19, 20] lead to physics similar to fermions. This is due to the fact that in one dimension statistics cannot be separated from interactions. Interacting bosons can thus be represented by a bosonization representation quite similar to the one of fermions [39]. The density can be written as

$$\rho(r) = \rho_0 - \rho_0 \partial_x \phi + \sum_n e^{in(2\pi\rho_0 x - 2\phi(r))}, \tag{4.2}$$

very similar to the fermionic form, while the single-particle operator is now

$$\psi(r) \sim \rho_0^{1/2} e^{i\theta} \tag{4.3}$$

(note the difference with the fermionic operator). The Hamiltonian is still described by the quadratic form (3.1). Now $K = \infty$ for noninteracting bosons and $K = 1$ for hard-core ones. There is thus a transition [19, 20] between a superfluid state (for $K > \frac{3}{2}$) to a Bose glass state, where the bosons are localized by the random potential at $K = \frac{3}{2}$.

In a similar way, one can of course treat fermions with spins. Equations are more complicated since they involve the charge and spin sectors, and we will not discuss the full physics here but refer the reader to [20]. Let us just draw attention to one interesting consequence of the renormalization equation of the disorder for the problem with spins, which reads

$$\frac{d\mathcal{D}}{dl} = (3 - K_\rho - K_\sigma - g_{1\perp})\mathcal{D}, \tag{4.4}$$

where $g_{1\perp}$ is the back-scattering interaction between opposite spins. For a Hubbard type interaction $g_{1\perp} = U$. For spin isotropic systems, $K_\sigma = 1 + g_{1\perp}/(2\pi v_{\mathrm{F}})$ and $g_{1\perp}$ is marginal with a flow

$$\frac{dg_{1\perp}}{dl} = -g_{1\perp}^2. \tag{4.5}$$

For more general spin couplings either $g_{1\perp} \to 0$ and $K_\sigma \to K_\sigma^*$, or $g_{1\perp}$ is relevant and a spin gap opens. The physics of the localization transition depends thus on the sign of interactions with special physical consequences. The transition point moves from $K_\rho = 2$ (for infinitesimal disorder) for repulsive interactions to $K_\rho = 3$ for attractive ones. For a Hubbard-type interaction for which $\frac{1}{2} < K_\rho < 2$, the delocalization point can never be reached and the system is localized regardless of the strength of interaction and disorder. Another consequence is that in the presence of spin degrees of freedom, the divergence of the localization length at the transition is not BKT like any more. One could naively think that in the repulsive case, the $g_{1\perp}$ term could be omitted and that the renormalization of the disorder could be written $(3 - K_\rho - K_\sigma^*)$. For spin isotropic interactions, this misses an important part of the physics. Indeed, for a Hubbard interaction $K_\rho = 1 - U/(2\pi v_{\mathrm{F}})$. Substituting in (4.4) leads (for the initial steps of the flow) to

$$\frac{d\mathcal{D}}{dl} = \left(1 - \frac{U}{\pi v_{\mathrm{F}}}\right)\mathcal{D}, \tag{4.6}$$

whereas the incorrect substitution at the fixed point would lead to $(1 + U/(\pi v_{\mathrm{F}}))$, leading to quite different physics. Equation (4.6) implies that for Hubbard-type interactions repulsive interactions make the system *less* localized [40] than for attractive interactions, i.e., that the $L_{\mathrm{loc}}^{U>0} > L_{\mathrm{loc}}^{U<0}$. Similar effects exist for the charge stiffness and the persistent currents; i.e., for a system with spins, the persistent currents are in fact enhanced by repulsive interactions. This counterintuitive statement can be explained physically. Interactions have two effects: (i) they tend to reinforce, when attractive, the superconducting fluctuations in the system. This screens disorder and makes it less effective. This is the only effect occurring for spinless fermions. (ii) When spin degrees of freedom exist, repulsive interactions also tend to make the density more uniform by spreading the charge. This makes it more difficult to couple to disorder. These two effects compete. This has several consequences, in particular for mesoscopic systems. Of course, for fermions, true delocalization can only be achieved with attractive interactions reaching at least the nearest neighbor.

Many other systems have been treated by such method. There are one-dimensional systems with long-range $1/r$ interactions that lead to a pinned Wigner crystal [41], doped spin 1 chains [42], fermionic [43] and bosonic [44] ladders, spin 1 chains in a magnetic disorder [45], and spin ladders [46]. Since we want to focus here on the methods we refer the reader to the above references for a detailed discussion of the physics of such systems.

5 A Zest for Numerics

Although we are mainly concerned about analytical method in these notes, let us mention some numerical results and methods that have been used in connection with the RG predictions. Although numerical studies have become very powerful in one dimension for pure systems, the presence of disorder complicates matters. Three main methods have been used.

Exact diagonalizations have been used to study both the phase diagram and the charge stiffness of both spinless fermions [47, 48] (or equivalently XXZ spin chains) and fermions with spins [49,50] with short-range or long-range interactions [51]. Using the finite size scaling of the spin stiffness $\rho_s = (1/L)\partial^2 E(\theta)/\partial\varphi^2 = f(L/L_{\mathrm{loc}})$, where $E(\theta)$ is the ground-state energy of the disordered XXZ chain with boundary conditions $S_L^+ = e^{i\varphi} S_1^+$, the localization length L_{loc} can be obtained [47], and is in good agreement with the RG results of Section 3.2. The behavior of the correlation length close to the transition point appeared consistent with the predicted BKT-like behavior. The results also suggested that a finite disorder was needed to disorder the ground state for $K > \frac{3}{2}$. A similar study with systems sizes up to $L = 18$ sites was also made for XXZ spin chains with a random exchange in [52]. Analysis of persistent currents was also in agreement with the RG prediction of Section 4. Unfortunately the exact diagonalization approach of the last section is limited to zero temperature and small system size.

In order to consider bigger system sizes, one can use Quantum Monte Carlo methods. In [53], such a study was performed for disordered bosons. The superfluid density was obtained as a function of interaction for a given disorder strength. It was shown that for not-too-repulsive equations, there was a phase with a finite superfluid density. For more repulsive interactions, a phase with finite compressibility by zero superfluid density was obtained, in agreement with the Bose Glass theory of Section 4.

The most promising recent method is the Density Matrix Renormalization Group. It been introduced in the recent years as a method specially designed to calculate the ground state of correlated one-dimensional systems [54]. This method has been also applied to the problem of the XXZ chain in a random magnetic field parallel to the z axis by Schmitteckert et al. [55]. The authors of [55] have been able to consider system sizes up to $L = 60$ sites, and average over several hundred realizations of the disorder. Localization and phase diagram were also in good agreement with the RG predictions.

Clearly, various numerical checks confirm the predictions of the RG. Unfortunately, so far only the phase diagram, stiffness, and localization length have been computed. This is clearly related to the complexity of the problem at hand. What would be extremely useful would be information on quantities deep in the localized phase, such as the single-particle Green function, the AC or DC conductivity. Analysis of such quantities would nicely complement the RG analytical study and allow for comparison with other analytical techniques more suited for the localized phase such as the variational method we analyze in the next section.

6 Variational Method

Let us now study this problem using a completely different and at first sight more formal method. As usual it is very convenient to get rid of the disorder from the start. Given the nonquadratic nature of (3.1), supersymmetric methods are unapplicable and we have to turn to replicas. The idea of the replica method in itself is quite simple. If we want to compute an observable O we have to do both average over disorder and thermodynamic average:

$$\overline{\langle O \rangle} = \int \mathcal{D}V p(V) \langle O \rangle_V = \int DV p(V) \frac{\int \mathcal{D}\phi O[\phi] e^{-S_V[\phi]}}{\int D\phi e^{-S_V[\phi]}}. \tag{6.1}$$

The action is usually linear in disorder and for Gaussian disorder the distribution of random potential is $p(V) \propto e^{-\int dx V(x)^2}$, so the average would be quite trivial without the denominator in (6.1). The idea is thus to introduce

n fields and to compute

$$\int D\phi_1 D\phi_2 \dots D\phi_n O[\phi_1] e^{-\sum_{i=1}^{n} S_V[\phi_i]} = \int D\phi O[\phi] e^{-S_V[\phi]} \left[\int D\phi e^{-S_V[\phi]} \right]^{n-1}, \quad (6.2)$$

which is exactly the quantity we want to average over disorder in (6.1) if one takes the formal limit $n \to 0$. Since (6.2) has no denominator, averaging over disorder is trivial. Of course there is a price to pay: before the averaging the replicas are all independent fields but the averaging introduces an interaction between them. We have thus traded a theory depending on a random variable V but a single field for a theory without disorder but with n coupled fields. Usually this is still a situation we are better equipped to solve because of the large number of field theoretic method dealing with "normal" (i.e., transitionally invariant actions). For the particular case (3.1) the replicated action is

$$S/\hbar = \int dx\, d\tau \, \frac{1}{2\pi K} \sum_a \left[\frac{1}{v} (\partial_\tau \phi_a)^2 + v (\partial_x \phi_a)^2 \right] - \frac{\mathcal{D}}{(2\pi\alpha)^2 \hbar} \sum_{ab} \int dx\, d\tau\, d\tau' \, \cos\big(\phi_a(x,\tau) - \phi_b(x,\tau')\big), \quad (6.3)$$

where $a = 1, \dots, n$ is the replica index. Disorder averaging has coupled the replicas via the cosine term. Because the disorder is time independent, this coupling contains two fields that can be at arbitrary time and is thus highly nonlocal. For fermions one usually prefers to go to frequency space, where this implies conservation of the frequency for each replica index, but this would not simplify things here because of the cosine.

This is up to now a totally formal procedure and nothing has been accomplished. Equation (6.3) is totally equivalent to (3.1) and the difficulty is of course to solve it. Based on the RG equation (3.18) one could think naively that since the localized phase corresponds to $\mathcal{D} \to \infty$ it would be safe to expand the cosine term in (6.3). Unfortunately, it is easy to check that it fails seriously *when* $n \to 0$ *is taken* (it of course works perfectly for a finite number of field $n \geq 2$). In order to circumvent this problem let us try to improve over this simple-minded expansion of the cosine with a variational ansatz. We introduce a trial action S_0:

$$S_0/\hbar = \frac{1}{2\beta L \hbar} \sum_{q,\omega_n} \sum_{ab} \phi_a(q,\omega_n) G_{ab}^{-1}(q,\omega_n) \phi_b(-q,-\omega_n), \quad (6.4)$$

where the propagators G^{-1} are our variational "parameters." As usual $(1/L)\sum_q \to \int dq/(2\pi)$. If we introduce

$$Z = \int \mathcal{D}\phi e^{-S/\hbar}, \quad (6.5)$$

we then have the variational theorem for the free energy $F = -\hbar \log(Z)$:

$$F \le F_{\text{tr}} = F_0 + \langle S - S_0 \rangle_{S_0}. \tag{6.6}$$

Since S_0 is quadratic, (6.6) can be in general computed quite explicitly as a function of the (unknown) propagators G. The "best" quadratic action S_0 is thus the one that satisfies the saddle point equation

$$\frac{\partial F_{\text{tr}}}{\partial G_{ab}(q, \omega_n)} = 0, \tag{6.7}$$

which gives a set of integral equations allowing us to determine the unknown functions G.

The observables are simply defined by quantities diagonal in replica indices, as can be seen from (6.2). For some quantities such as the compressibility, it is necessary to be more careful since one has to substract the average, which is usually zero in a pure system or after averaging over disorder but nonzero for a *specific* realization of the disorder. Let us introduce the various propagators (time ordering in τ is always implied):

$$\begin{aligned} B_{ab}(x,\tau) &= \langle [\phi_a(x,\tau) - \phi_b(0,0)]^2 \rangle \\ &= \big(G_{aa}(0,0) + G_{bb}(0,0) - 2G_{ab}(x,\tau)\big) \qquad (6.8) \\ G_{ab}(q_x, \omega_n) &= \langle \phi_a(q_x, \omega_n) \phi_b(-q_x, -\omega_n) \rangle . \qquad (6.9) \end{aligned}$$

The compressibility is given by

$$\chi(q,\omega_n) = \frac{1}{\hbar} \int dx \int_0^{\beta\hbar} d\tau \, e^{-\imath(qx - \omega_n \tau)} \times \overline{\langle T_\tau (n(x,\tau) - \langle n(x,\tau)\rangle)(n(0,0) - \langle n(0,0)\rangle)\rangle}, \tag{6.10}$$

which leads to the average static compressibility $\chi_s = \lim_{q\to 0}(\lim_{\omega \to 0} \chi(q,\omega))$ (see (3.4)). When expressed in terms of the replicated bosonized operators, (6.10) gives

$$\chi_s = \lim_{q\to 0} \lim_{\omega \to 0} q^2 G_c(q,\omega), \tag{6.11}$$

where we introduced an important propagator: the connected one defined as $G_c^{-1}(q) = \sum_b G_{ab}^{-1}(q)$.

Without the replicas, this method is nothing but the well-known Self-Consistent Harmonic Approximation (SCHA), which is known to work very well for sine-Gordon-type Hamiltonians. Such a method correctly gives in particular the two phases (massless and massive). Extension of this method to disordered systems was done in the context of classical elastic systems such as interfaces [56]. In quantum problems, another level of complexity occurs because of the aforementioned nonlocality of the interaction in time. But before going to these problems, specific to the quantum systems, let us illustrate the aspects of this variational method when applied to disordered systems, on a technically simpler example (for which this method was extremely fruitful [57]): the case of classical periodic systems.

6.1 A Classical Example

Let us take the action (6.3) but with only a *single* time integral for the disorder term. Such action would be the result of the average on a disorder both dependent on space *and* time. Of course, such a disorder would be quite unrealistic for quantum problems. However (6.12) would be a perfectly natural Hamiltonian for a *classical* problem where $z = v\tau$ is now just one of the spatial dimensions [57,58]. To make the analogy more transparent, let us use $z = v\tau$, and replace the integral over x by an integral in $d-1$ dimensions. If denote by r the d-dimensional space variable $r = (x, z)$, the starting action is

$$S/\hbar = \int d^d r \, \frac{1}{2\pi K} \sum_a (\partial_r \phi_a)^2 - \frac{\mathcal{D}}{(2\pi\alpha)^2 \hbar v} \sum_{ab} \int d^d r \, \cos\Big(2\big(\phi_a(r) - \phi_b(r)\big)\Big). \qquad (6.12)$$

One can see that (6.12) is exactly the Hamiltonian describing an elastic system such as a vortex lattice or a classical CDW in the presence of point like defects in d dimensions. $\hbar$ plays the role of the temperature for the classical system, the elastic constant c is given by $c = 1/(\pi \overline{K})$ with $\overline{K} = K/\hbar$ and $\rho_0^2 \Delta/2 = \mathcal{D}\hbar/(2\pi\alpha)^2$ would be the correlator of the classical disorder [58].

If we call $q = (q_x, \omega)$ the d-dimensional momentum, without loss of generality, the matrix $G^{-1}_{ab}(q)$ can be chosen of the form $G^{-1}_{ab} = cq^2\delta_{ab} - \sigma_{ab}$. We obtain by minimization of the variational free energy the saddle point equations

$$G_c^{-1}(q) = cq^2, \quad \sigma_{a\neq b} = \frac{2\mathcal{D}}{(\pi\alpha)^2} e^{-2B_{ab}(r=0)}. \qquad (6.13)$$

Using (6.8)–(6.9) one obtains

$$B_{ab}(r) = \hbar \int \frac{d^d q}{(2\pi)^d} \big(G_{aa}(q) + G_{bb}(q) - 2\cos(qr) G_{ab}(q)\big). \qquad (6.14)$$

For this particular problem, the connected part is not affected by disorder. This is the consequence of a hidden symmetry (statistical tilt symmetry) of (6.12), whose disorder part is not affected by any *local* shift of $\phi_a(r)$ such as $\phi_a(r) \to \phi_a(r) + f(r)$, where f is an arbitrary function. Such a symmetry does not exist for the time-correlated disorder natural in a quantum problem, with important physical consequences to which we will come back. The only interesting equation here is thus the equation for the off-diagonal part $\sigma_{a\neq b}$. Because of the locality of the interaction term between replicas in (6.12), the self-energy σ_{ab} is simply a matrix of constants.

Given the symmetry of the original action/Hamiltonian (6.12) by permutation of the replica indices, it is very natural to look for a variational ansatz with the same symmetry. This would mean that the G^{-1} matrix would have only (for each value of q) *two* independent values: the diagonal one $G_{aa} = \tilde{G}$ and the off diagonal one $G_{a\neq b}$. Such a matrix can easily be inverted for any n, and the analytic continuation for $n \to 0$ gives

$$G_c = \tilde{G} - G_{a\neq b} = \frac{1}{G_c^{-1}} \tag{6.15}$$

$$G_{a\neq b} = -\frac{G^{-1}_{a\neq b}}{(G_c^{-1})^2}. \tag{6.16}$$

Using these inversion formulas is it easy to solve for (6.13). In $d > 2$, B depends on G_c only and thus $\sigma_{a\neq b}$ is simply a constant proportional to disorder. Given the Gaussian nature of the trial action (6.4), the correlation functions such as the density-density can easily be computed:

$$\overline{\langle \rho(r)\rho(0)\rangle} = \langle \rho_a(r)\rho_a(0)\rangle \propto e^{-2B_{aa}(r)}. \tag{6.17}$$

Using (6.15) shows that $G_{aa}(q) \sim \mathcal{D}/q^4$ leading to a growth

$$B_{aa}(r) \sim \mathcal{D} r^{4-d}. \tag{6.18}$$

Although this solution is perfectly well behaved, a stability analysis of the replica symmetric saddle point shows that it is unstable. This can be checked from the eigenvalue λ of the replicon mode [56,58].

$$\lambda = 1 - \frac{8\hbar\mathcal{D}}{(\pi\alpha)^2} e^{-4\hbar \int d^d p\, G_c(p)/(2\pi)^d} \int \frac{d^d q}{(2\pi)^d} G_c^2(q). \tag{6.19}$$

A negative eigenvalue λ indicates an instability of the replica symmetric solution. We introduce a small regularizing mass in G_c: $G_c(q)^{-1} = cq^2 + \mu^2$ and take the limit $\mu \to 0$. It is easy to see from (6.19) that for $d < 2$ the replica symmetric solution is always stable. In that case disorder is in fact irrelevant, due to the strong quantum (or thermal for the associated classical system) fluctuations. For $d = 2$, the condition becomes $\mu^{2(\overline{K}-1)} < 1$ for small μ. Thus there is a transition at $\overline{K} = 1$ between a replica symmetric stable high-temperature phase where disorder is irrelevant and a low-temperature (glassy) phase where the symmetric saddle point is unstable. For the classical system this is the well-known Cardy–Ostlund transition [59], with very interesting physical aspects of its own. This transition is the equivalent for *time-dependent* disorder of the localization transition studied in Section 3.2. For time-dependent disorder, one time integral drops in (3.17) and (3.18) would become $(2 - 2K)\mathcal{D}$, giving the transition at $K = 1$. More details can be found in [57,58,60].

For $2 < d < 4$, the replica symmetric solution is *always unstable.*

One should thus look for another way of inverting the 0×0 matrices than the replica symmetric one. Fortunately, such a scheme was invented

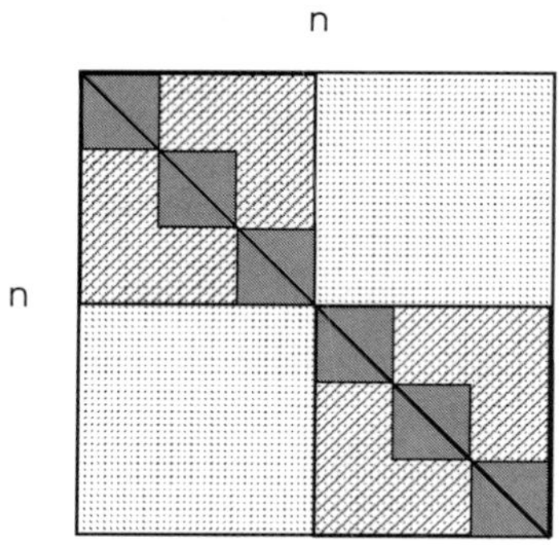

FIGURE 5.4. Replica symmetry broken matrices (here for a 2-step RSB). Each pattern correspond to a different value σ_i (see text), and the diagonal one (black line) is $\tilde{\sigma}$.

in the context of spin glasses. Instead of having a single value σ for the off-diagonal term, one introduces a whole set of values. Let us briefly illustrate the procedure here, refering the reader to [61] for details. Let us introduce a set of integers $m_0 = n, m_1, \ldots, m_{k+1} = 1$ such that m_i/m_{i+1} is an integer. One cuts the matrix in blocks of size m_1 as illustrated on Fig. 5.4.

Elements outside the blocks have the value σ_0. The procedure is then recursively applied for the inner blocks. At the last step, the value on the block of size m_k is σ_k and a diagonal value $\tilde{\sigma}$. Quite generally, the matrix is thus now parametrized by a diagonal element and a whole function (in the limit $n \to 0$) $\sigma(u)$, where $u \in [0,1]$ (notice the range of variation of the m_i when $n \to 0$). Such matrices can also be inverted in the limit $n \to 0$), albeit with more complicated inversion rules than for the RS solution. When one has a continuous function (see Fig. 5.5), the RSB is said to be continuous. Simpler cases are, of course, a constant function (a single off-diagonal value), which is simply the RS solution or a function continuous by steps. We have represented in Fig. 5.4 (see also Fig. 5.5) the case of a two-step RSB. We denote $\tilde{G}(q) = G_{aa}(q)$, similarly $\tilde{B}(x) = B_{aa}(x)$, and parametrize $G_{ab}(q)$ by $G(q,v)$, where $0 < v < 1$, and $B_{ab}(x)$ by $B(x,v)$. Physically, v parametrizes pairs of low-lying states, in the hierarchy of states, $v = 0$ corresponding to states further apart. The saddle point equation becomes

$$\sigma(v) = \frac{2\mathcal{D}}{(\pi\alpha)^2} e^{-2B(0,v)}, \tag{6.20}$$

where

$$B(0,v) = 2\hbar \int \frac{d^d q}{(2\pi)^d} \left(\tilde{G}(q) - G(q,v)\right). \tag{6.21}$$

$B(0,v)$ corresponds physically to the mean squared phase fluctuations at the same point in space ($r = 0$) (for the associated classical system this would be mean squared relative displacements of the same object) but in two replica states, or more physically in two different low lying metastable states. The large distance behavior of disorder-averaged correlators is determined by the small v behavior of $B(0,v)$. We look for a solution such that $\sigma(v)$ is constant for $v > v_c$, v_c itself being a variational parameter, and has an arbitrary functional form below v_c. This corresponds to full RSB (see Fig. 5.5). The algebraic rules for inversion of hierarchical matrices [56] give

$$B(0,v) = B(0,v_c) + \int_v^{v_c} dw \int \frac{d^d q}{(2\pi)^d} \frac{2\hbar\sigma'(w)}{(G_c(q)^{-1} + [\sigma](w))^2}, \tag{6.22}$$

where $[\sigma](v) = u\sigma(v) - \int_0^v dw\, \sigma(w)$ and

$$B(0,v_c) = \int \frac{d^d q}{(2\pi)^d} \frac{2\hbar}{G_c(q)^{-1} + [\sigma](v_c)} \tag{6.23}$$

This is a simple number. Taking the derivative of (6.20) with respect to v, using $[\sigma]'(v) = v\sigma'(v)$, (6.22), and (6.20), again one finds

$$1 = \sigma(v) \int \frac{d^d q}{(2\pi)^d} \frac{4\hbar}{(cq^2 + [\sigma](v))^2} \simeq \sigma(v) \left(\frac{4\hbar c_d}{c^{d/2}}\right) [\sigma(v)]^{(d-4)/2}. \tag{6.24}$$

Since the integral is ultraviolet convergent, we have taken the short-distance momentum cutoff to infinity. c_d is a simple number,

$$c_d = \int \frac{d^d q}{(2\pi)^d} \left(\frac{1}{q^2+1}\right)^2 = \frac{(2-d)\pi^{1-d/2}}{2^{d+1}\sin(d\pi/2)\Gamma(d/2)}, \tag{6.25}$$

with $c_{d=3} = 1/(8\pi)$, $c_{d=2} = 1/(4\pi)$. Derivating one more time yields the effective self-energy:

$$[\sigma](v) = (u/u_0)^{2/\theta}, \tag{6.26}$$

where $\theta = (d-2)$ and $v_0 = 8\hbar c_d c^{-d/2}/(4-d)$. The shape of $[\sigma](u)$ is shown in Fig. 5.5.

The solution (6.26), is a priori valid up to a breakpoint u_c, above which $[\sigma]$ is constant, since $\sigma'(u) = 0$ is also a solution of the variational equations. u_c can also be extracted from the saddle point equations and we refer the reader to [58] for details. The precise value of u_c is unimportant for our purpose but the existence of the two distinct regimes in $[\sigma](u)$ has a simple physical interpretation that we examine in Section 6.2. Using (6.26), one can now compute the correlation functions. Larger distances correspond to less massive modes and are dominated by the small u behavior of (6.26). One obtains

$$\overline{\langle(\phi(r) - \phi(0))^2\rangle} = 2\hbar \int \frac{d^d q}{(2\pi)^d} (1 - \cos(qr)) \tilde{G}(q) \tag{6.27}$$

$$\tilde{G}(q) = \frac{1}{cq^2}\left(1 + \int_0^1 \frac{dv}{v^2} \frac{[\sigma](v)}{cq^2 + [\sigma](v)}\right) \sim \frac{Z_d}{q^d}, \tag{6.28}$$

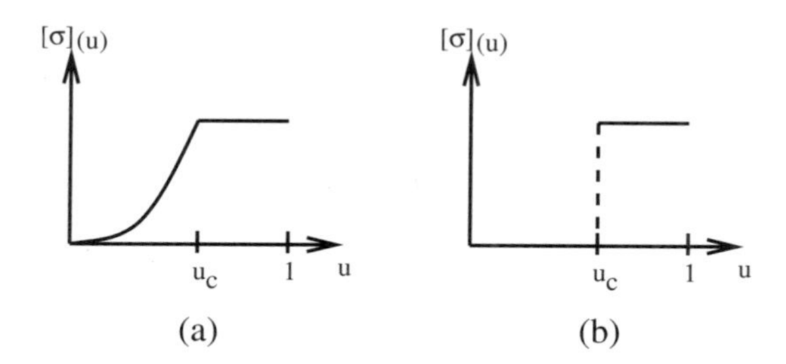

FIGURE 5.5. Shape of the self-energy σ as a function of u. (a) is the full RSB solution occurring for $2 < d < 4$. (b) a one step RSB ($d = 2$).

with $Z_d = (4-d)/(4\hbar S_d)$ and $1/S_d = 2^{d-1}\pi^{d/2}\Gamma[d/2]$. Thus for $2 < d < 4$, this leads to a logarithmic growth

$$\overline{\langle (\phi(x) - \phi(0))^2 \rangle} = \tfrac{1}{2} A_d \log|x|, \tag{6.29}$$

with $A_d = 4 - d$, instead of the power law growth (6.18) of the replica symmetric solution. Note that the amplitude is independent of disorder.

6.2 If It Ain't Broken ...

It is thus necessary to break the replica symmetry to get the correct asymptotic physics ($1/q^d$ propagator) instead of the $1/q^4$ given by the replica symmetric solution. From (6.28), one easily sees that at large enough q (i.e., for short distances), one recovers the replica symmetric solution. Thus u_c and $[\sigma(u_c)]$ (see Fig. 5.5) define a lengthscale $L \sim (c/[\sigma(u_c)])^{1/2}$ above which the RS solution does not describe the physics. This lengthscale corresponds to the pinning length similar to (3.12) obtained by balancing the elastic energy with the disorder one. Here

$$E_{\text{el}} \sim L^{d-2} \tag{6.30}$$

$$E_{\text{dis}} \sim \mathcal{D}^{1/2} L^{d/2} \tag{6.31}$$

leading to the famous Larkin [62, 63] length $L_{\text{loc}} \sim (1/\mathcal{D})^{1/(4-d)}$ which corresponds to the distance for which relative displacements are of the order of the lattice spacing[1] (or for which the phase here is of the order 2π).[2] Below this lengthscale, the system has a single equilibrium state. It

[1]There are some subtleties for systems for which the size of the particles differs from the lattice spacing. This leads to the definitions of two characteristic lengths. We ignore these differences here since they do not occur for classical CDW and 1D fermionic systems. See [57] for consequences for classical systems and [70] for the Wigner crystal.

[2]Such a length was also obtained by Imry and Ma for the random field Ising model. So a proper acronym for this length might be the Loflim length (for Larkin–Ovchinikov, Fukuyama–Lee, Imry–Ma).

can be described by simply expanding in the displacements in (3.1),

$$H_{\text{dis}} = \int d^d x\, f(x)\phi(x), \tag{6.32}$$

where f is a random force, derivative of the random potential V. It is easy to see that this model gives the $1/q^4$ propagator. This breaks down when the displacements cannot be expanded, i.e., for distances larger than L_{loc}. As anticipated by Larkin and Ovchinikov [63], pinning occurs that tends to keep a domain of size L_{loc} in place despite the thermal fluctuations.

This helps us understand the physics of the RSB solution. To illustrate it let us focus for simplicity on the case when $d = 2$. As can be seen from (6.26), when $d \to 2$, the RSB solution is a one-step breaking, as shown in Fig. 5.5. There are some additional complications but they are unimportant for our discussion here. A normal self-consistent approximation would approximate the energy by a simple Gaussian centered in $\phi = 0$. The only way to incorporate the pinning is thus to put a mass term in the propagator

$$q^2 \to q^2 + L_{\text{loc}}^{-2}. \tag{6.33}$$

As can be immediately inferred from the discussion of Section 3.1, this is a much too crude approximation. The RSB solution is smarter and approximates effectively the distribution of displacements by a hierarchical superposition of Gaussians centered at different randomly located points in space. A pictorial view of these two cases is shown in Fig. 5.6.

The double distribution over environment and thermal (or quantum) fluctuations is approximated as follows [57]. In each environment there are effective "pinning centers," corresponding to the low-lying metastable states (preferred configurations). Since all q modes are in effect decoupled within this approximation, for each q mode a preferred configuration (a state) is $\phi_\alpha(q)$. They are distributed according to

$$P(\{\phi_\alpha(q)\}) \sim \prod_\alpha e^{-cq^2|\phi_\alpha(q)|^2/2T_g}. \tag{6.34}$$

Each is endowed with a free energy f_α distributed according to an exponential distribution $P(f) \sim \exp(u_c f/T)$ (here $u_c = T/T_g$). Once these seed states are constructed, the full thermal distribution of the q mode ϕ_q is obtained by letting it fluctuate thermally around one of the states:

$$P(\phi_q) \sim \sum_\alpha W_\alpha e^{-c(q^2+L_{\text{loc}}^{-2})|\phi_q-\phi_\alpha(q)|^2/2T}, \tag{6.35}$$

where each state is weighted with probability $W_\alpha = e^{-f_\alpha/T}/\sum_\beta e^{-f_\beta/T}$. One thus recovers qualitatively the picture of Larkin Ovchinikov as the solution of the problem with the replica variational method. The Larkin length naturally appears as setting the (internal) size of the elastically

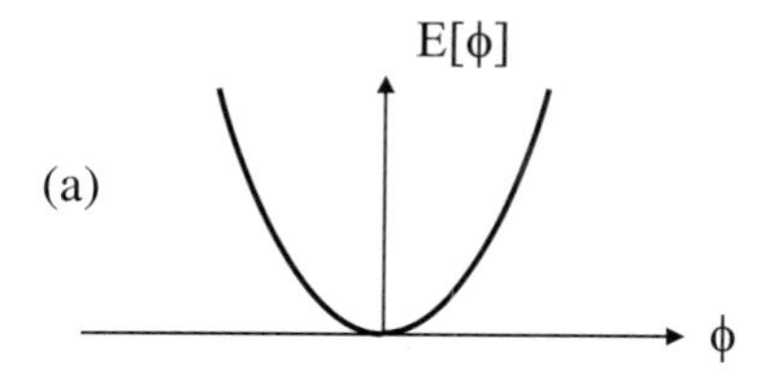

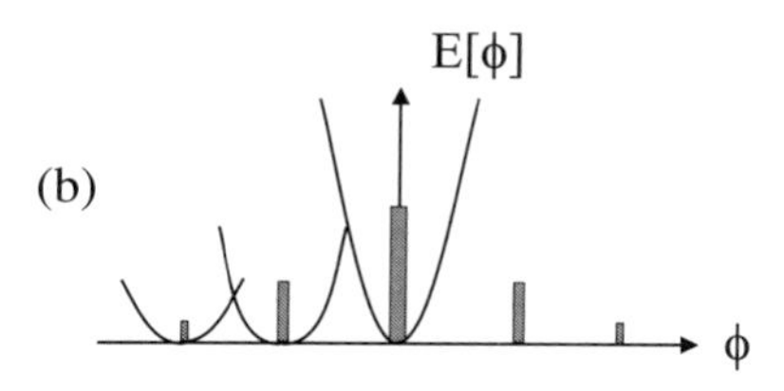

FIGURE 5.6. (a) The RS solution describes fluctuations around a single state, with a single spring constant. (b) The RSB solution takes correctly into account the fact that different regions of space correspond to different equilibrium values of ϕ, with a priori different spring constants. The weight of a given configuration has been represented by the height of the shaded region.

correlated domains. The full RSB case corresponds to more levels in this hierarchy of Larkin domains (in some sense there are clusters of domains of size larger than L_{loc}) and the way this hierarchy scales with distance reproduces the exponents for displacements and energy fluctuations. We refer the reader to [57] for more details on the use of the variational method in this context and for the physical properties of such systems.

6.3 Quantum Problems

We now apply the same method to the quantum problem. Because of the nonlocality in time of the interaction in (6.3) the solution will have quite different properties. We again use the trial action (6.4), with the parametrization

$$vG_{ab}^{-1}(q, \omega_n) = \frac{((vq)^2 + \omega_n^2)}{\pi K}\delta_{ab} - \sigma_{ab}(\omega_n). \tag{6.36}$$

Although, as in Section 6.1, σ_{ab} is still independent of q because of the locality in space, it is now dependent on ω_n. This would render the solution extremely complicated if it were not for a remarkable property of quantum systems. Off-diagonal replica terms such as $\sigma_{a\neq b}$ *only* exist for the mode $\omega_n = 0$. The general argument [64, 65] is that in each realization of the random potential V, the disorder does not depend on τ. Therefore before averaging over disorder,

$$G_{ab,V} = \langle \phi_a(x,\tau)\phi_b(0,0)\rangle = \langle \phi_a(x,\tau)\rangle\langle \phi_b(0,0)\rangle \\ = \langle \phi_a(x,0)\rangle\langle \phi_b(0,0)\rangle. \tag{6.37}$$

It is important to note that such a property crucially depends on the assumption that the Hamiltonian is τ-independent and on the fact that equilibrium has being attained. This is the case considered here.

The static mode $\omega_n = 0$ thus plays a special role. This is quite natural in a time-independent disorder. Quite naively, one sees already that the properties of the variational solution will thus be very similar to the ones of *point like* (i.e., totally uncorrelated) disorder in d spatial dimensions (here $d = 1$). This suggests strongly that the variational method will pull out a static solution, reminiscent of the one introduced in Section 3.1 and treat the fluctuations around this static solution. We will come back to this point later. The solution can be obtained quite readily in all dimensions and we refer the reader to [65] for details. We focus here on the case $d = 1$. In this case, two types of solution exist: (i) A simple RS solution with $\sigma_{ab} = 0$. This solution is stable for $K > \frac{3}{2}$. It corresponds of course to the delocalized regime where the cosine term in (6.3) is irrelevant. The variational method correctly reproduces the (Gaussian) delocalized regime, and the correct transition point, but of course misses the renormalization of the Luttinger parameters given by the RG. (ii) For $K < \frac{3}{2}$, although an RS solution still exists, it is unstable and physically obviously incorrect [65]. One should look for an RSB solution. In that case, the correct solution is a one-step RSB solution (it would be full RSB for $d > 2$) of the type shown in Fig. 5.5.

$$G_c^{-1}(q,\omega_n) = \frac{\hbar}{\pi K}(vq^2 + \frac{\omega_n^2}{v}) + \frac{2\mathcal{D}}{\hbar(\pi\alpha)^2} \\ \times \int_0^{\beta\hbar} d\tau\,(1-\cos(\omega_n\tau))\Big[\exp\big(-2\hbar\tilde{B}(x=0,\tau)\big) \\ - \int_0^1 du\,\exp\big(-2\hbar B(u)\big)\Big] \tag{6.38}$$

with

$$\sigma(q,\omega_n,u) = \frac{2\mathcal{D}v}{(\pi\alpha)^2}\beta\exp(-\hbar 2B(u))\delta_{\omega_n,0}. \tag{6.39}$$

These equations are still formidable to solve. A simple parametrization is

$$vG_c^{-1}(q,\omega_n) = \frac{1}{\pi \overline{\overline{K}}}((vq)^2 + \omega_n^2) + \Sigma_1(1-\delta_{n,0}) + I(\omega_n) \tag{6.40}$$

$$I(\omega_n) = \frac{2\mathcal{D}v}{(\pi\alpha)^2\hbar}\int_0^{\beta\hbar}[e^{-2\hbar\tilde{B}(\tau)} - e^{-2\hbar B(u>u_c)}](1-\cos(\omega_n\tau))d\tau \tag{6.41}$$

$$\Sigma_1 = u_c(\sigma(u>u_c) - \sigma(u<u_c)) = [\sigma](u>u_c) \tag{6.42}$$

$$\sigma(u) = \frac{2\mathcal{D}v}{(\pi\alpha)^2} e^{-\hbar 2B(u)} \beta\delta_{n,0}. \tag{6.43}$$

The parameters Σ_1, the breakpoint u_c and the function $I(\omega_n)$ have to be determined self-consistently. Let us examine first the general properties of the solution.

Since $I(\omega_n = 0) = 0$, it is easy to check from (3.4) that the compressibility is unchanged by the disorder, since the "mass" term Σ_1 goes also away *at* $\omega_n = 0$. The variational method thus correctly reproduces that the compressibility of an Anderson insulator is still finite, and practically unchanged (for free electrons where we can compute it) from the value without disorder. Correlation functions are also easy to obtain. We just give here the important physical point without the explicit derivation [65]. Because of the presence of Σ_1 in the propagator (6.40) they will be massive. This leads to

$$B(x \to \infty, \tau = 0) \to x/L_{\text{loc}}. \tag{6.44}$$

A full calculation of the correlations shows that $\tilde{B}(\tau)$ grows until $\tau \sim L_{\text{loc}}$, when it saturates:

$$B(x = 0, \tau \to \infty) \to \text{Cst} = (\rho_0^{-1} l_\perp)^2. \tag{6.45}$$

Since the variational action is Gaussian, one has for the correlation of the $2k_{\text{F}}$ part of the density

$$\chi_\rho(x, \tau) = e^{-2B(x,\tau)}. \tag{6.46}$$

Two different physical effects are described by the above correlation functions. In the absence of disorder, $\chi_\rho(\tau \to \infty)$ would go to zero as a power law, a sign of the wandering of the particles. Equation (6.45) expresses the fact that the time-independent potential localizes the particles at a given point in space, instead of letting them fluctuate (unboundedly) due to quantum fluctuations in the absence of disorder. This leads to density correlations in time going to a constant up to a Debye Waller like factor. The behavior (6.45) thus shows that a *static* solution $\phi_0(x)$ exists. The variational approach thus provides a good justification for the static solution ϕ_0 on which the method of Section 3 is built. The correlation length of this static solution is given by (6.44). Since this length controls through (6.46) the exponential decay of the spatial correlations of density, it is thus related to the standard localization length. Note that here both L_{loc} and *the full static solution* are determined by the variational method. Simple dimensional analysis on (6.40) shows that

$$L_{\text{loc}} \propto 1/\sqrt{\Sigma_1}. \tag{6.47}$$

The solution of the variational equations [65] leads back to the expression (3.16) for L_{loc}. Quite interestingly, (6.45) defines a length if one writes the

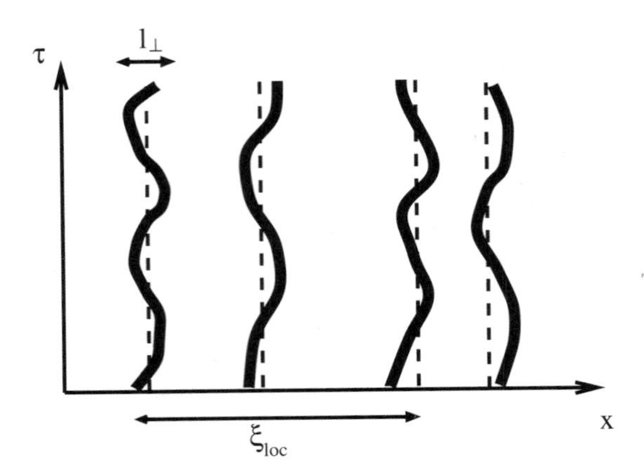

FIGURE 5.7. Localization.

constant in units of the fermion spacing. This length is the "width" of the fluctuations of the particles, as shown in Fig. 5.7.

Note the difference between the two lengths $l_\perp$ and L_{loc}. They are related through [65]:

$$l_\perp^2 = \frac{\alpha^2}{\pi^2}(\hbar\overline{K})\ln(L_{\text{loc}}/a). \tag{6.48}$$

At the transition, one expects that L_{loc} diverges as $L_{\text{loc}} \sim \exp(b/(K_c - K)^\alpha)$ with $\alpha = \frac{1}{2}$ from the RG (see (3.23)). Thus relation (6.48) predicts that

$$l_\perp \sim \frac{1}{(K_c - K)^{\alpha/2}} \tag{6.49}$$

diverges as a power law, which could be measured in numerical simulations.

Let us now look at the conductivity, for which we need $I(\omega_n)$. If we call $I'(\omega)$ and $I''(\omega)$ the real and imaginary parts of the analytic continuation of $I(\omega_n)$, then the conductivity is given by

$$\sigma(\omega) = \frac{\omega I'' + \omega(-\omega^2 + I' + \Sigma_1)}{(-\omega^2 + \Sigma_1 + I')^2 + (I'')^2}. \tag{6.50}$$

It is easy to see that $\Re\sigma(\omega)$ goes to zero at zero frequency and thus the phase is indeed localized. But *because* of the analytic continuation, the existence or not of a gap in the optical conductivity is *not* linked to the existence of a "mass" Σ_1 but to whether I'' is nonzero at small frequencies. The equation for $I(\omega_n)$ takes a particularly simple form in the limit $K, \hbar \to 0$ while keeping $\overline{K}$ fixed. In this limit, we can write

$$I(\omega_n) = \Sigma_1 f\left(\frac{\omega_n}{\sqrt{\pi\overline{K}\Sigma_1}}\right), \tag{6.51}$$

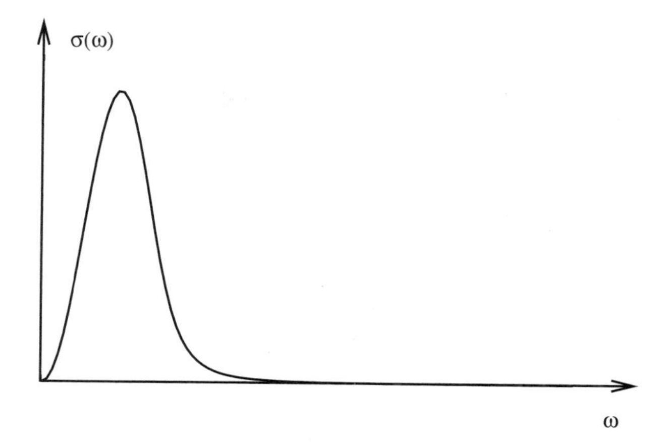

FIGURE 5.8. The conductivity of the one-dimensional Anderson insulator in the limit $K \to 0, \hbar \to 0, \overline{K} = K/\hbar$ fixed.

where the scaling function f satisfies

$$f(x) = 2\left[1 - \frac{1}{\sqrt{1 + x^2 + f(x)}}\right]. \tag{6.52}$$

It can be shown that for $\omega \to 0$

$$\Re\sigma(\omega) \sim \omega^2. \tag{6.53}$$

In a similar way (6.50) gives

$$\Im\sigma(\omega) \sim \omega/\Sigma_1. \tag{6.54}$$

Such behavior is in agreement with exact results in one dimension [2, 3] up to logarithmic prefactors. At high frequency, one can show that $\sigma(\omega) \sim \omega^{2K-4}$, in agreement with the RG result. The resulting conductivity is plotted in Fig. 5.8.

6.4 The Fine Prints

The variational method is thus an extremely efficient method for this type of disordered problem. It allows us to get most of the physical properties in the localized phase, and give the qualitative features of the transition and the delocalized phase. Its physics is very similar that described for point-like disorder in Section 6.1. The variational method determines the "static" solution for the mode $\omega_n = 0$ (but by taking into account the effects of all modes) without being restricted to a single Gaussian. It can then correctly compute the fluctuations (both in q and ω_n) around this solution, contrarily to the approximate method of Section 3.1, which was impeded by the lack of knowledge of the static solution.

In $d = 1$, an additional remarkable property can be seen. Although most of the properties are independent of the value of u_c, the conductivity is strongly dependent on it. As for the static solution, one has continuous replica symmetry breaking for $d > 2$, which goes continuously to the one-step solution in $d = 2$. In these dimensions, taking the value of u_c out of the variational equations gives the correct conductivity (no gap in $\sigma(\omega)$). In $d = 1$, the u_c obtained by minimizing the free energy would give an *incorrect* (gapped) conductivity. Another way to determine u_c is to use the marginality condition [65] that corresponds to the instability of the replicon mode, similar to the condition (6.19). This condition, more dynamical in nature, coincides with the free-energy value of u_c for $d \geq 2$ but is different in $d = 1$. One can check that the marginality condition always gives the correct conductivity. Some arguments for why it is so were given in [65], but this point is not yet fully understood. This phenomenon has since been found to occur in other systems [66].

The success of the variational method is obviously linked to the fact that we are looking at small "quadratic" fluctuations around a certain (in our case highly disordered) solution. In this case the replacement

$$-\cos(\phi) \to \tfrac{1}{2}\phi^2 \qquad (6.55)$$

is very reasonable. Such an approximation is very good to compute correlation functions of the variable ϕ. This is the case for the density-density correlation and the optical conductivity. What is missing in the approximation (6.55) are the solitons that go from one minimum of the potential to another minimum. The energy cost of such excitations is grossly overestimated by the variational method, as is shown in Fig. 5.9.

This has important consequences for the calculation of correlation functions that involve the soliton creation operator $e^{i\theta}$. These correlation functions are found incorrectly to decay exponentially with distance at equal time and to be zero at unequal time. As a result, the variational method does not allow the calculation of Fermion Green's function (they involve the operators $e^{i(\theta \pm \phi)}$) nor superconducting correlations (which involve the operator $e^{2i\theta}$). In the case of a disordered XXX spin chain, the situation is even worse, since $S_x \propto \cos(\theta)$, whereas $S_z \sim -\partial_x \phi/\pi + (-)^{x/a} \cos 2\phi$. Therefore, even in the presence of a disorder that preserves SU(2) symmetry (such as a random bond disorder), the variational method would lead to a spurious breaking of rotational symmetry. It would also give poor results for systems that include both the θ and the ϕ field in the Hamiltonian. Examples of such theories include disordered Hubbard ladders [43] , disordered spin ladders [46], or XXX spin chain in a random fields [22].

The mishandling of soliton excitations also limits our knowledge of the transport properties at finite temperatures. Indeed, the optical conductivity does not correspond to transport of charge but charge oscillations around the equilibrium positions, it is thus well described by our harmonic approximation, as shown in Fig. 5.9. On the contrary, transport at finite T involves

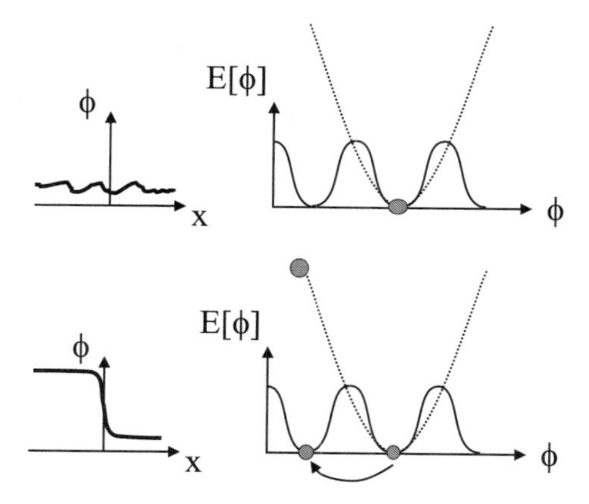

FIGURE 5.9. The Gaussian approximation correctly treats the excitations where the phase stays in one of the minima of the cosine. They contribute to the AC conductivity at $T = 0$. Soliton excitations bring the phase from one minimum to another. With the $\cos(\phi)$ term these excitations only cost energy at the kink. When the potential is replaced by the quadratic approximation, the energy cost of the "wrong" minimum is very high, hence a poor result for the physical results. These excitations dominate the DC conductivity for $T > 0$.

real charge displacements. Since $\rho(x) \sim \nabla\phi$, displacing a charge amounts to making a solitonic excitation in the field ϕ, as shown in Fig. 5.10. Indeed, to compute the conductivity of noninteracting electrons in the presence of phonons $\sigma(T) \sim \exp[-(T_0/T)^{1/(d+1)}]$, according to Mott's arguments, is strongly reminiscent of an instanton calculation, as shown in Fig. 5.10.

A similar argument applies to Efros and Schklovskii's calculations [67]. An extension of the variational approach to finite temperatures would indeed still lead to $\sigma(T, \omega = 0) = 0$, which proves that it misses the excitations that are important at finite temperature. Unfortunately, despite some attempts [68], no way to treat such solitons has been found at present.

Despite these limitations, the variational method is up to now the only analytical method that gives information for such localized systems in the localized phase. As with all variational approaches, some physical insight in the properties of the system under consideration is needed to determine whether the method has any chance of success. Clearly, one must apply this method only to systems that can be reasonably well understood qualitatively from their classical action. Fortunately, many systems fall in this category, and we examine some of those in the following.

6.5 Higher Dimension: Electronic Crystals and Classical Systems

First, the GVM can be used to study classical systems using the standard mapping $\tau \to z$. The action (6.3) and its extension to higher spatial di-

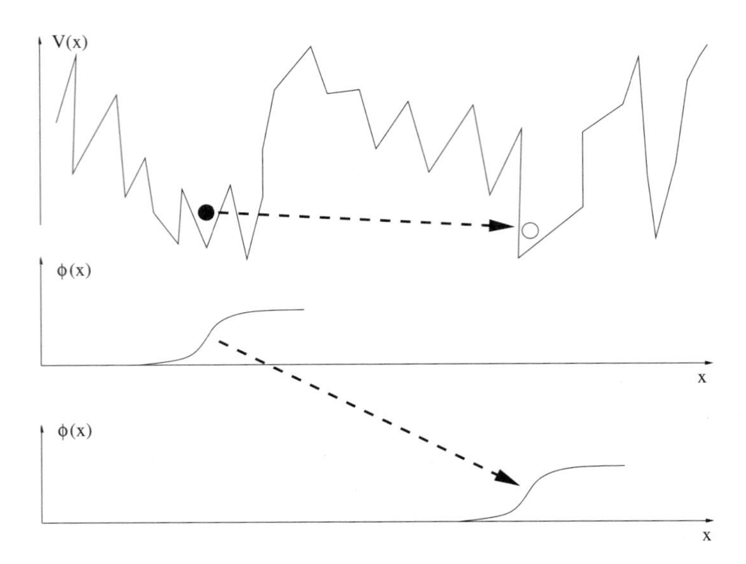

FIGURE 5.10. Hopping of charges that leads to a finite d.c. conductivity. This corresponds to soliton-type excitations in the phase variable ϕ.

mensions describe elastic objects (lines in this case, as shown in Fig. 5.11) pinned by columnar (i.e., time- or z-independent) defects. This situation is realized, for example, in vortex in type II superconductors irradiated by heavy ions (creating the linear track of disorder). This system in $2+1$ dimension is equivalent to a $d=2$ quantum Bose system in presence of pins.

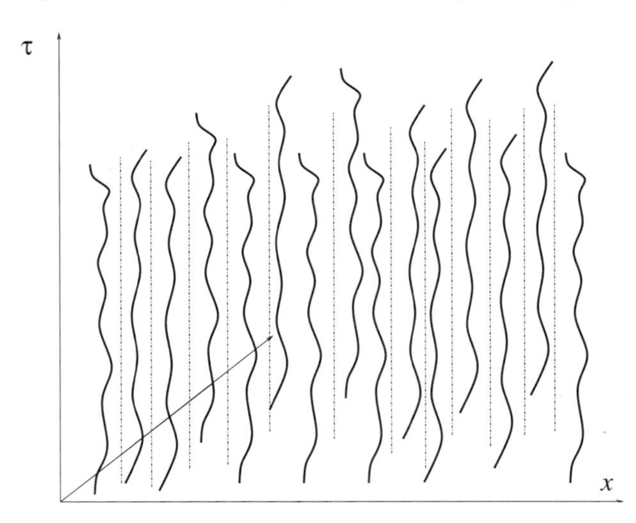

FIGURE 5.11. The action (6.3) in $d=2$ describes elastic lines in the presence of columnar pins.

In a similar way to $d = 1$ (see Section 6.3), such a system has a pinned phase (the Bose glass) [21, 69]. The variational method can be used to describe the Bose glass phase [65]. However, contrary to $d = 1$, it cannot be used to go to the superfluid regime since to describe a two-dimensional "melting" of the Bose glass phase, dislocations are important (no dislocations exist in $d = 1$) and for reasons explained above, the GVM overestimates the energy cost of topological excitations. Another way to say it is that in $d > 1$ we loose the elastic description (4.2) of the fermion or boson operators. The GVM can thus only be used in phases where the particles are localized so that some elastic description can again be used.

We can thus use the variational method in higher dimensions to study electronic crystals. This includes charge density wave, but also the two-dimensional Wigner crystal of electrons. In such a phase, the electrons are confined by their repulsion (and in some systems a magnetic field). An elastic description can be used. Some level of quantumness is hidden in the elastic parameters ("size" of the particles, quantization of the phonon modes of the crystal). For such systems, the calculation of the optical conductivity is particularly useful since it is one of the few probes of such systems. Since the physics of such systems would deserve a review of its own, we will not dwell further on it here but refer the reader to [70, 71] for details.

7 Commensurate Systems

When the filling of the fermion system is commensurate, the physics discussed above is modified in various ways since the backward scattering on disorder becomes real. If the forward scattering still exists, not much is changed. Two special cases will thus occur: (i) the forward scattering is absent. This occurs because of a symmetry of the system. This is the case, for example, for fermions at half filling with a random hopping or for spin chains with random exchange. (ii) The commensurate potential (either due to the lattice or due to electron-electron interaction) would open a gap. There will be a competition between Mott physics wanting to get a commensurate (gapped) insulator and the disorder that would like to destroy such a gap (push the system locally away from commensurability).

7.1 The Peculiar Random Exchange

For electrons at half filling with a random exchange, the forward scattering does not exist and the disorder term is simply

$$
\begin{aligned}
H &= \int dx\, V(x) i[\psi_+^\dagger \psi_- - \text{H. c.}] \\
&= -\int dx\, \frac{V(x)}{(2\pi\alpha)} \sin\big(2\phi(x)\big). \qquad (7.1)
\end{aligned}
$$

Although this seems very similar to (3.1) one easily sees the difference in Fig. 5.12. Contrary to normal disorder where ϕ follows the random *phase* of the random potential, here $\phi = \pm\pi/4$, depending on the sign of the potential. Thus ϕ is nearly gapped but for its kinks. The low-energy properties will thus be dominated by the kinks in ϕ. Such a kink structure between the doubly degenerate minima makes it unlikely that the GVM can be used for this problem.

In order to get an idea of the physics, let use examine the case of free fermions $K = 1$. For $E = 0$, one can easily construct the eigenstates by solving

$$\partial\psi_+ - V(x)\psi_-(x) = 0 \qquad (7.2)$$

$$-\partial\psi_- + iV(x)\psi_+(x) = 0 \qquad (7.3)$$

to obtain the (unnormalized)

$$\psi_+(x) \sim \psi_-(x) \sim e^{-\int_0^x dy\, V(y)}, \qquad (7.4)$$

which obviously decays as

$$\psi_\pm(x) \sim e^{-\sqrt{x}}. \qquad (7.5)$$

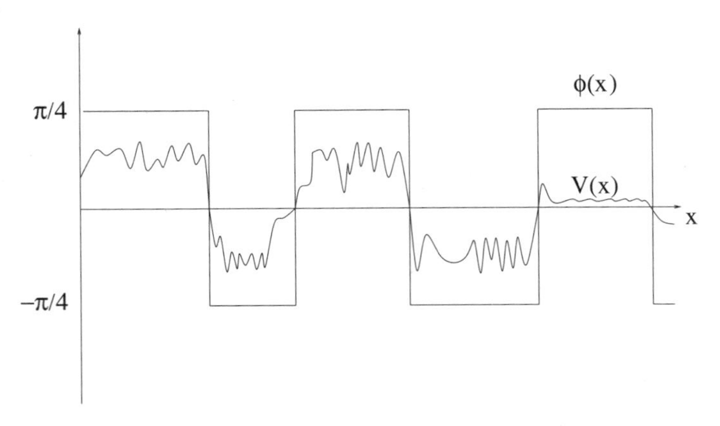

FIGURE 5.12. The profile of $\phi(x)$ for a given realization of disorder $V(x)$. As one can see, $\phi(x)$ is only sensitive to the *sign* of $V(x)$ and not to its amplitude. As a result, ϕ has kinks each time $V(x)$ goes to zero.

If one wants to view such a state as an exponentially localized state, this means that the localization length diverges when $E \to 0$. The divergence of the localization length is of course the signature that the physical properties will be quite different than the ones for the "normal" disorder studied in the previous sections. For the noninteracting case, they have been studied using a variety of techniques such as replicas, Berezinskii methods, supersymmetry, etc. [72–75]. The density of states diverge at the Fermi level as [76]

$$\rho(\epsilon) \sim \frac{1}{\epsilon \log(1/\epsilon)^3}. \tag{7.6}$$

Quite astonishingly, if one considers that all states are still localized, the DC conductivity is now *finite* [72,77]. This is surprising since it seems to violate the Mott phenomenological derivation. The conductivity is proportional to the absorbed power. To make a transition between two localized states one needs one occupied state and an empty one separated by an energy $\hbar\omega$. The number of such states is $\rho(\epsilon = 0)\hbar\omega$. The absorbed power (and the conductivity) is thus (up to log correction)

$$\sigma(\omega) \propto \rho(\epsilon = 0)(\hbar\omega)^2, \tag{7.7}$$

correctly giving back for a constant density of states at the Fermi level the ω^2 dependence of Section 6.3. One would thus naively have expected a $\sigma(\omega) \sim \omega$ in the case of a singular density of states (7.6). A qualitative understanding of the behavior of the DC conductivity is thus still lacking.

When interactions are included, the problem becomes more difficult to tackle. Going back to the spin chain version of (7.1), a real space renormalization procedure has been introduced. This procedure works beautifully and allows the calculation of most correlation functions. We refer the reader to [78, 79] for details. As for the normal disorder (see Section 3.2), interaction wads found to be irrelevant at the disordered fixed point. The divergence of the localization length also drastically changes the correlation functions. In particular the spin-spin correlation functions now decay as power law instead of being exponential. It would be extremely interesting to have an equivalent derivation of this real space RG directly in the boson representation.

7.2 Mott Versus Anderson

A particularly interesting situation occurs when the nondisordered system possesses a gap. In that case the competition between this gap and the disorder is nontrivial. This arises in a large number of systems, such as disordered Mott insulators [80–83], systems with external (Peierls or spin-Peierls systems [84]) or internal commensurate potential (ladders or spin ladders [85–88], disordered spin 1 chains [42, 89]). It is physically simple to see that in order to destroy the gap one needs a disorder comparable to the gap (although in some specific cases, Imry–Ma effects can destroy the gap for infinitesimal disorder [80]). This makes the complete description of the gap closure and of the physics of the resulting phases extremely difficult with the usual analytic techniques such as the perturbative renormalization group described in section 3.2, due to the absence of a weak coupling fixed point. Indeed both the commensurate Mott insulator (which can be described by a sine-Gordon Hamiltonian [90]) and the disordered Anderson insulator correspond to strong coupling fixed points.

To be concrete, but keep the technical details to a minimum, let us consider again spinless fermions. The commensurability can be described by adding to the Hamiltonian (2.9) a periodic potential at the Fermi level

$$H_{\rm com} = g \int dx (\psi_+^\dagger \psi_- + \psi_-^\dagger \psi_+), \tag{7.8}$$

which becomes in the boson representation

$$H_{\rm com} = \int dx \frac{g}{\pi \alpha \hbar} \cos 2\phi. \tag{7.9}$$

The reason to consider here a periodic potential is specific to pathologies associated with the Mott gap for spinless fermions [80]. Similar results are expected for a Mott insulator (which is due to the $4k_{\rm F}$ potential of the lattice).

If this system is studied by RG then one is faced with the competition of two strong coupling fixed points since

$$\frac{dg}{dl} = (2 - K) g \tag{7.10}$$

$$\frac{d\mathcal{D}}{dl} = (3 - 2K)\mathcal{D}. \tag{7.11}$$

Using the usual qualitative argument consisting in taking the most divergent operator, it was concluded [81] that if disorder reaches a strong coupling ($\mathcal{D}(l^*) = 1 > g(l^*)$) first, we will be in the Anderson phase, with a localization length $l_0 = ae^{l^*} \sim (1/\mathcal{D})^{1/(3-2K)}$, whereas if the commensurate potential reaches strong coupling first, we will be in the Mott phase with a correlation length (or soliton size) $d \sim (1/g)^{1/(2-K)}$. The phase transition between the two phases occurs for $l_0 \sim d$. This picture relies on the important assumption that there is no other stable fixed point than the Mott insulator and the Anderson insulator. Even if it was so, it would not be possible to determine the phase boundaries, nor determine what type of critical point separate the Mott and the Anderson insulator. In order to make progress on these issues and obtain a more complete picture of dirty one-dimensional Mott insulators, one needs to solve the problem nonperturbatively [91]. The methods of Section 6.3 are well adapted since the problem can be cast in the sine-Gordon form.

7.3 Variational Approach

For the Mott versus Anderson problem, the variational action reads

$$\frac{S_{\text{rep.}}}{\hbar} = \sum_a \left[\int \frac{dx\, d\tau}{2\pi K} \left(v(\partial_x \phi_a)^2 + \frac{(\partial_\tau \phi_a)^2}{v} \right) - \frac{g}{\pi\alpha\hbar} \int dx\, d\tau \, \cos 2\phi_a \right]$$
$$- \frac{\mathcal{D}}{(2\pi\alpha\hbar)^2} \sum_{a,b} \int dx \int_0^\beta d\tau\, d\tau' \, \cos\Big(2\big(\phi_a(x,\tau) - \phi_b(x,\tau')\big)\Big) \quad (7.12)$$

The term $\cos 2\phi_a$ in (7.12) is responsible for the opening of a gap. We search for a saddle point with a form of the variational connected Green function slightly generalized with respect to (6.40):

$$vG_c^{-1}(q,\omega_n) = \frac{1}{\pi\overline{K}}(\omega_n^2 + v^2q^2) + m^2 + \Sigma_1(1-\delta_{n,0}) + I(\omega_n), \quad (7.13)$$

where the parameter Σ_1 and the function $I(\omega_n)$ satisfy the equations (6.42) and (6.41). We give here the main steps of the solution and refer the reader to [91,92] for more details. The variational self-energy satisfies the equation (6.43). Finally, m satisfies the equation

$$m^2 = \frac{4gv}{\pi\alpha} e^{-2\hbar\tilde{G}(0,0)}. \quad (7.14)$$

The important physical quantities are simply given such as conductivity and compressibility are simply given by (3.3) and (6.11).

Since we expect the physics to be continuous for small enough K (i.e., repulsive enough interactions), one can gain considerable insight by considering (see Section 3.1) the classical limit $\hbar \to 0$, $K \to 0$ keeping $\overline{K} = K/\hbar$ fixed. In this limit, one can solve analytically the saddle point equations (6.40)–(6.43), (7.13)–(7.14) and compute m, Σ_1 and $I(\omega_n)$. The resulting phase diagram is parameterized with two physical lengths (for $K \to 0$): The correlation length (or soliton size) of the pure gapped phase,

$$d = \left(\frac{4g\overline{K}}{\alpha v} \right)^{-1/2}, \quad (7.15)$$

and the localization (or pinning) in the absence of commensurability length,

$$l_0 = \left(\frac{(\alpha v)^2}{16\mathcal{D}\overline{K}^2} \right)^{1/3}. \quad (7.16)$$

Contrary to the naive direct transition predicted by the extrapolation of the RG, we find *three* phases [91], as shown in Fig. 5.13.

Their main characteristics in term of conductivity and compressibility are summarized in Fig. 5.14.

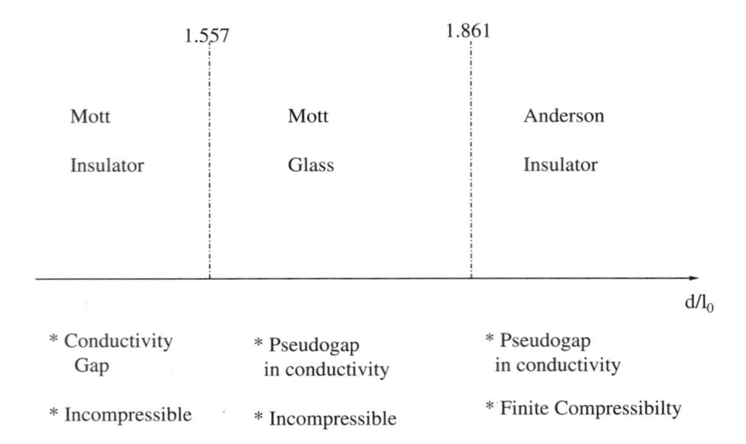

FIGURE 5.13. The phase diagram of a system with commensurability and disorder. d is the soliton size, l_0 the localization (or pinning length)

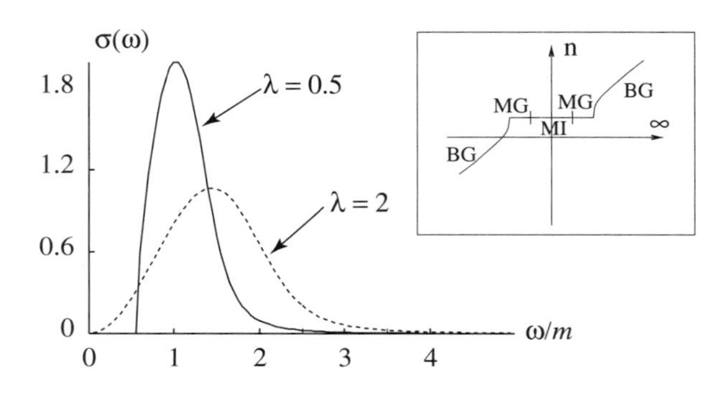

FIGURE 5.14. Conductivity in the Mott Insulator (solid line) and Mott Glass phases (dashed line). Insert: density n versus the chemical potential μ. λ is defined in the text.

Mott Insulator

At weak disorder we find a replica symmetric solution with $\Sigma_1 = 0$ but with $m \neq 0$ (m depends on the disorder). m is given by the simple equation

$$m^2 = \frac{4gv}{\pi\alpha} \exp\left[-\frac{\mathcal{D}\overline{K}^{1/2}}{\alpha^2 \pi^{3/2} v^{1/2} m^3} \right]. \tag{7.17}$$

It is convenient to work with the physical lengths d and l_0 and the correlation length ξ, where

$$\xi^2 = \frac{v^2}{(\pi \overline{K} m^2)}. \tag{7.18}$$

One can then rewrite (7.17) as

$$\frac{1}{\xi^2} = \frac{1}{d^2} \exp\left[-\frac{1}{16} \left(\frac{\xi}{l_0} \right)^3 \right]. \tag{7.19}$$

For $l_0/d > \frac{1}{2}(3e/4)^{1/3}$, (7.19) has a single physical solution. For $l_0/d < \frac{1}{2}(3e/4)^{1/3}$, (7.19) has no solution, which means that the Mott insulator becomes unstable. In fact, due to another constraint (see below), the Mott Insulator becomes unstable at an even smaller disorder.

Having $m \neq 0$ leads to zero compressibility $\kappa = 0$. Disorder reduces the gap created by the commensurate potential and thus increases ξ compared to the pure case. Let us now examine the equation for $I(\omega_n)$ giving the transport properties. An expansion around $\hbar = 0$ in (6.41) gives the self-consistent equation for $I(\omega_n)$:

$$I(\omega_n) = \frac{8\mathcal{D}v}{(\pi\alpha)^2} \int_0^{\beta\hbar} G_c(x=0,\tau)\big(1 - \cos(\omega_n \tau)\big)\, d\tau \tag{7.20}$$

Introducing the scaling form (to be contrasted with (6.51)):

$$I(\omega_n) = m^2 f\left(\frac{\omega_n}{\sqrt{\pi \overline{K}} m} \right), \tag{7.21}$$

(7.20) can be recast in the form:

$$f(x) = \lambda \left[1 - \frac{1}{\sqrt{1 + x^2 + f(x)}} \right], \tag{7.22}$$

where

$$\lambda = \frac{4\mathcal{D}\overline{K}^{1/2} v}{\pi^{3/2} \alpha^2 m^3} = \frac{1}{4} \left(\frac{\xi}{l_0} \right)^3 \tag{7.23}$$

Let us note that for $\lambda = 2$, the equation (7.22) reduces to the equation (6.52). For $\lambda > 2$ (7.22) has no physical solution. Using (7.23), this condition becomes

$$\frac{l_0}{d} < \frac{1}{2} e^{1/4}. \tag{7.24}$$

For $\lambda < 2$, there is a physical solution of (7.22) such that $\lim_{x \to \pm\infty} f(x) = 1 + \lambda$ and for $x \ll 1$, $f(x) = 1 + \alpha x^2 + o(x^2)$ with $\alpha = \lambda/(2 - \lambda)$. The conductivity of the Mott insulator can be obtained from f in the form

$$\sigma(\omega) = \frac{\xi \overline{K}}{\pi} \frac{\imath x}{(1 + f(\imath x) - x^2)}, \tag{7.25}$$

where $x = \omega/\omega^*$ and $\omega^* = v/\xi$ is the characteristic frequency associated with the correlation length ξ. The conductivity $\sigma(\omega)$ is zero if

$$\omega < \omega_c = \omega^* \sqrt{1 + \lambda - 3\left(\frac{\lambda}{2} \right)^{2/3}}. \tag{7.26}$$

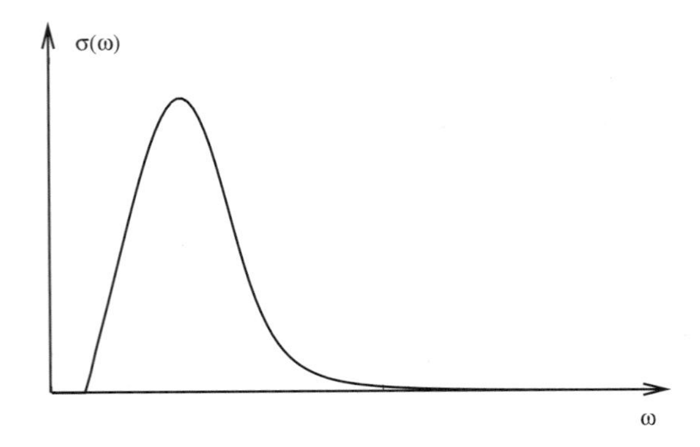

FIGURE 5.15. The frequency-dependent conductivity of the Mott Insulator

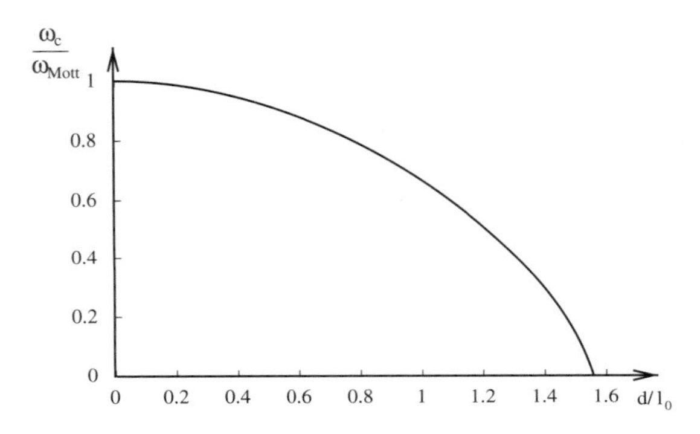

FIGURE 5.16. The dependence of the conductivity gap with disorder strength in the Mott insulator

The replica symmetric phase with $m \neq 0$ has thus a conductivity gap $\hbar\omega_c$ and can be assimilated to a Mott insulator (MI). The behavior of the conductivity is plotted in Fig. 5.15. Close to the threshold, one has $\sigma(\omega) \sim (\omega - \omega_c)^{1/2}$.

However the gap in the conductivity $\hbar\omega_c$ decreases when disorder increases, and closes for $d/l_0 = 2e^{-1/4} \simeq 1.557$ (see Fig. 5.16). For $d/l_0 \to 2e^{-1/4}$, the conductivity gap vanishes linearly with d/l_0.

For $d/l_0 > 2e^{-1/4}$, the MI solution becomes unphysical *even though* the mass m remains finite at this transition point, i.e., the system remains incompressible. For stronger disorder one must break replica symmetry, as for the pure disorder case of Section 6.3. Here, however *two* possibilities arise, depending on whether the saddle point allows for $m \neq 0$ or not. In the presence of a breaking of replica symmetry, one extra equation is needed to determine the breakpoint. As in the case of the Anderson insulator, such an equation is provided by the marginality of the replicon condition discussed in Section 6.4. Two phases exist, as described below.

The Anderson Glass

For large disorder compared to the commensurate potential $d/l_0 > 1.861$, $m = 0$ is the only saddle point solution. The saddle point equations then reduce to (6.40)–(6.43). Thus, we recover the Anderson glass with interactions of Section 6.3. As we have seen in Section 6.3, in such a phase the conductivity starts as $\sigma(\omega) \sim \omega^2$, showing no gap, and the compressibility is finite.

In the Anderson glass phase, the disorder completely washes out the commensurate potential. The MI and the AG were the only two phases accessible by renormalization techniques [81]. Within the replica variational formalism, however, we find that an intermediate phase exists between them.

The Mott Glass

For intermediate disorder $2e^{-1/4} < d/l_0 < 1.861$, a phase with *both* $\Sigma_1 \neq 0$ and $m \neq 0$ is obtained. We shall call this phase the *Mott glass* (MG). We shall not discuss in full detail the one-step solution of the saddle point equation here. We will rather stress the salient features of our solution. First, as a result of the marginality of replicon mode condition, $m^2 + \Sigma_1$ remains constant in the MI and MG as disorder strength is increased [91, 92]. In the MG phase, $I(\omega_n)$ is still of the form (7.21) but m is replaced by $\sqrt{m^2 + \Sigma_1}$. The reduced self-energy $f(x)$ satisfies (7.22) but with $\lambda = 2$ in the whole Mott Glass phase. This implies that (see (6.52)) the AC conductivity of the Mott glass is identical to the one of an Anderson glass. However, since $m \neq 0$ in the MG, the system is *incompressible* ($\kappa = 0$) like a Mott insulator. Thus, the Mott Glass is a new glassy phase (since it has Replica Symmetry Breaking) with characteristics intermediate between those of an Anderson insulator and those of a Mott insulator.

7.4 Physical Discussion

The existence of a phase with a compressibility gap but no conductivity gap is quite remarkable since, by analogy with noninteracting electrons [82], one is tempted to associate a zero compressibility to the absence of available states at the Fermi level and hence to a gap in the conductivity as well. Our solution shows this is not the case, when interactions are turned on the excitations that consists in adding one particle (the important ones for the compressibility) become quite different from the particle-hole excitations that dominate the conductivity. A similar situation is obtained in the case of the one-dimensional Wigner crystal [93], which has the conductivity of a perfect 1d metal, $\sigma(\omega) \propto \delta(\omega)$ but a zero compressibility since $\chi = \lim_{q \to 0} 1/\ln q$. This argument suggests that the difference in one-particle and two-particle properties is a consequence of the strong repulsion in the system.

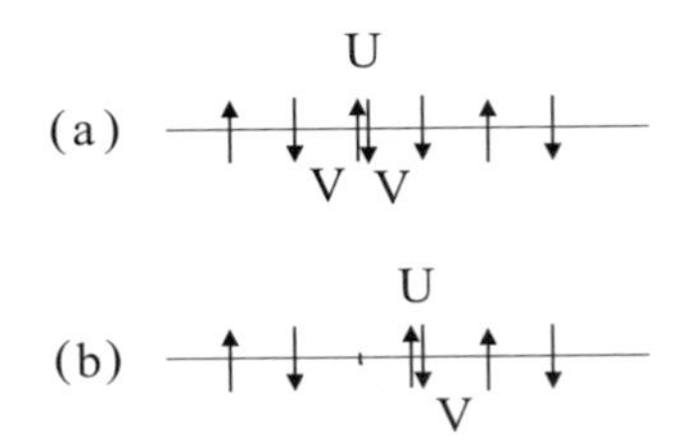

FIGURE 5.17. (a) Energy needed to add one particle. (b) Energy needed to make the particle hole excitations entering in the optical conductivity. Because of excitonic processes, when disorder is added the gap in the optical conductivity can close before the single particle gap. This phenomenon, leading to the MG phase, should occur regardless of the dimension.

In addition to the variational method itself, the Mott glass phase can also be obtained by two other independent methods. Higher-dimensional extensions of the present problem, similar to the one made in Section 6.1, can be treated around four spatial dimensions using a $d = 4 - \epsilon$ functional renormalization group method (totally different from the $d = 2$ RG). Such study confirms [91, 92] the existence of the intermediate Mott glass phase. One can also analyze (7.12) for zero kinetic energy and obtain the MG phase [92]. Although we have done the derivation of the Mott phase for fermions in one dimension we expect its physics to survive into higher dimensions. This can be seen by looking at the atomic limit (zero hopping) of an interacting fermionic system (in any dimension). If the repulsion extends over at least one interparticle distance, leading to small values of K, particle-hole excitations are lowered in energy by excitonic effects. For example, for fermions with spins with both an on site U and a nearest-neighbor V, the gap to add one particle is $\Delta = U/2$. On the other hand, the minimal particle-hole excitations would be to have the particle and hole on neighboring sites (excitons) and cost $\Delta_{\text{p.h.}} = U - V$, as shown in Fig. 5.17.

When disorder is added, the gaps decrease respectively[3] as $\Delta \to \Delta - W$ and $\Delta_{\text{p.h.}} \to \Delta_{\text{p.h.}} - 2W$. Thus the conductivity gap closes, the compressibility remaining zero (for bounded disorder). According to this physical picture of the MG, the low frequency behavior of conductivity is dominated by excitons (involving neighboring sites). This is at variance from the AG, where the particle and the hole are created on distant sites. This may have consequences on the precise low-frequency form of the conductivity such as logarithmic corrections. When hopping is restored, we expect the excitons to dissociate and the MG to disappear above a critical value $K > K^*$. Since finite range is needed for the interactions, in all cases (fermions or bosons) $K^* < 1$. In addition, we expect $K^* < \frac{1}{2}$ for fermions with spins.

[3]For small bounded disorder the gap of the Mott insulator phase is robust.

One interesting question is the one of DC transport in the three phases, and whether the Mott glass has a DC transport closer to the Anderson or the Mott phase. Since the excitons are neutral, one simple guess would be that such excitations would not contribute to the DC transport. The DC conductivity in the Mott glass phase would thus be still exponentially activated just as in the Mott insulator. Of course, more detailed studies would be needed to confirm this point.

8 Conclusions

Many disordered fermionic systems can thus be successfully described by an elastic disordered theory. In one dimension, this situation is ubiquitous due to the importance of collective excitations. Most physical systems, whether one starts with fermions, bosons, or spins, can be represented in terms of bosonic excitations. In higher dimensions such a description is valid in crystalline phases such as a Wigner crystal in which the quantum particles are strongly localized due to their interactions or charge density waves. Disorder then leads to rich physical phenomena coming from the competition between the elasticity, wanting a well-ordered structure, and the disorder that distorts the structure to gain pinning energy. This leads to the existence of many metastable configurations and glassy properties. Using the various methods described in these notes, we now have a good description of the low-energy excitations of such structures. This gives access to a host of physical properties such as the a.c. transport properties.

Clearly, one of the most important open questions is the issue of topological defects in such structures. Indeed, such defects are needed to describe the melting of these crystalline phases and will be necessary to go to more "liquid" phases in which the statistics (fermionic or bosonic) of the particles will play a much more crucial role. In addition, DC transport is obviously dominated by such excitations. Unfortunately, so far the methods able to tackle the properties in the localized phase such as the Gaussian Variational Method cannot handle such solitonic excitations, so radically new methods must be designed to handle them.

Acknowledgments: The work presented in these notes results from many fruitful and enjoyable collaborations. We would like to thank R. Chitra, H. Maurey, B.S. Shastry, and specially P. Le Doussal. Most importantly, nothing would have started without an initial collaboration with H.J. Schulz, to whose memory we would like to dedicate these lecture notes.

9 References

[1] N.F. Mott and W.D. Twose, Adv. Phys. **10**, 107 (1961).

[2] V.L. Berezinskii, Sov. Phys. JETP **38**, 620 (1974).

[3] A.A. Abrikosov and J.A. Rhyzkin, Adv. Phys. **27**, 147 (1978).

[4] E. Abrahams, P.W. Anderson, D.C. Licciardello, and T.V. Ramakrishnan, Phys. Rev. Lett. **42**, 673 (1979).

[5] F. Wegner, Z. Phys. B **35**, 207 (1979).

[6] K.B. Efetov, A.I. Larkin, and D.E. Khmel'nitskii, Sov. Phys. JETP **52**, 568 (1980).

[7] K.B. Efetov, Adv. Phys. **32**, 53 (1983).

[8] P. Nozières, *Theory of Interacting Fermi Systems* (W.A. Benjamin, New York, 1961).

[9] B.L. Altshuler and A.G. Aronov, in *Electron-Electron Interactions in Disordered Systems*, eds. A.L. Efros and M. Pollak (North-Holland, Amsterdam, 1985).

[10] A.M. Finkelstein, Z. Phys. B **56**, 189 (1984).

[11] P.A. Lee and T.V. Ramakhrishnan, Rev. Mod. Phys. **57**, 287 (1985).

[12] For a review see: D. Belitz and T. R. Kirkpatrick, Rev. Mod. Phys. **66** 261 (1994).

[13] J. Sólyom, Adv. Phys. **28**, 209 (1979).

[14] V.J. Emery, in *Highly Conducting One-Dimensional Solids*, eds. J.T. Devreese and al. (Plenum, New York, 1979), p. 327.

[15] H.J. Schulz, in *Mesoscopic Quantum Physics*, Les Houches LXI, eds. E. Akkermans, G. Montambaux, J.L. Pichard, and J. Zinn-Justin (Elsevier, Amsterdam, 1995).

[16] J. Voit, Rep. Prog. Phys. **58**, 977 (1995).

[17] D. Sénéchal, this volume.

[18] P. Jordan and E. Wigner, Z. Phys. **47**, 631 (1928).

[19] T. Giamarchi and H.J. Schulz, Europhys. Lett. **3**, 1287 (1987).

[20] T. Giamarchi and H.J. Schulz, Phys. Rev. B **37**, 325 (1988).

[21] M.P.A. Fisher, P.B. Weichman, G. Grinstein, and D.S. Fisher, Phys. Rev. B **40**, 546 (1989).

[22] C. Doty and D.S. Fisher, Phys. Rev. B **45**, 2167 (1992).

[23] G. Grüner, Rev. Mod. Phys. **60**, 1129 (1988).

[24] H. Fukuyama, J. Phys. Soc. Jpn. **41**, 513 (1976).

[25] H. Fukuyama and P.A. Lee, Phys. Rev. B **17**, 535 (1978).

[26] M.V. Feigelmann and V.M. Vinokur, Phys. Lett. A **87**, 53 (1981).

[27] Y. Suzumura and H. Fukuyama, J. Phys. Soc. Jpn. **52**, 2870 (1983).

[28] L.P. Gorkov and I.E. Dzyaloshinski, JETP Lett. **18**, 401 (1973).

[29] D.C. Mattis, J. Math. Phys. **15**, 609 (1974).

[30] A. Luther and I. Peschel, Phys. Rev. Lett. **32**, 992 (1974).

[31] W. Apel, J. Phys. C **15**, 1973 (1982).

[32] W. Apel and T.M. Rice, J. Phys. C **16**, L271 (1982).

[33] W. Apel and T.M. Rice, Phys. Rev. B **26**, 7063 (1982).

[34] J.M. Kosterlitz and D.J. Thouless, J. Phys. C **6**, 1181 (1973).

[35] J.M. Kosterlitz, J. Phys. C **7**, 1046 (1974).

[36] C. Kane and M.P.A. Fisher, Phys. Rev. Lett. **68**, 1220 (1992).

[37] C. Kane and M.P.A. Fisher, Phys. Rev. B **46**, 15233 (1992).

[38] T. Giamarchi and H. Maurey, in *Correlated Fermions and Transport in Mesoscopic Systems*, eds. T. Martin, G. Montambaux, and J. Tran Thanh Van (Editions Frontières, Gif-sur-Yvette, France, 1996), cond-mat/9608006.

[39] F.D.M. Haldane, Phys. Rev. Lett. **47**, 1840 (1981).

[40] T. Giamarchi and B.S. Shastry, Phys. Rev. B **51**, 10915 (1995).

[41] H. Maurey and T. Giamarchi, Phys. Rev. B **51**, 10833 (1995).

[42] N. Kawakami and S. Fujimoto, Phys. Rev. B **52**, 6189 (1995).

[43] E. Orignac and T. Giamarchi, Phys. Rev. B **56**, 7167 (1997).

[44] E. Orignac and T. Giamarchi, Phys. Rev. B **57**, 11713 (1998).

[45] V. Brunel and T. Jolicoeur, Phys. Rev. B **58**, 8481 (1998).

[46] E. Orignac and T. Giamarchi, Phys. Rev. B **57**, 5812 (1998).

[47] K.J. Runge and G.T. Zimanyi, Phys. Rev. B **49**, 15212 (1994).

[48] G. Bouzerar, D. Poilblanc, and G. Montambaux, Phys. Rev. B **49**, 8258 (1994).

[49] G. Bouzerar and D. Poilblanc, Phys. Rev. B **52**, 10772 (1995).

[50] R.A. Römer and A. Punnoose, Phys. Rev. B **52**, 14809 (1995).

[51] R. Berkovits and Y. Avishai, Europhys. Lett. **29**, 475 (1995).

[52] S. Haas, J. Riera, and E. Dagotto, Phys. Rev. B **48**, 13174 (1993).

[53] R.T. Scalettar, G.G. Batrouni, and G.T. Zimanyi, Phys. Rev. Lett. **66**, 3144 (1991).

[54] S.R. White, Phys. Rev. B **48**, 10345 (1993).

[55] P. Schmitteckert et al., Phys. Rev. Lett. **80**, 560 (1998).

[56] M. Mézard and G. Parisi, J. de Phys. I **1**, 809 (1991).

[57] T. Giamarchi and P. Le Doussal, in *Statics and Dynamics of Disordered Elastic Systems*, ed. A.P. Young (World Scientific, Singapore, 1998), p. 321, cond-mat/9705096.

[58] T. Giamarchi and P. Le Doussal, Phys. Rev. B **52**, 1242 (1995).

[59] J.L. Cardy and S. Ostlund, Phys. Rev. B **25**, 6899 (1982).

[60] P. Le Doussal and T. Giamarchi, Phys. Rev. Lett. **74**, 606 (1995).

[61] M. Mézard, G. Parisi, and M.A. Virasoro, *Spin Glass Theory and beyond* (World Scientific, Singapore, 1987).

[62] A.I. Larkin, Sov. Phys. JETP **31**, 784 (1970).

[63] A.I. Larkin and Y.N. Ovchinnikov, J. Low Temp. Phys **34**, 409 (1979).

[64] This remark is originally due to S. Sachdev.

[65] T. Giamarchi and P. Le Doussal, Phys. Rev. B **53**, 15206 (1996).

[66] A. Georges, O. Parcollet and S. Sachdev, cond-mat/9909239.

[67] A.L. Efros and B.I. Shklovskii, J. Phys. C **8**, L49 (1975).

[68] P.A. Lee and A.I. Larkin, Phys. Rev. B **17**, 1596 (1978).

[69] D.R. Nelson and V.M. Vinokur, Phys. Rev. B **48**, 13060 (1993).

[70] R. Chitra, T. Giamarchi, and P. Le Doussal, Phys. Rev. Lett. **80**, 3827 (1998).

[71] R. Chitra, T. Giamarchi and P. Le Doussal, to be published.

[72] A.A. Gogolin, Phys. Rep. **86**, 1 (1982).

[73] A. Comtet, A. Georges, and P. Le Doussal, Phys. Lett. B **208**, 487 (1988).

[74] A.O. Gogolin, A.A. Nersesyan, A.M. Tsvelik, and L. Yu, Nucl. Phys. B **540**, 705 (1997).

[75] L. Balents and M.P.A. Fisher, Phys. Rev. B **56**, 12970 (1997).

[76] T.P. Eggarter and R. Riedinger, Phys. Rev. B **18**, 569 (1978).

[77] K. Damle, O. Motrunich, and D.A. Huse, Phys. Rev. Lett. **84**, 3434 (2000).

[78] D.S. Fisher, Phys. Rev. B **50**, 3799 (1994).

[79] D.S. Fisher, Phys. Rev. B **51**, 6411 (1995).

[80] R. Shankar, Int. J. Mod. Phys. B **4**, 2371 (1990).

[81] S. Fujimoto and N. Kawakami, Phys. Rev. B **54**, R11018 (1996).

[82] M. Mori and H. Fukuyama, J. Phys. Soc. Jpn. **65**, 3604 (1996).

[83] M.A. Paalanen, J.E. Graebner, R.N. Bhatt, and S. Sachdev, Phys. Rev. Lett. **61**, 597 (1988).

[84] B. Grenier et al., Phys. Rev. B **57**, 3444 (1998).

[85] E. Orignac and T. Giamarchi, Phys. Rev. B **56** 7167 (1997); **57** 5812 (1998).

[86] M. Azuma et al., Phys. Rev. B **55**, 8658 (1997).

[87] S.A. Carter et al., Phys. Rev. Lett. **77**, 1378 (1996).

[88] S. Fujimoto and N. Kawakami, Phys. Rev. B **56**, 9360 (1997).

[89] C. Monthus, O. Golinelli, and T. Jolicoeur, Phys. Rev. B **58**, 805 (1998).

[90] T. Giamarchi, Physica B **230-232**, 975 (1997).

[91] E. Orignac, T. Giamarchi, and P.L. Doussal, Phys. Rev. Lett. **83**, 2378 (1999).

[92] T. Giamarchi, P. Le Doussal and E. Orignac, to be published.

[93] H.J. Schulz, Phys. Rev. Lett. **71**, 1864 (1993).

Part III

Functional Derivatives, Mean-Field, Self-Consistent Methods, Slave-Bosons, and Extensions

6

Self-Consistent Many-Body Theory for Condensed Matter Systems

N.E. Bickers

ABSTRACT Self-consistent field techniques for the many-electron problem are examined using the modern formalism of functional methods. Baym–Kadanoff, or Φ-derivable, approximations are introduced first. After a brief review of functional integration results, the connection between conventional mean-field theory and higher-order Baym–Kadanoff approximations is established through the concept of the action functional. The Φ-derivability criterion for thermodynamic consistency is discussed, along with the calculation of free-energy derivatives. Parquet, or crossing-symmetric, approximations are introduced next. The principal advantages of the parquet approach and its relationship to Baym–Kadanoff theory are outlined. A linear eigenvalue equation is derived to study instabilities of the electronic normal state within Baym–Kadanoff or parquet theory. Finally, numerical techniques for the solution of self-consistent field approximations are reviewed, with particular emphasis on renormalization group methods for frequency and momentum space.

1 Introduction

Interacting many-electron systems have been analyzed by a diverse set of techniques including variational approximations, direct numerical simulation, and diagonalization in small systems. In this chapter we examine techniques which combine aspects of two of the earliest approaches to the many-electron problem, mean-field, or Hartree–Fock, theory and perturbative expansions. These techniques have lost favor in recent years because their quantitative (and in some cases qualitative) accuracy is constrained to the extreme weak-coupling limit. Despite this fact, *extensions* of the mean-field approach based on diagrammatic perturbation theory have recently become a subject of renewed interest. This interest stems largely from increases in computing power, which allow the self-consistent solution of complex approximations previously considered intractable. We shall refer to such approximations generically as self-consistent field (SCF) theories. SCF theories include the "Φ-derivable approximations" first studied by Baym and Kadanoff [1, 2], which follow from functional differentiation of an approximate free energy; and the "parquet approximations" developed by de Dominicis and Martin [3], which simultaneously renormalize two-particle vertex functions and one-particle self-energies.

Our intent is to provide a self-contained introduction to Baym–Kadanoff and parquet approximations which (a) places these techniques in a current context and (b) treats them as the logical extensions of mean-field theory, rather than as infinite-order perturbative expansions. We begin in Section 2 with a brief review of mean-field theory, emphasizing the role of the one-body mean-field Hamiltonian. In order to introduce action functionals into the discussion, a short summary of functional integration results is provided in Section 3. In Section 4 the connection between mean-field theory and all higher-order Baym–Kadanoff approximations is established through the concept of the SCF action functional, the nonlocal generalization of the mean-field Hamiltonian. In Section 5 the criterion of Φ-derivability [2], which insures the existence of a self-consistent thermodynamics and conserving correlation functions, is discussed. The method for calculating first and second derivatives of the SCF free energy by functional differentiation of the Φ functional is outlined in Section 6. In Section 7 we begin our discussion of parquet theory by introducing the concept of crossing symmetry and illustrating its violation within Baym–Kadanoff approximations. The general parquet equations for normal-state Fermi systems are derived in Section 8, then rewritten in a spin-diagonalized form in Section 9. In Section 10 the relationship between parquet theory and the so-called fluctuation exchange approximation [4] is established, and a "pseudopotential parquet" approximation is worked out. Linear eigenvalue equations for the investigation of normal-state ordering instabilities are derived in Section 11. In Section 12 we discuss a numerical renormalization group approach for solving SCF equations efficiently at a hierarchy of temperature scales. We briefly survey representative results from SCF solutions for several correlated electron models in Section 13, then make some concluding remarks in Section 14.

2 Review of Mean-Field Theory

A traditional starting point for the analysis of many-electron systems is the mean-field, or Hartree–Fock, approximation. Each electron in an N-electron system experiences a dynamically readjusting interaction with the other $N - 1$ electrons. Mean-field theory reduces this generally unsolvable problem to a set of one-body problems which must be solved self-consistently: in the simplest "Hartree" applications each electron moves in a time-independent *local* field determined by the average dynamics of all other electrons. (Considerations of exchange symmetry introduce spatially nonlocal "Fock" terms into the one-electron fields. Such terms are a special case of general nonlocal interactions, which are most simply discussed after the introduction of the action functional. For this reason we concentrate here on local mean-field approximations.)

Hartree and Hartree–Fock approximations are common in atomic physics, where self-consistent one-electron orbitals are used to evaluate matrix elements of the electron-electron interaction. It is convenient to choose an example instead from the prototypical model for interacting electrons on a periodic lattice, the Hubbard model. In the simplest Hubbard model, electrons are assumed to move by near-neighbor hopping on a cubic (or square) array of identical orbitals. The orbitals themselves may describe the solution of a complex atomic physics problem. Only the intra-orbital Coulomb matrix element is retained, so that the Hamiltonian takes the form

$$\mathcal{H} - \mu N = \sum_{\mathbf{k}\sigma} (\varepsilon_{\mathbf{k}} - \mu) n_{\mathbf{k}\sigma} + U \sum_i n_{i\uparrow} n_{i\downarrow}, \tag{2.1}$$

with $\varepsilon_{\mathbf{k}}$ the one-electron band energy and U the Coulomb interaction.

If the system retains translational and spin rotational invariance (symmetries of the underlying Hamiltonian), and fluctuations in the single-orbital occupancy,

$$n_i = \sum_\sigma n_{i\sigma}, \tag{2.2}$$

are small, it is reasonable to replace the Coulomb term in (2.1) with

$$\tfrac{1}{2} U \langle n \rangle \sum_{i\sigma} n_{i\sigma}, \tag{2.3}$$

where $\langle n \rangle$ is the average orbital occupancy:

$$\langle n \rangle = \frac{1}{N} \sum_i \langle n_i \rangle. \tag{2.4}$$

This can be done formally by writing

$$n_{i\sigma} = \tfrac{1}{2} \langle n \rangle + \left[n_{i\sigma} - \tfrac{1}{2} \langle n \rangle \right] \tag{2.5}$$

and separating out the fluctuation correction from the mean-field Hamiltonian:

$$\begin{aligned} \mathcal{H} &= \mathcal{H}_{\mathrm{MF}} + \Delta\mathcal{H} \\ \mathcal{H}_{\mathrm{MF}} &= \sum_{\mathbf{k}\sigma} (\widetilde{\varepsilon}_{\mathbf{k}} - \mu) n_{\mathbf{k}\sigma} - \tfrac{1}{4} U \langle n \rangle^2 \cdot N \\ \Delta\mathcal{H} &= U \sum_i \left[n_{i\uparrow} - \tfrac{1}{2} \langle n \rangle \right] \left[n_{i\downarrow} - \tfrac{1}{2} \langle n \rangle \right] \\ \widetilde{\varepsilon}_{\mathbf{k}} &= \varepsilon_{\mathbf{k}} + \tfrac{1}{2} U \langle n \rangle. \end{aligned} \tag{2.6}$$

The self-consistency condition determining $\langle n \rangle$ is just

$$\langle n \rangle = 2 \sum_{\mathbf{k}} f_{\mathbf{k}}$$
$$f_{\mathbf{k}} = \frac{1}{e^{\beta(\tilde{\varepsilon}_{\mathbf{k}} - \mu)} + 1}. \tag{2.7}$$

The local mean field in (2.3) induces a uniform shift in the electron dispersion curve, since it is equivalent to a renormalization of the chemical potential μ. Such a field is called *diagonal*, since it respects all symmetries of the underlying Hamiltonian. (Note that a diagonal field may be local or nonlocal; e.g., the field $\sum_{\mathbf{rr}'} v(\mathbf{r} - \mathbf{r}') c_{\mathbf{r}}^{\dagger} c_{\mathbf{r}'}$ is spatially nonlocal, but translationally invariant.) *Off-diagonal* fields, which violate one or more symmetries of the Hamiltonian, may be introduced as well: self-consistency requirements dictate whether nonzero values are allowed. Such fields imply the existence of *order parameters*, anomalous expectation values which vanish identically in the symmetric state. Examples of possible order parameters for the Hubbard model include the following:

$$\langle m \rangle = \frac{1}{N} \sum_{\ell} \langle n_{\ell\uparrow} - n_{\ell\downarrow} \rangle$$
$$\langle m_{\Pi} \rangle = \frac{1}{N} \sum_{\ell} (-1)^{\ell} \langle n_{\ell\uparrow} - n_{\ell\downarrow} \rangle$$
$$\langle n_{\Pi} \rangle = \frac{1}{N} \sum_{\ell} (-1)^{\ell} \left[\langle n_{\ell} \rangle - \langle n \rangle \right] \tag{2.8}$$
$$\langle \Delta^{\dagger} \rangle = \frac{1}{N} \sum_{\ell} \langle c_{\ell\uparrow}^{\dagger} c_{\ell\downarrow}^{\dagger} \rangle.$$

These fields respectively violate spin rotational invariance, spin rotational and translational invariance, translational invariance, and U(1) gauge symmetry. The corresponding ordered states are a ferromagnet, antiferromagnet, charge density wave solid, and singlet superconductor.

In some cases more than one mean-field solution may be possible. (For example, the diagonal solution in (2.6) is always possible.) In such cases the optimum mean-field solution is that which minimizes the free energy. This implies both local and global stability: a system is locally stable if all susceptibilities (second derivatives with respect to external symmetry-breaking fields) are finite and positive; when a particular susceptibility diverges, the electronic normal state becomes unstable to introduction of an off-diagonal mean field, and a second-order (or continuous) phase transition occurs. As an example, the half-filled ($\langle n \rangle = 1$) positive-U Hubbard model has a finite-temperature mean-field instability to antiferromagnetism, while the negative-U model has simultaneous instabilities to charge density wave and superconducting order. Global stability considerations become important when two or more ordered states with competing free energies are possible. It may happen in this case that the system's order parameter changes discontinuously in a first-order phase transition. As an example, the nearly half-filled positive-U Hubbard model exhibits two mean-field phase transitions as the temperature is reduced: a second-order transition from normal state to commensurate ($\mathbf{Q} = \boldsymbol{\pi}/a$) antiferromagnet, then a first-order transition from commensurate to incommensurate ($\mathbf{Q} \neq \boldsymbol{\pi}/a$) antiferromagnet.

The neglect of fluctuations in mean-field theory, at first sight a reasonable approximation, is in fact highly suspect due to correlation effects: in the Hubbard example, the presence of an up-spin electron at site i drastically reduces the probability for finding a down-spin electron at the same site. Furthermore, despite the fact that the Hubbard interaction is instantaneous and on-site, the induced correlations have long-range components in space and time. This makes purely local extensions of the mean-field treatment impossible. The quantum mechanical action functional S, introduced in the next section, provides the necessary generalization of the Hamiltonian for incorporating fluctuation effects, while retaining the general SCF framework.

3 Basics of Functional Integration

In order to generalize the mean-field analysis of Section 2, it is necessary to develop a systematic approach for treating nonlocal interactions. Nonlocality may be either (a) spatial or (b) temporal. The kinetic energy operator for lattice electrons

$$\widehat{T} = \sum_{\langle ij \rangle \sigma} t_{ij}(c^{\dagger}_{i\sigma} c_{j\sigma} + HC) \tag{3.1}$$

is a familiar example of an operator with spatial nonlocality. Temporal nonlocality, or retardation, effects arise whenever particles interact through the exchange of Bose-like field quanta. Such effects become important when the velocity of the quanta is comparable to or smaller than the velocity of the particles themselves. In condensed matter physics, examples of such quanta include phonons (i.e., lattice vibrations); electronic spin fluctuations ($S = 1$ electron-hole pairs); and singlet-pair fluctuations ($S = 0$ electron-electron pairs).

While spatial nonlocality can be incorporated in a conventional time-independent Hamiltonian, temporal nonlocality cannot. The natural generalization of the Hamiltonian for such cases is the so-called *action functional S*. In the Lagrangian formulation of quantum statistical mechanics, the partition function of a Fermi, Bose or spin system is written as an integral of the statistical weighting factor e^{-S} over an appropriately defined space of field variables ϕ:

$$Z = \int_\phi e^{-S(\phi)}. \tag{3.2}$$

The same basic techniques are used to treat Bose and Fermi systems. Since the Fermi case requires the introduction of anticommuting variable spaces, it is simplest to introduce basic concepts first in the Bose context (Section 3.1), then to discuss the modifications necessary for Fermi systems (Section 3.2) [5].

3.1 Bose Systems

As an introductory example, it is useful to consider at length the single-level Bose Hamiltonian

$$\widehat{\mathcal{H}} = \varepsilon b^\dagger b. \tag{3.3}$$

The partition function takes the form

$$Z_0 = (1 - e^{-\beta\varepsilon})^{-1}, \tag{3.4}$$

with

$$\beta = 1/kT. \tag{3.5}$$

This partition function may be rewritten as a functional integral over a space of complex fields $\phi(\tau) = \mathrm{Re}\,\phi(\tau) + i\,\mathrm{Im}\,\phi(\tau)$:

$$S(\phi) = \int_0^\beta d\tau \left[\bar{\phi}(\tau)\frac{d\phi(\tau)}{d\tau} + \varepsilon\bar{\phi}(\tau)\phi(\tau)\right] \tag{3.6}$$

$$\begin{aligned} \int_\phi \equiv \int D\bar{\phi}D\phi &= \int \prod_\tau d\bar{\phi}(\tau)d\phi(\tau) \\ &= \int \prod_\tau \frac{d[\mathrm{Re}\,\phi(\tau)]d[\mathrm{Im}\,\phi(\tau)]}{\pi}. \end{aligned} \tag{3.7}$$

The action and integration measure above are really shorthand notation for bulkier expressions involving a discretely indexed variable set:

$$S = \lim_{N\to\infty} \sum_{\ell=1}^{N} \Delta\tau_\ell \left[\bar{\phi}_\ell \left(\frac{\phi_\ell - \phi_{\ell-1}}{\Delta\tau_\ell}\right) + \varepsilon\bar{\phi}_\ell\phi_{\ell-1}\right] \tag{3.8}$$

$$\int_\phi \equiv \lim_{N\to\infty} \int \prod_{j=1}^{N} d\bar{\phi}_j d\phi_j = \lim_{N\to\infty} \int \prod_{j=1}^{N} \frac{d(\mathrm{Re}\,\bar{\phi}_j)d(\mathrm{Im}\,\phi_j)}{\pi}, \tag{3.9}$$

subject to

$$\sum_{\ell=1}^{N} \Delta\tau_\ell = \beta \tag{3.10}$$

and the "boundary condition"

$$\phi_0 = \phi_N. \tag{3.11}$$

Since the integral in (3.2) assumes a Gaussian form in this case, it may be verified by explicit calculation (using the discrete variable set) that the result of (3.4) is recovered. It is useful to rewrite the shorthand form for the action in (3.6) as a vector product,

$$S(\phi) = \bar{\phi}_\ell (-G_0^{-1})_{\ell\ell'} \phi_{\ell'}, \tag{3.12}$$

where

$$(-G_0^{-1})_{\ell\ell'} = \beta\left(\frac{\partial}{\partial\tau} + \varepsilon\right)_{\ell\ell'} = \beta\left(\frac{\partial}{\partial\tau}\right)_{\ell\ell'} + (\beta\varepsilon)\delta_{\ell\ell'} \tag{3.13}$$

and

$$\bar{u}_\ell v_\ell \equiv \frac{1}{\beta}\int_0^\beta d\tau\, \bar{u}(\tau) v(\tau) \tag{3.14}$$

for arbitrary "vectors" $\bar{u}_\ell$ and v_ℓ. The dimensionless matrix

$$G_0 = -\beta^{-1}\left(\frac{\partial}{\partial\tau} + \varepsilon\right)^{-1} \tag{3.15}$$

is called the bare *propagator*, or *Green's function*.[1] This terminology is consistent with the common usage of "propagator" for the two-point correlation function $-\langle \mathrm{T}_\tau\, b(\tau) b^\dagger \rangle$. As discussed below, the matrix elements of G_0 are exactly the values of the correlation function in a noninteracting system.

The general complex Gaussian integration formula

$$\int \prod_\ell d\bar{\phi}_\ell\, d\phi_\ell\, e^{-\bar{\phi}A\phi + \bar{u}\phi + \bar{\phi}v} = (\det A)^{-1} e^{\bar{u}A^{-1}v} \tag{3.16}$$

implies the shorthand

$$\begin{aligned} Z_0 &= \left[\det(-G_0^{-1})\right]^{-1} \equiv e^{-\beta\mathcal{F}_0} \\ \mathcal{F}_0 &= \frac{1}{\beta}\ln\det(-G_0^{-1}) = \frac{1}{\beta}\operatorname{Tr}\ln(-G_0^{-1}). \end{aligned} \tag{3.17}$$

If one introduces formal Fourier transform variables

$$\begin{aligned} \phi(\tau) &= \sum_{m=-\infty}^{\infty} e^{-i\nu_m\tau}\phi_m \\ \bar{\phi}(\tau) &= \sum_{m=-\infty}^{\infty} e^{i\nu_m\tau}\bar{\phi}_m, \end{aligned} \tag{3.18}$$

[1] Our convention is that all one-particle propagators are dimensionless, so that equations take the same form in a time- or frequency-space representation. This accounts for the factor of $1/\beta$ in (3.15).

it is clear that the quadratic form $\beta\bar{\phi}_\ell(\partial/\partial\tau + \varepsilon)_{\ell\ell'}\phi_{\ell'}$ can be diagonalized as $\beta\bar{\phi}_m(-i\nu_m + \varepsilon)\phi_m$, so that in turn

$$\mathcal{F}_0 \text{ "="} \; \frac{1}{\beta} \sum_{m=-\infty}^{\infty} \ln \beta(-i\nu_m + \varepsilon). \tag{3.19}$$

The "equality" in (3.19) requires additional comment. The Bose periodicity condition $\phi_0 = \phi_N$ implies that the set of frequencies ν_m consists of the *even Matsubara frequencies:*

$$\nu_m = (2m)\pi/\beta, \quad m = 0, \pm 1, \pm 2, \ldots . \tag{3.20}$$

Using this fact, it may be demonstrated that the right-hand side of (3.19) is *divergent:* the problem is that the discretized expression in (3.8) must be more carefully Fourier-transformed to obtain a well-defined limit. For this purpose it is useful to define a reference system with ε replaced by ε_{ref}. It may then be shown rigorously that

$$Z_0/Z_{\text{ref}} = e^{-\beta(\mathcal{F}_0 - \mathcal{F}_{\text{ref}})} \tag{3.21}$$

where

$$\mathcal{F}_0 = \frac{1}{\beta} \sum_{m=-\infty}^{\infty} e^{i\nu_m 0^+} \ln \beta(-i\nu_m + \varepsilon) \tag{3.22}$$

and $\mathcal{F}_{\text{ref}}$ takes the same form with $\varepsilon \to \varepsilon_{\text{ref}}$. Note that the "regularizations" in this result arise naturally from a careful treatment of the discretized functional integral. If these regularizations are taken into account, the formal notation of (3.17) and (3.19) may be applied without error.

The extension of this treatment to spatially extended Bose systems with one-body Hamiltonians is straightforward: it is only necessary to enlarge the label space of the matrix propagator G_0 to include space, as well as time, labels. If a natural lattice is present or if continuous space is discretized, matrix traces acquire an additional discrete summation over spatial labels in addition to the summation over Matsubara frequencies. The inverse propagator for a Bose Hamiltonian

$$\widehat{\mathcal{H}} = \sum_{ij} b_i^\dagger (\mathcal{H}_0)_{ij} b_j \tag{3.23}$$

is just

$$-G_0^{-1} = \beta\left(\frac{\partial}{\partial\tau} + \mathcal{H}_0\right), \tag{3.24}$$

with $\mathcal{H}_0$ diagonal in time, but possibly off-diagonal in spatial labels. External one-body perturbations $\mathcal{U}$ may be incorporated in the inverse propagator in similar fashion; nonlocal contributions appear as off-diagonal matrix elements. For example, if the action contains a nonlocal perturbation

$\beta\bar{\phi}_{i\ell}\mathcal{U}_{i\ell,j\ell'}\phi_{j\ell'}$, then

$$-G_0^{-1} = \beta\left(\frac{\partial}{\partial\tau} + \mathcal{H}_0 + \mathcal{U}\right). \tag{3.25}$$

Two-body Bose interactions may be transcribed to the action formalism as simply as one-body terms: each boson operator $b_i^\dagger(\tau)$ in the normal-ordered interaction is replaced by the corresponding complex field $\bar{\phi}_{i\ell}$; the resulting expression is multiplied by β and summed over all indices to obtain the interaction contribution to the action. Note that this prescription holds for an instantaneous or retarded interaction. As an example, the term

$$\widehat{V} = V_0 b_1^\dagger b_2^\dagger b_2 b_1 \tag{3.26}$$

gives rise to

$$S_V = (\beta V_0)\bar{\phi}_{1\ell}\bar{\phi}_{2\ell}\phi_{2\ell}\phi_{1\ell}, \tag{3.27}$$

with a sum on ℓ implied. Such terms make the action nonquadratic and prevent the exact calculation of the partition function as a Gaussian integral. In this case additional approximation techniques (such as the Baym–Kadanoff technique developed in Section 4 for fermions) become necessary.

Before moving on to the treatment of Fermi systems, it is useful to mention at this point how Bose correlation functions are evaluated in the functional integral formalism. It may be shown [5] that a general time-ordered correlation function has the representation

$$\begin{aligned}\langle \mathrm{T}_\tau\, b_{i_1}(\tau_1) b_{i_2}(\tau_2)\cdots b_{i_n}^\dagger(\tau_n)\rangle \\ = Z^{-1}\int D\bar{\phi}\, D\phi\, \phi_{i_1\ell_1}\phi_{i_2\ell_2}\cdots\bar{\phi}_{i_n\ell_n} e^{-S};\end{aligned} \tag{3.28}$$

i.e., boson operators are simply replaced by their complex-field counterparts and inserted in the field-space integral with statistical factor e^{-S}. The properties of the functional integral automatically lead to enforcement of time ordering. Note that for a quadratic action all correlation functions reduce to Gaussian integrals and may be evaluated exactly. For example, in the absence of interactions the two-point correlation function is just

$$\begin{aligned}G_0(xx') &\equiv -\langle \mathrm{T}_\tau\, b(x) b^\dagger(x')\rangle_0 = (G_0)_{xx'} \\ x &\equiv (\mathbf{x}, \tau).\end{aligned} \tag{3.29}$$

3.2 Fermi Systems

The preceding results must be modified in order to treat Fermi, rather than Bose, systems. The principal modification is the introduction of anticommuting c-numbers (or "Grassmann variables") to replace the complex variables $\phi(\tau)$ used previously. Any two anticommuting c-numbers η and ξ satisfy the relation

$$\eta\xi + \xi\eta = 0; \tag{3.30}$$

in particular, the "square" of any such number is identically zero:

$$\eta\eta = 0. \tag{3.31}$$

Formal operations of differentiation and integration may be defined for functions built up from anticommuting c-numbers, and the resulting "Grassmann integrals" used to construct a representation for Fermi partition functions. We will not at this point review the properties of Grassmann integration in detail, but will instead state without proof the required results [5], each of which has a close analog in the treatment of Bose systems. In general, for invertible matrices A and anticommuting c-number "functions" $\{\bar{\eta}_n, \eta_n\}$ and $\{\bar{\xi}_n, \xi_n\}$,

$$\int \prod_\ell d\bar{\eta}_\ell \, d\eta_\ell \, e^{-\bar{\eta}A\eta + \bar{\eta}\xi + \bar{\xi}\eta} = (\det A) e^{\bar{\xi}A^{-1}\xi}. \tag{3.32}$$

Note that this expression differs from (3.16) in that $\det A$, rather than $(\det A)^{-1}$, appears.

The representation of the partition function for noninteracting fermions with Hamiltonian $\widehat{\mathcal{H}}_0$ subject to a nonlocal one-body perturbation $\mathcal{U}$ is

$$Z_0 = \int_c e^{-S(c)} \tag{3.33}$$

$$S(c) = \bar{c}_{i\ell}(-G_0^{-1})_{i\ell,j\ell'} c_{j\ell'} \tag{3.34}$$

$$(-G_0^{-1})_{i\ell,j\ell'} = \beta \left(\frac{\partial}{\partial \tau} + \mathcal{H}_0 + \mathcal{U} \right)_{i\ell,j\ell'} \tag{3.35}$$

$$\int_c \equiv \int D\bar{c}\, Dc = \int \prod_{i\ell} d\bar{c}_{i\ell} \, dc_{i\ell}, \tag{3.36}$$

subject in this case to

$$c_0 = -c_N \tag{3.37}$$

(antiperiodic, rather than periodic, boundary conditions). Because of this antiperiodicity, the Fourier transform variables are now indexed by *odd Matsubara frequencies:*

$$\omega_m = (2m+1)\pi/\beta, \quad m = 0, \pm 1, \pm 2, \ldots . \tag{3.38}$$

In analogy with the Bose case, the Fermi free energy for a noninteracting system may be written

$$\mathcal{F}_0 \text{ ``='' } -\frac{1}{\beta} \operatorname{Tr} \ln(-G_0^{-1}). \tag{3.39}$$

As before, regularizations are necessary to recover a proper limit when a frequency space representation is employed.

Two-body interactions generate nonquadratic contributions to the action exactly as in the Bose case; as previously, the interaction operator must be normal-ordered before its transcription to functional notation. Also in analogy with the Bose case, time-ordered correlation functions are evaluated as integrals of anticommuting c-numbers weighted by the statistical factor e^{-S}. For example, the two-point correlation function (or one-electron propagator) takes the form

$$\begin{aligned} G(xx') &\equiv -\langle \mathrm{T}_\tau\, c(x)c^\dagger(x')\rangle = -Z^{-1}\int D\bar{c}\,Dc\, c(x)\bar{c}(x')e^{-S} \\ x &\equiv (\mathbf{x},\tau), \end{aligned} \tag{3.40}$$

and in the absence of interactions,

$$G_0(xx') = (G_0)_{xx'}. \tag{3.41}$$

4 Self-Consistent Approximations for the Action Functional

Having reviewed briefly the formalism for describing many-particle systems in terms of a nonlocal action functional, we turn to the use of this formalism in deriving approximations for interacting systems. The partition function for a system of spin-$\frac{1}{2}$ particles with spin-independent interactions takes the form

$$Z = e^{-\beta\mathcal{F}} = \int D\bar{c}\,Dc\, e^{-S}, \tag{4.1}$$

where

$$S = S_0 + S_V \tag{4.2}$$

$$S_0 = \beta \sum_{xx'\sigma} \bar{c}_\sigma(x)(\partial/\partial\tau + \mathcal{H}_0)_{xx'} c_\sigma(x') \tag{4.3}$$

$$S_V = \tfrac{1}{2}\beta \sum_{xx'\sigma\sigma'} v(x-x')\bar{c}_\sigma(x)\bar{c}_{\sigma'}(x')c_{\sigma'}(x')c_\sigma(x), \tag{4.4}$$

with

$$\begin{aligned} x &= (\mathbf{x},\tau) \\ \sum_x &= \frac{1}{\beta}\int_0^\beta d\tau \sum_{\mathbf{x}}. \end{aligned} \tag{4.5}$$

Here $\mathbf{x}$ represents a lattice site or a discretized chunk of continuous space. Note that $v(x-x')$ is a (possibly time-retarded) two-body interaction. In the simplest case, the Hubbard model, the interaction is point-like in space and time:

$$v(x-x') = U\delta_{xx'}. \tag{4.6}$$

After Fourier transforming, the action may be conveniently rewritten

$$S_0 = \beta \sum_{k\sigma} (-i\omega_n + \varepsilon_{\mathbf{k}}) \bar{c}_\sigma(k) c_\sigma(k) \tag{4.7}$$

$$S_V = \tfrac{1}{2}\beta \sum_{kk'q\sigma\sigma'} v(q) \bar{c}_\sigma(k+q) \bar{c}_{\sigma'}(k'-q) c_{\sigma'}(k') c_\sigma(k), \tag{4.8}$$

where

$$\begin{aligned} k &= (\mathbf{k}, i\omega_n) \\ q &= (\mathbf{q}, i\nu_m), \end{aligned} \tag{4.9}$$

$$\begin{aligned} \omega_n &= (2n+1)\pi/\beta \\ \nu_m &= (2m)\pi/\beta, \end{aligned} \tag{4.10}$$

and $\mathbf{k}$ and $\mathbf{q}$ are restricted to an appropriate Brillouin zone.

In order to make progress (short of calculating correlation functions "exactly" by a statistical sampling, or Monte Carlo, technique), it is necessary to replace S_V with a quadratic SCF approximation. In analogy with the mean-field treatment of Section 2 for a Hamiltonian, the action may be rewritten as

$$S = S^{\mathrm{SCF}} + \Delta S \tag{4.11}$$

$$S^{\mathrm{SCF}} = S_0 + \beta \sum_{kk'\sigma\sigma'} \Sigma_{\sigma\sigma'}(kk') \bar{c}_\sigma(k) c_{\sigma'}(k') + \beta\overline{\mathcal{F}} \tag{4.12}$$

$$\Delta S = S - S^{\mathrm{SCF}}, \tag{4.13}$$

where Σ is a nonlocal self-consistent field and $\overline{\mathcal{F}}$ is a c-number. In the simplest case Σ is a diagonal field, which respects all symmetries of the system, i.e.,

$$\Sigma_{\sigma\sigma'}(kk') = \Sigma(k) \delta_{kk'} \delta_{\sigma\sigma'}. \tag{4.14}$$

As in the mean-field analysis, there is no reason *a priori* to restrict consideration to diagonal fields. For example, the choice

$$\begin{aligned} \Sigma_{\sigma\sigma'}(kk') &= \Sigma_\sigma(k) \delta_{kk'} \delta_{\sigma\sigma'} \\ \Sigma_\sigma(k) &\neq \Sigma_{-\sigma}(k) \end{aligned} \tag{4.15}$$

implies broken spin rotational symmetry, i.e., ferromagnetism.

In cases where the off-diagonal order is relatively simple, it is convenient to generalize the quantity $\Sigma(k)$ in (4.14) to a low-dimensional matrix, which incorporates diagonal and off-diagonal fields. For example, in the presence of antiferromagnetism, the mean-field action may be written

$$\begin{aligned} S^{\mathrm{SCF}} &= S_0 + \beta\overline{\mathcal{F}} + \tfrac{1}{2}\beta \sum_{k\sigma} \begin{bmatrix} \bar{c}_{k\sigma} & \bar{c}_{k+\Pi,\sigma} \end{bmatrix} \begin{bmatrix} \Sigma(k) & \sigma\Delta(k) \\ \sigma\Delta(k+\Pi) & \Sigma(k+\Pi) \end{bmatrix} \begin{bmatrix} c_{k\sigma} \\ c_{k+\Pi,\sigma} \end{bmatrix} \\ \Pi &= (\boldsymbol{\pi}, 0). \end{aligned} \tag{4.16}$$

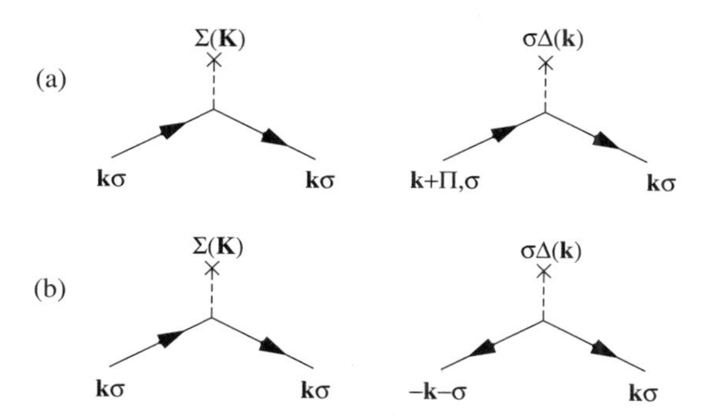

FIGURE 6.1. Diagonal and off-diagonal SCF's for (a) commensurate antiferromagnetism and (b) singlet superconductivity.

Alternately, in the presence of singlet superconductivity, the corresponding expression is

$$S^{\mathrm{SCF}} = S_0 + \beta\overline{\mathcal{F}} + \tfrac{1}{2}\beta\sum_{k\sigma}\begin{bmatrix}\bar{c}_{k\sigma} & c_{-k,-\sigma}\end{bmatrix}\begin{bmatrix}\Sigma(k) & \sigma\Delta(k)\\ \sigma\Delta^*(k) & -\Sigma^*(k)\end{bmatrix}\begin{bmatrix}c_{k\sigma}\\ \bar{c}_{-k,-\sigma}\end{bmatrix}. \quad (4.17)$$

The diagonal and off-diagonal fields in these examples have the diagrammatic representations shown in Fig. 6.1.

The optimal SCF solution is that with the smallest approximate free energy. (Clearly the discussion is not yet complete, since the choice of $\overline{\mathcal{F}}$ remains arbitrary at this point.) A search of several possibilities may be necessary to assure the attainment of a global minimum. In most cases it is sufficient to consider SCF's with a definite translational period, either commensurate or incommensurate with the underlying lattice, though nonuniform solutions are in principle also possible. Likewise, it is usual to restrict attention to states with a well-defined spin symmetry. This means that order parameters may be grouped into four classes: $S = 0$ (charge) and $S = 1$ (spin) in the particle-hole channel; and $S = 0$ (singlet) and $S = 1$ (triplet) in the particle-particle channel. Modifications are necessary in the presence of strong spin-orbit coupling [6], but we will not discuss that case here.

The principal question is how to choose S^{SCF} (and $\overline{\mathcal{F}}$) to make the correction ΔS, which still contains the dynamical variables $c_{k\sigma}$, "small." In the discussion which follows a diagonal SCF is assumed for simplicity. The SCF action then contains a dimensionless propagator matrix which, in the absence of external symmetry-breaking fields, is diagonal in momentum space:

$$G^{\mathrm{SCF}}(k) = \frac{T}{i\omega_n - (\varepsilon_{\mathbf{k}} - \mu) - \Sigma(k)}. \quad (4.18)$$

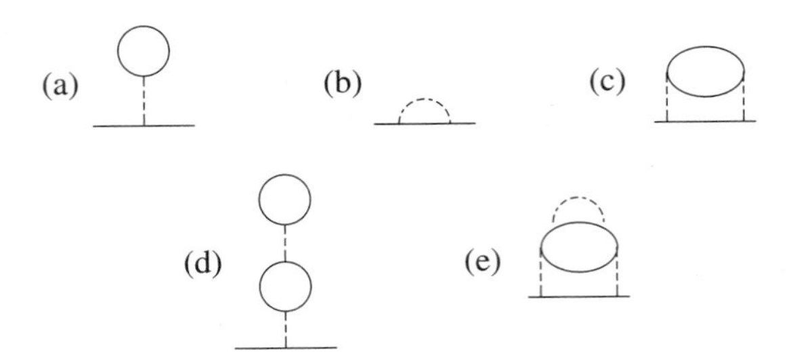

FIGURE 6.2. Low-order self-energy corrections for a general spin-independent action $S = S_0 + S_V$. (a–c) Skeleton diagrams. (d–e) Non-skeleton diagrams.

A perturbative expansion in ΔS of the two-point correlation function requires expanding e^{-S}, retaining S^{SCF} in the exponent. The proper self-energy corrections which arise are expressed in terms of G^{SCF}, rather than G_0. It is convenient to label these corrections $\Sigma_\Delta(k)$ to distinguish them from the SCF itself. Since $\Delta S = S - S^{\mathrm{SCF}}$, the self-energy corrections have two components: an infinite set of diagrammatic contributions from S_V; and a "mass counterterm" $-\Sigma(k)$. If a set of S_V contributions (evaluated with self-consistent propagators) can be exactly offset by the counterterm, the corresponding diagrams are eliminated from perturbation theory in ΔS.

By manipulating the counterterm in this way, the effect of ΔS may be made arbitrarily "small" (at least in a formal sense). It is important to note that when a set of S_V contributions is canceled, that set disappears *completely* from the theory: the diagrams in question no longer appear explicitly as terms in the self-energy or implicitly as self-energy corrections on internal lines in more complicated diagrams. This means that only *skeleton* S_V diagrams should be considered when deducing approximations for the counterterm. Skeleton diagrams are just those without secondary self-energy corrections on internal lines. As an example, diagrams (a–c) in Fig. 6.2 are skeleton, while diagrams (d), (e) are not.

Each skeleton diagram may be used to define a different SCF approximation. Note that since the skeleton diagrams are evaluated with SCF propagators, the assumption of an off-diagonal SCF leads to the possibility of diagrams which vanish in the fully symmetric system. For example, in the case of antiferromagnetism, the SCF propagator in Fig. 6.3a and the skeleton diagram in Fig. 6.3b are both nonzero.

The SCF condition determining Σ is that the chosen set of S_V skeleton diagrams, evaluated using G^{SCF}, exactly reproduce Σ. For Hartree approximations this is just the requirement that "tadpole" diagrams (Fig. 6.4) are eliminated. The accuracy of higher-order approximations depends on a judicious choice of skeleton diagrams. In particular, it may happen that an approximation which respects all symmetries of the action S is unstable with respect to the development of an infinitesimal broken-symmetry so-

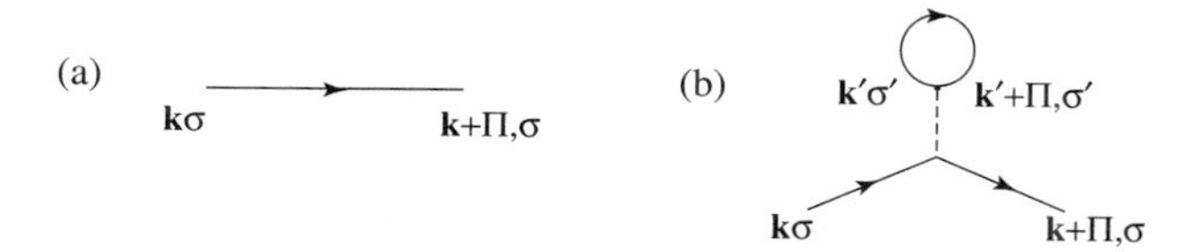

FIGURE 6.3. Off-diagonal propagator (a) and skeleton self-energy diagram (b) for a commensurate antiferromagnet.

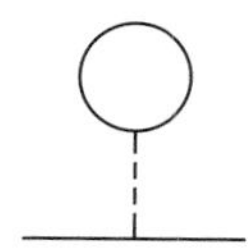

FIGURE 6.4. Hartree self–energy diagram.

lution. We return to this point in Section 7 after discussing the analytical form of the SCF free energy and the additional conditions which must be imposed to generate a thermodynamically consistent solution.

5 Φ-Derivability and Thermodynamic Self-Consistency

The SCF free energy $\mathcal{F}_{\text{SCF}}$ is defined by the equation

$$Z_{\text{SCF}} \equiv e^{-\beta\mathcal{F}_{\text{SCF}}} = \int D\bar{c}\, Dc\, e^{-S^{\text{SCF}}}. \tag{5.1}$$

Hence,

$$\mathcal{F}_{\text{SCF}} = -\frac{1}{\beta}\,\text{Tr}\ln\big[-(G^{\text{SCF}})^{-1}\big] + \overline{\mathcal{F}}, \tag{5.2}$$

where, in the absence of symmetry breaking,

$$G^{\text{SCF}}_{k\sigma,k'\sigma'} = G^{\text{SCF}}(k)\delta_{kk'}\delta_{\sigma\sigma'} \tag{5.3}$$

with $G^{\text{SCF}}(k)$ as in (4.18).

At this stage the contribution $\beta\overline{\mathcal{F}}$, the c-number part of the SCF action, is still arbitrary (since it can be subtracted back out in ΔS). This arbitrariness disappears and the allowable forms for Σ are further restricted if it is required that the free energy possess a number of properties obeyed by the exact solution. These properties include (a) consistency of thermodynamic derivatives and (b) respect of local conservation laws (or continuity equations) for particle number, momentum, and energy.

It is convenient at this point to introduce the concept of *generalized thermodynamic derivatives*. In simple cases differentiation with respect to a uniform field (such as a chemical potential or magnetic field) produces the corresponding thermodynamic variable (the uniform density or magnetization). A second differentiation produces a uniform susceptibility, which indicates sensitivity to changes in the field. Generalized thermodynamic derivatives measure the response to an arbitrary (possibly nonuniform or nonlocal) external field. A general number-conserving perturbation takes the form

$$S_{\text{ext}} = \beta \sum_{xx'\sigma\sigma'} \mathcal{U}_{\sigma\sigma'}(xx')\bar{c}_\sigma(x)c_{\sigma'}(x'). \tag{5.4}$$

Perturbations which violate number conservation (i.e., pair creation and annihilation fields) are also possible. To simplify notation in the discussion which follows, only number-conserving fields are considered explicitly; furthermore, spin labels are everywhere omitted. No additional conceptual complications are introduced by the presence of fields off-diagonal in particle number or spin.

Since S_{ext} is a one-body perturbation, it can be incorporated exactly into SCF theory. Note that S_{ext} changes $\mathcal{F}_{\text{SCF}}$ directly (through the change in S_0) and indirectly (through the change in $\Sigma(G^{\text{SCF}})$). It can be demonstrated[2] that a sufficient condition for obtaining a fully self-consistent and conserving approximation is that

$$\left.\frac{\delta \mathcal{F}}{\delta \mathcal{U}_{\sigma'\sigma}(x'x)}\right|_{\mathcal{U}=0} = G_{\sigma\sigma'}(xx'); \tag{5.5}$$

i.e., the perturbation acts as a "source" for the one-electron propagator. In this equation $\mathcal{F}$ and G are exact or SCF. For the exact solution, (5.5) is clearly satisfied since

$$\begin{aligned}\frac{\delta \mathcal{F}}{\delta \mathcal{U}_{\sigma'\sigma}(x'x)} &= (-\beta Z)^{-1} \int D\bar{c}\, Dc \left[-\beta \bar{c}_{\sigma'}(x')c_\sigma(x)\right] e^{-S} \\ &= -\langle c_\sigma(x)\bar{c}_{\sigma'}(x')\rangle = G_{\sigma\sigma'}(xx'). \end{aligned} \tag{5.6}$$

For an SCF approximation based on skeleton diagrams, (5.5) can be satisfied by choosing an appropriate $\overline{\mathcal{F}}$: First of all,

$$\mathcal{F}_{\text{SCF}}(\mathcal{U}) = -\frac{1}{\beta}\,\text{Tr}\ln\beta\left[\partial/\partial\tau + \mathcal{H}_0 + \mathcal{U} + \Sigma(G^{\text{SCF}}(\mathcal{U}))\right] + \overline{\mathcal{F}}. \tag{5.7}$$

Hence the total variation in $\mathcal{F}_{\text{SCF}}$ for given $\delta\mathcal{U}$ (viewed as a matrix in space-time and spin space) is

$$\begin{aligned}\delta\mathcal{F}_{\text{SCF}} &= -\frac{1}{\beta}\,\text{Tr}\left[(\partial/\partial\tau + \mathcal{H}_0 + \mathcal{U} + \Sigma)^{-1}(\delta\mathcal{U} + \delta\Sigma)\right] + \delta\overline{\mathcal{F}} \\ &= \text{Tr}\big(G^{\text{SCF}}\delta\mathcal{U}\big) + \text{Tr}\big(G^{\text{SCF}}\delta\Sigma\big) + \delta\overline{\mathcal{F}}. \end{aligned} \tag{5.8}$$

[2] See [2] for a thorough discussion of this point. We do not present a derivation of continuity equations or a proof of thermodynamic self-consistency in the present article.

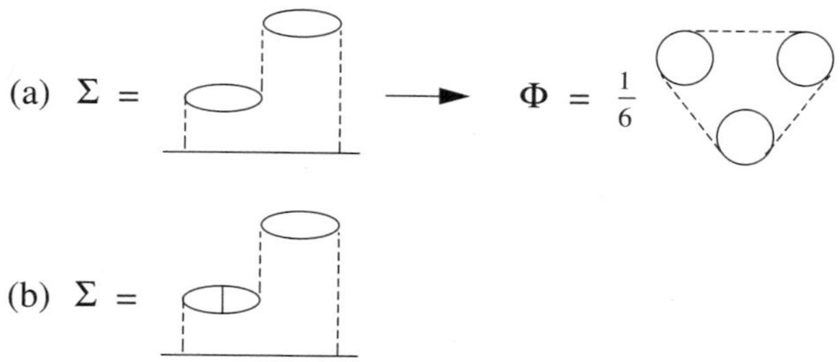

FIGURE 6.5. Illustration of Φ derivability. (a) Sample Φ-derivable self-energy and corresponding term in Φ. (b) Sample non-Φ-derivable skeleton diagram.

In order to satisfy (5.5), the last relation demands

$$\mathrm{Tr}\big(G^{\mathrm{SCF}}\delta\Sigma\big) + \delta\overline{\mathcal{F}} = 0. \tag{5.9}$$

A solution $\overline{\mathcal{F}}$ may be written

$$\overline{\mathcal{F}} = -\,\mathrm{Tr}\big(\Sigma G^{\mathrm{SCF}}\big) + \Phi, \tag{5.10}$$

where

$$\frac{\delta\Phi}{\delta G^{\mathrm{SCF}}_{\sigma\sigma'}(xx')} = \Sigma_{\sigma'\sigma}(x'x). \tag{5.11}$$

Equation (5.10) provides the required expression for $\overline{\mathcal{F}}$, the previously undetermined part of the SCF action. Such an expression can only be written down when the set of skeleton diagrams determining Σ is Φ-derivable, i.e., can be obtained from some functional $\Phi(G_{\mathrm{SCF}})$ by functional differentiation. In such cases the entire approximation is thermodynamically self-consistent and conserving. Examples of skeleton diagrams which do and do not satisfy the criterion of Φ-derivability are shown in Fig. 6.5. In the first case, the symmetry factor of $\frac{1}{6}$ in Φ accounts for the number of equivalent propagator lines which may be broken by functional differentiation. In contrast, since the two bubbles in the self-energy diagram of Fig. 6.5(b) are not equivalent, this contribution cannot be expressed as an exact functional derivative. It should be viewed instead as one of several diagrams of $O(V^4)$ which taken together form a Φ-derivable contribution to Σ.

Physically well-motivated normal-state approximations are in most cases automatically Φ-derivable. As an example, for the Hubbard model mean-field approximation considered in Section 2,

$$\begin{aligned}\Sigma_{\sigma\sigma'}(xx') &= U\langle n_{-\sigma}\rangle\delta_{xx'}\delta_{\sigma\sigma'}\\ \Phi &= NU\langle n_\uparrow\rangle\langle n_\downarrow\rangle\\ \mathrm{Tr}\big(\Sigma G^{\mathrm{SCF}}\big) &= 2\Phi\\ \overline{\mathcal{F}} = -\Phi &= -\tfrac{1}{4}NU\langle n\rangle^2.\end{aligned} \tag{5.12}$$

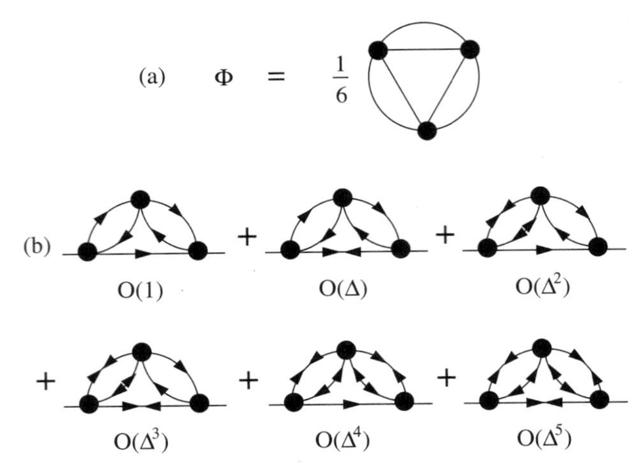

FIGURE 6.6. Self-energy in the presence of a pairing field Δ. (a) $\mathrm{O}(V^3)$ contribution to Φ. The solid dot represents a particle-hole interaction (which may be expanded in two ways, corresponding to direct and exchange scattering). The solid lines may be diagonal or off-diagonal. Additional diagrams appear when particle-particle scattering is included. (b) Contributions to the proper self-energy from functional differentiation of Φ. The diagrams of even order in Δ are diagonal; the diagrams of odd order, off-diagonal. For the terms of $\mathrm{O}(\Delta^2)$ and $\mathrm{O}(\Delta^3)$, only one of two distinct contributions is shown.

Note that $\overline{\mathcal{F}}$ takes the form found previously by explicitly separating the Hamiltonian into mean-field and fluctuation contributions. The "2" in $\mathrm{Tr}\big(\Sigma G^{\mathrm{SCF}}\big)$ arises from the summation on spins.

Some extra care must be exercised in broken-symmetry situations, since every line in Φ diagrams represents a full propagator, which has both diagonal and off-diagonal parts. As an example, consider the self-energy diagrams which arise in the presence of an off-diagonal particle-particle pairing field from the single $\mathrm{O}(V^3)$ Φ diagram in Fig. 6.6. For simplicity, assume that the heavy dot represents a particle-hole scattering vertex, i.e., that the two lines on each side of the dot correspond to a particle-hole pair, rather than a particle-particle pair. The off-diagonal propagator is denoted by a line with double arrowheads. With this convention, the particle-hole diagonal (Σ) and off-diagonal (Δ) self-energies each correspond to four terms (of varying order in Δ). Near the pairing transition temperature, where a power series expansion in Δ is permissible, only the first terms in each self-energy need be retained. (The picture is considerably complicated if both particle-hole and particle-particle vertices are retained.)

6 Thermodynamic Derivatives

The evaluation of generalized thermodynamic derivatives proceeds naturally from the choice of the generating functional Φ. Since only SCF and bare propagators appear in this section, the superscript "SCF" is dropped without loss of clarity. Local first derivatives of $\mathcal{F}^{\mathrm{SCF}}$ are related to propagators in the same way as in the exact solution. For example,

$$-\frac{\partial \mathcal{F}^{\mathrm{SCF}}}{\partial \mu} = \langle N \rangle = \mathrm{Tr}\, G = \sum_{\mathbf{x}\sigma} G_{\sigma\sigma}(\mathbf{x}\tau, \mathbf{x}\tau^{+}) \tag{6.1}$$

with μ a chemical potential; and

$$-\frac{\partial \mathcal{F}_{\mathrm{SCF}}}{\partial h} = \langle M_z \rangle = \langle N_{\uparrow} \rangle - \langle N_{\downarrow} \rangle = \mathrm{Tr}\, \sigma G, \tag{6.2}$$

where

$$h = \tfrac{1}{2} g \mu_B H, \tag{6.3}$$

with $\mathbf{H} = H\hat{z}$ a z-axis magnetic field. Note the origin of the minus signs in (6.1) and (6.2): the respective perturbations are

$$\mathcal{U}_{\sigma\sigma'}(xx') = -\mu \delta_{\mathbf{x}\mathbf{x}'} \delta_{\tau\tau'^{+}} \delta_{\sigma\sigma'} \tag{6.4}$$

and

$$\mathcal{U}_{\sigma\sigma'}(xx') = -\sigma h \delta_{\mathbf{x}\mathbf{x}'} \delta_{\tau\tau'^{+}} \delta_{\sigma\sigma'}. \tag{6.5}$$

Continuity equations for charge, spin, and energy currents (all expressible as first derivatives of the free energy in the presence of an external nonequilibrium field) may be written down using the same techniques applicable to the exact solution.

Second derivatives of the free energy, or generalized susceptibilities, are of particular interest for the detection of ordering instabilities. The divergence of a susceptibility signals that an infinitesimal field is sufficient to produce a finite response: in other words, the system can spontaneously lower its free energy by the formation of a symmetry-breaking order parameter.

As an example, the susceptibility to a uniform static magnetic field along the z-axis is just

$$\chi_{\mathrm{m}} = \left.\frac{\partial \langle M_z \rangle}{\partial h}\right|_{h=0} = \mathrm{Tr}\, \sigma \left.\frac{\partial G}{\partial h}\right|_{h=0}, \tag{6.6}$$

using the result of (6.2) for the magnetization. More generally, one may consider a nonlocal z-axis field, represented in the action S_0 by a term

$$\beta \sum_{\sigma 11'} (-\sigma) h(11') \bar{c}_{\sigma}(1) c_{\sigma}(1'), \tag{6.7}$$

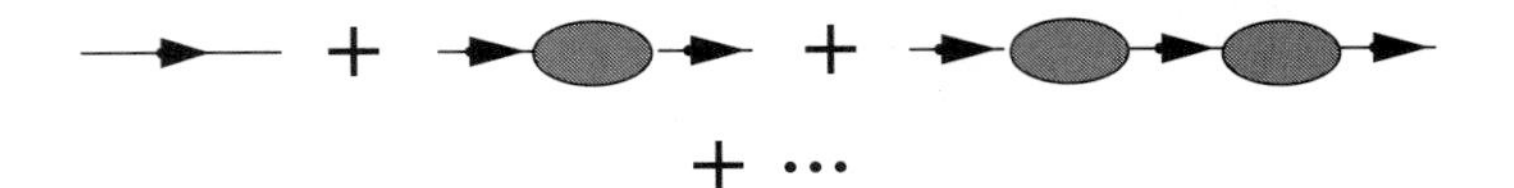

FIGURE 6.7. Symbolic expansion of the SCF propagator $G(h)$. The shaded bubble represents a factor of Σ.

FIGURE 6.8. Differentiation of the bare propagator $G^0_{\sigma\sigma}(h)$ with respect to the external field.

where the compact notation

$$1 \equiv (\mathbf{x}_1, \tau_1) \tag{6.8}$$

has been adopted. The *dimensionless* thermodynamic derivative of interest in this case is

$$\frac{1}{\beta}\frac{\delta G_{\sigma\sigma}(11')}{\delta h(22')}. \tag{6.9}$$

The object in (6.9) is commonly called a *particle-hole propagator*, since it plays the same role for a particle-hole pair (in this case with total spin $S = 1$ and $S^z = 0$) as the propagator G plays for one electron; differentiation of off-diagonal elements of G with respect to a pairing field generates a *particle-particle propagator* in like manner.

The SCF propagator $G_{\sigma\sigma}(11'; h)$ takes the form

$$G_{\sigma\sigma}(11'; h) = -\beta^{-1}\big[\partial/\partial\tau + \mathcal{H}_0 - \sigma h + \Sigma_\sigma\big(G(h)\big)\big]^{-1}{}_{11',\sigma\sigma}. \tag{6.10}$$

This form is illustrated diagrammatically in Fig. 6.7. The solid lines in this figure represent the bare propagator G^0 and incorporate only the explicit magnetic field dependence:

$$G^0_{\sigma\sigma}(11'; h) = -\beta^{-1}\big(\partial/\partial\tau + \mathcal{H}_0 - \sigma h\big)^{-1}{}_{11',\sigma\sigma}. \tag{6.11}$$

(Note that a superscript, rather than subscript, "0" is used throughout this section for notational convenience.) Each derivative with respect to the field splits a solid line in two and produces a factor of $-\sigma$:

$$\frac{1}{\beta}\frac{\delta G^0_{\sigma\sigma}(12; h)}{\delta h(43)} = (-\sigma) G^0_{\sigma\sigma}(14; h) G^0_{\sigma\sigma}(32; h). \tag{6.12}$$

The operation is illustrated diagrammatically in Fig. 6.8. It follows that the total contribution to $\beta^{-1}\delta G/\delta h$ from direct dependence on h is

$$(-\sigma) G_{\sigma\sigma}(14; h) G_{\sigma\sigma}(32; h), \tag{6.13}$$

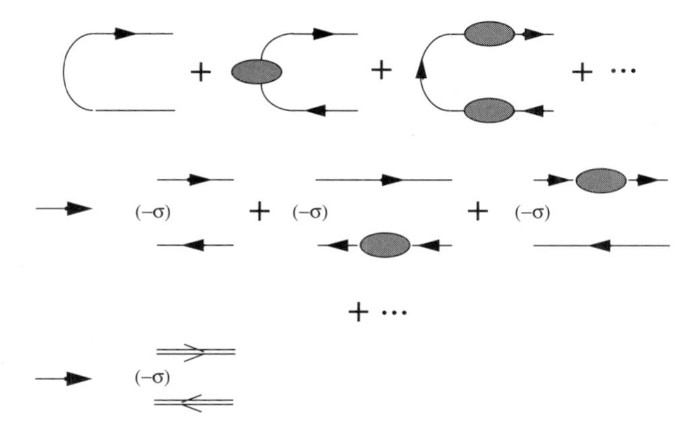

FIGURE 6.9. Differentiation of the SCF propagator $G_{\sigma\sigma}$ with respect to explicit field dependence.

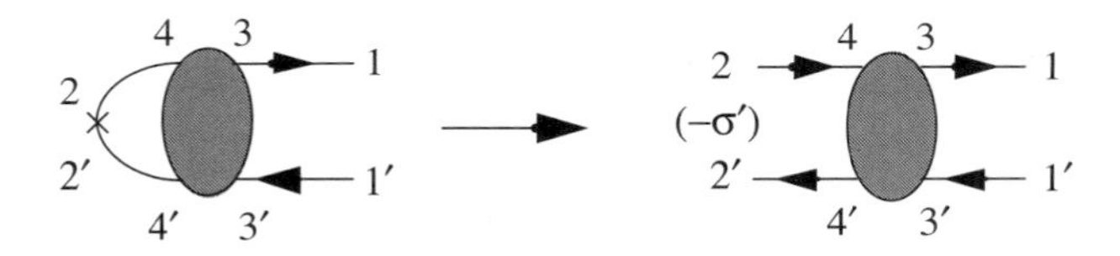

FIGURE 6.10. Differentiation of the lowest-order self-energy correction to $G_{\sigma\sigma}$ with respect to implicit field dependence.

as shown in Fig. 6.9. The double lines indicate SCF propagators containing Σ to all orders.

Explicit differentiation does not give the complete answer, however, since the indirect dependence of G on h through the SCF itself has not yet been taken into account. This requires differentiation *within* the SCF bubbles. It is helpful to concentrate first on the contribution to G of lowest order in Σ. If $\Sigma(G;V)$ were instead a functional Σ^0 depending on G^0 and V, the new term generated in $\beta^{-1}\delta G/\delta h$ would be (see Fig. 6.10)

$$
\begin{aligned}
G^0_{\sigma\sigma}(13)&\frac{\delta\Sigma^0_{\sigma\sigma}(33')}{\delta h(22')}G^0_{\sigma\sigma}(3'1') \\
&= G^0_{\sigma\sigma}(13)\left[\frac{\delta\Sigma^0_{\sigma\sigma}(33')}{\delta G^0_{\sigma'\sigma'}(44')}\frac{\delta G^0_{\sigma'\sigma'}(44')}{\delta h(22')}\right]G^0_{\sigma\sigma}(3'1') \qquad (6.14)\\
&= \beta(-\sigma')G^0_{\sigma\sigma}(13)G^0_{\sigma\sigma}(3'1')\Gamma^{\mathrm{ph}}_{\sigma\sigma,\sigma'\sigma'}(33';44')G^0_{\sigma'\sigma'}(42)G^0_{\sigma'\sigma'}(2'4')\\
&\qquad \Gamma^{\mathrm{ph}}_{\sigma\sigma,\sigma'\sigma'}(33';44') \equiv \frac{\delta\Sigma^0_{\sigma\sigma}(33')}{\delta G^0_{\sigma'\sigma'}(44')}.
\end{aligned}
$$

(Note that a sum on internal indices is implied.) Diagrammatically, $\Gamma^{\mathrm{ph}}_{\sigma\sigma,\sigma'\sigma'}$ is obtained by removing one internal line from the skeleton self-energy in all possible ways. The resulting object functions as a particle-hole scattering

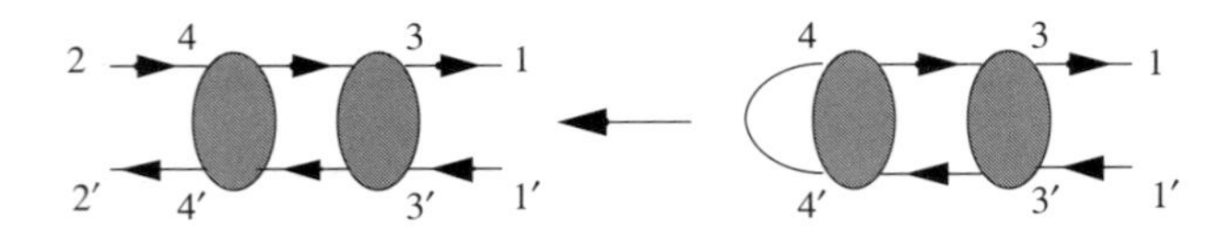

FIGURE 6.11. Correspondence between particle-hole reducible terms in the particle-hole propagator and non-skeleton self-energy corrections.

vertex. Since the self-energy is of skeleton form, the scattering vertex is particle-hole irreducible (in the horizontal channel): this means it cannot be divided into two disjoint portions containing the space-time points $(33')$ and $(44')$ by cutting through a single pair of propagator lines. If this were possible, the parent diagram in Σ would necessarily be non-skeleton (see Fig. 6.11).

Reinstating $\Sigma(G;V)$ in place of $\Sigma^0(G^0;V)$ and retaining terms of all orders in Σ in the original expansion of G leads to three effects in $\beta^{-1}\delta G/\delta h$: (i) incoming and outgoing factors of G^0G^0 are replaced by factors of GG; (ii) the irreducible vertex assumes the form

$$\Gamma^{\text{ph}}_{\sigma\sigma,\sigma'\sigma'}(33';44') = \frac{\delta\Sigma_{\sigma\sigma}(33')}{\delta G_{\sigma'\sigma'}(44')}; \tag{6.15}$$

and (iii) multiple scattering processes based on the irreducible vertex are introduced. The first two effects follow by the same chain of reasoning leading from (6.12) to (6.13). The last effect comes about because differentiation with respect to h may take place within *nested* self-energy insertions when SCF propagators are employed in the diagrammatic representation of Σ; see again the example in Fig. 6.11. Note that the notation for Γ^{ph} has a natural interpretation in terms of the scattering of particle-hole pairs: the compound index $33'$ specifies the space-time coordinates of a final-state particle and hole, while the index $44'$ specifies the same for an initial-state particle and hole.

When all effects are incorporated, the complete particle-hole propagator takes the form

$$\begin{aligned}\frac{\delta G_{\sigma\sigma}(11';h)}{\delta h(22')} &= (-\sigma')\big[\delta_{\sigma\sigma'}G^{\text{ph}}_{\sigma\sigma,\sigma\sigma}(11';22') \\ &\quad + G^{\text{ph}}_{\sigma\sigma,\sigma\sigma}(11';33')\Gamma_{\sigma\sigma,\sigma'\sigma'}(33';44')G^{\text{ph}}_{\sigma'\sigma',\sigma'\sigma'}(44';22')\big],\end{aligned} \tag{6.16}$$

where

$$\begin{aligned}\Gamma_{\sigma\sigma,\sigma'\sigma'}(33';44') &\equiv \Gamma^{\text{ph}}_{\sigma\sigma,\sigma'\sigma'}(33';44') \\ &\quad + \Gamma_{\sigma\sigma,\sigma''\sigma''}(33';55')G^{\text{ph}}_{\sigma''\sigma'',\sigma''\sigma''}(55';66')\Gamma^{\text{ph}}_{\sigma''\sigma'',\sigma'\sigma'}(66';44').\end{aligned} \tag{6.17}$$

and

$$G^{\text{ph}}_{\sigma\sigma,\sigma\sigma}(11';22') \equiv \beta G_{\sigma\sigma}(12)G_{\sigma\sigma}(2'1'). \tag{6.18}$$

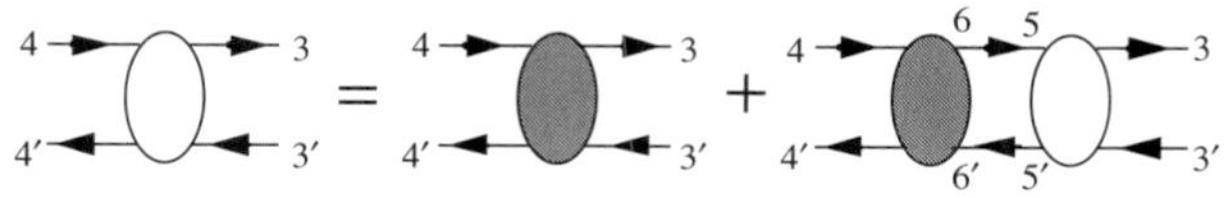

FIGURE 6.12. Symbolic representation of the Bethe–Salpeter equation connecting $\Gamma^{\rm ph}$ and Γ. The irreducible vertex $\Gamma^{\rm ph}$ is represented by a shaded bubble and the complete vertex Γ by an open bubble.

Equation (6.17) relates the irreducible particle-hole vertex $\Gamma^{\rm ph}$ to the complete (or reducible) vertex Γ (see Fig. 6.12). The equation may be recast in an abbreviated matrix notation as

$$\Gamma = \Gamma^{\rm ph} + \Gamma G^{\rm ph}\Gamma^{\rm ph}. \tag{6.19}$$

The solution of this equation sums a formal geometric series, or "ladder," based on the irreducible vertex. Such *Bethe–Salpeter equations* for vertex functions are the analog of Dyson's equation for one-particle self-energies. (Bethe–Salpeter equations are discussed in a much more general context in the sections on parquet theory which follow.)

The uniform static magnetic susceptibility (the object which motivated this discussion) may finally be written out explicitly by multiplying (6.16) by σ, setting $1 = 1'$ and $2 = 2'$, then summing over all variables:

$$\begin{aligned}\chi_{\rm m} = -\sigma\sigma'[&\delta_{\sigma\sigma'}G^{\rm ph}_{\sigma\sigma,\sigma\sigma}(11;22)\\ &+ G^{\rm ph}_{\sigma\sigma,\sigma\sigma}(11;33')\Gamma_{\sigma\sigma,\sigma'\sigma'}(33';44')G^{\rm ph}_{\sigma'\sigma',\sigma'\sigma'}(44';22)],\end{aligned} \tag{6.20}$$

with a summation on all variables understood. It is instructive to check for the Hubbard model that the result of mean-field theory (or the so-called random phase approximation) is recovered. Introducing frequency and momentum variables, one finds in this case

$$\Gamma^{\rm ph}_{\sigma\sigma,\sigma'\sigma'}(kk;k'k') = (U/N)\delta_{\sigma,-\sigma'} \tag{6.21}$$

$$\Gamma_{\sigma\sigma,\sigma'\sigma'}(kk;k'k') = \frac{U}{N}\left[\frac{1}{1-U^2\chi_0^2}\delta_{\sigma,-\sigma'} - \frac{U\chi_0}{1-U^2\chi_0^2}\delta_{\sigma\sigma'}\right], \tag{6.22}$$

where

$$k \equiv (\mathbf{k}, i\omega) \tag{6.23}$$

and

$$\chi_0 = \frac{1}{N}\sum_{\mathbf{k}}(-\partial f_{\mathbf{k}}/\partial\tilde{\varepsilon}_{\mathbf{k}}) \tag{6.24}$$

is the bare particle-hole susceptibility. The quantities $\tilde{\varepsilon}_{\mathbf{k}}$ and $f_{\mathbf{k}}$ are as in (2.6) and (2.7). The uniform static magnetic susceptibility becomes

$$\begin{aligned}\chi_{\rm m} &= 2\chi_0 + \frac{2U\chi_0^2}{1-U^2\chi_0^2} + \frac{2U^2\chi_0^3}{1-U^2\chi_0^2}\\ &= \frac{2\chi_0}{1-U\chi_0},\end{aligned} \tag{6.25}$$

as expected.

In more elaborate applications it is frequently useful to introduce vertices defined relative to the total spin S and spin projection S^z of a particle-hole (or particle-particle) pair, rather than relative to individual spins as in the example above. This is because the total spin is conserved in scattering processes whenever the underlying action respects spin rotational symmetry (the usual case for normal-state systems with $h = 0$). A procedure for carrying out this spin diagonalization in the context of general vertex functions is described in Section 9; see also the discussion in [4]. As an example, the magnetic susceptibility in the last example may be rewritten using a magnetic vertex function as

$$\begin{aligned} \chi_{\mathrm{m}} &= -2\big[G^{\mathrm{ph}}(11;22) + G^{\mathrm{ph}}(11;33')\Gamma_{\mathrm{m}}(33';44')G^{\mathrm{ph}}(44';22)\big] \\ G^{\mathrm{ph}}(12;34) &= \beta G(13)G(42) \\ \Gamma_{\mathrm{m}} &= \Gamma_{\sigma\sigma,\sigma\sigma} - \Gamma_{\sigma\sigma,-\sigma-\sigma}. \end{aligned} \tag{6.26}$$

Note that Γ_{m} may be obtained from an irreducible magnetic vertex

$$\Gamma_{\mathrm{m}}^{\mathrm{ph}} = \Gamma_{\sigma\sigma,\sigma\sigma}^{\mathrm{ph}} - \Gamma_{\sigma\sigma,-\sigma-\sigma}^{\mathrm{ph}} \tag{6.27}$$

using the same Bethe–Salpeter equation which relates $\Gamma_{\sigma\sigma,\sigma'\sigma'}^{\mathrm{ph}}$ and $\Gamma_{\sigma\sigma,\sigma'\sigma'}$.

Finally, as an example of an off-diagonal response function, let us consider the mean-field susceptibility of the Hubbard model to a uniform static singlet-pairing field Δ. Due to the presence of the pairing field in S_0, two Σ diagrams must be considered at Hartree–Fock level (see Fig. 6.13). It is convenient to adopt a Nambu notation based on spinors of anticommuting variables, as in (4.17). Functions dependent on "particle-hole" and spin labels may then be decomposed in terms of two sets of Pauli matrices, $\widehat{\tau}_i$ and $\widehat{\sigma}_i$. The dimensionless bare propagator matrix takes the form

$$G^0(\mathbf{k},\tau) = -\beta^{-1}\left[\frac{\partial}{\partial\tau}\hat{1} + \varepsilon_{\mathbf{k}}\hat{\tau}_3 + \Delta(i\hat{\sigma}_2)\hat{\tau}_1\right]^{-1}. \tag{6.28}$$

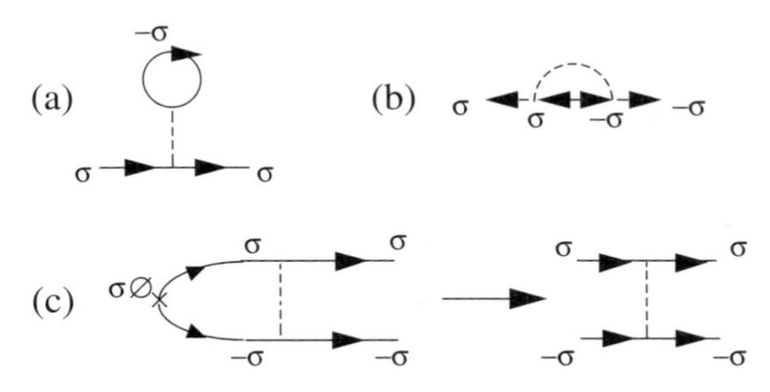

FIGURE 6.13. Calculation of the Hartree–Fock singlet-pair susceptibility for the Hubbard model. (a–b) Diagonal and off-diagonal Hartree–Fock self-energies. (c) Differentiation of the off-diagonal self-energy correction with respect to implicit field dependence.

The SCF propagator contains both a diagonal, or normal, SCF of the form $\Sigma_N \hat{\tau}_3$ (Fig. 6.13a) and an off-diagonal, or anomalous, SCF of the form $\Sigma_A (i\hat{\sigma}_2)\hat{\tau}_1$ (Fig. 6.13b). (For comparison, these fields correspond to Σ and Δ in (4.17).) The explicit form of G^{SCF} is then

$$G^{\mathrm{SCF}}(\mathbf{k},\tau) = -\beta^{-1}\left[\frac{\partial}{\partial\tau}\hat{1} + (\varepsilon_{\mathbf{k}} + \Sigma_N)\hat{\tau}_3 + (\Delta + \Sigma_A)(i\hat{\sigma}_2)\hat{\tau}_1\right]^{-1}$$
$$\equiv G_1 \hat{1} + G_2\hat{\tau}_3 + G_A(i\hat{\sigma}_2)\hat{\tau}_1. \tag{6.29}$$

Just as in the particle-hole examples treated previously, the irreducible singlet scattering vertex may be evaluated as a functional derivative. In this simple case one can write

$$\Gamma_{\mathrm{s}}^{\mathrm{pp}} = \Gamma^{\mathrm{pp}}_{\sigma,-\sigma;\sigma,-\sigma} - \Gamma^{\mathrm{pp}}_{\sigma,-\sigma;-\sigma,\sigma} = 2\frac{\delta\Sigma_A}{\delta G_A}. \tag{6.30}$$

(As discussed in more detail in subsequent sections, care must be taken to prevent double-counting in particle-particle calculations due to the constraint imposed by indistinguishability. We are consistent in associating a factor of $\frac{1}{2}$ with each intermediate-state particle-particle propagator; see, e.g., (7.17).) In the Fourier-transformed basis the single off-diagonal diagram (Fig. 6.13b) gives

$$\Gamma_{\mathrm{s}}^{\mathrm{pp}}(k+Q,-k;k'+Q,-k') = 2U/N, \tag{6.31}$$

with Q the center-of-mass momentum-energy. To find the complete singlet vertex Γ_{s} it is necessary to solve a Bethe–Salpeter equation analogous to (6.17); since $\Gamma_{\mathrm{s}}^{\mathrm{pp}}$ is momentum- and frequency-independent, a closed-form solution is possible as before. One finds for $\Delta = 0$

$$\Gamma_{\mathrm{s}}(Q) = \frac{2U}{N}\frac{1}{1+U\chi_0^{\mathrm{s}}(Q)}, \tag{6.32}$$

where

$$\chi_0^{\mathrm{s}}(Q) = \frac{1}{N}\sum_{\mathbf{k}} \frac{1 - f_{\mathbf{k}} - f_{\mathbf{k+Q}}}{(\tilde{\varepsilon}_{\mathbf{k}} + \tilde{\varepsilon}_{\mathbf{k+Q}} - 2\mu) - i\Omega_m} \tag{6.33}$$

is the bare singlet susceptibility. Finally, the SCF singlet susceptibility follows by substitution of (6.32) in the particle-particle analog of (6.26):

$$\chi^{\mathrm{s}} = \chi_0^{\mathrm{s}}\left[1 - \frac{U\chi_0^{\mathrm{s}}}{1+U\chi_0^{\mathrm{s}}}\right]$$
$$= \frac{\chi_0^{\mathrm{s}}}{1+U\chi_0^{\mathrm{s}}}. \tag{6.34}$$

At this point a framework has been set up for the discussion of Baym–Kadanoff generalizations of mean-field theory. The most natural infinite-order generalization is the so-called fluctuation exchange, or FLEX, approximation. This approximation shares many formal features with parquet theories. For this reason we proceed in the following sections to derive and discuss the general features of parquet theories, before returning to the FLEX approximation in Section 10.

7 Crossing Symmetry

Baym–Kadanoff approximations provide a general means to calculate correlation functions within a self-consistent one-particle framework. Their accuracy is limited, however, by the failure to incorporate self-consistency at the two-particle level. For example, vertex functions which appear in the calculation of the one-particle self-energy are not the same as those generated by functional differentiation in two-particle correlation functions. This deficiency is addressed by parquet theories, which were first studied in condensed matter physics by de Dominicis and Martin [3].

Before proceeding it is important to note the fundamental difference in philosophy which underlies the derivation of Baym–Kadanoff and parquet theories. Baym–Kadanoff theories attach primary importance to Σ, the one-particle SCF, or self-energy. So long as the SCF is Φ-derivable, the resulting one-particle propagators satisfy continuity equations, and the two-particle correlation functions which result from functional differentiation are guaranteed to be thermodynamically self-consistent. On the other hand, Baym–Kadanoff theories invariably produce two-particle correlation functions which violate the set of exact symmetries dictated by the Pauli Principle. In contrast, parquet theories are formulated to satisfy these symmetries exactly. Primary importance is attached to the two-body scattering vertex Γ, and the one-particle self-energy Σ assumes a subsidiary role as an integral over Γ. As a result, parquet theories guarantee the self-consistent treatment of *both* one-particle and two-particle propagators. They do not, on the other hand, guarantee the exact respect of conservation laws and thermodynamic self-consistency.

Since neither Baym–Kadanoff nor parquet theories lead to the respect of all conservation laws and symmetries, which is to be considered more fundamental? In some sense the answer depends on the application anticipated. For example, in some transport studies it may be essential to employ propagators which satisfy exact continuity equations; in this case, a Baym–Kadanoff treatment may be essential. More generally, however, if the principal criterion is quantitative accuracy, we venture the opinion that parquet theories are the more fundamental since they incorporate a higher level of self-consistency. This view is justified more explicitly in Section 10, in which the connection between an infinite-order Baym–Kadanoff approximation (the "fluctuation exchange" approximation [4] and parquet theory is established.

In deriving the equations of parquet theory [7] it is convenient to start with the expression for the partition function of a lattice electron system

$$\begin{array}{c} 3 \rightarrow \boxed{\Gamma} \rightarrow 1 \\ 4 \leftarrow \quad \leftarrow 2 \end{array}$$

FIGURE 6.14. Complete particle-hole vertex $\Gamma(12;34)$. The final-state lines are on the right.

introduced in (4.1). For notational convenience in this section and the next, we incorporate the electron spin with the space and time coordinates in a single variable:

$$1 \equiv (\mathbf{x}_1, \tau_1, \sigma_1), \tag{7.1}$$

so that

$$\sum_1 \equiv \sum_{\sigma_1} \sum_{\mathbf{x}_1} \int_0^\beta \frac{d\tau_1}{\beta}. \tag{7.2}$$

The *complete particle-hole vertex* takes the form

$$\Gamma = \tfrac{1}{2} \sum_i \Gamma(12;34) \bar{c}(1) c(2) \bar{c}(4) c(3), \tag{7.3}$$

represented diagrammatically in Fig. 6.14. As in Section 6, this is just the complete amplitude for transitions between the particle-hole states labeled by the compound indices 34 and 12. (Note that the notational convention for Γ used here differs from that in [4] and [7].)

The symmetry of the complete vertex under label permutations is implied by the anticommutation properties of c and $\bar{c}$. For example, since

$$\Gamma = -\tfrac{1}{2} \sum_i \Gamma(12;34) \bar{c}(4) c(2) \bar{c}(1) c(3), \tag{7.4}$$

it follows by relabeling that

$$\Gamma(12;34) = -\Gamma(42;31). \tag{7.5}$$

Likewise

$$\Gamma(12;34) = \Gamma(43;21). \tag{7.6}$$

Equation (7.5) is known as a *crossing relation*, since it may be obtained diagrammatically by crossing lines 1 and 4 in Fig. 6.14. Alternatively, the left-hand side of (7.5) describes the scattering process which takes place from left to right, while the right-hand side describes the scattering process which takes place from top to bottom. (7.6) is the more trivial statement that the amplitude remains the same if a scattering process is reversed.

The *complete particle-particle vertex* may be written

$$\Gamma_P = \tfrac{1}{2} \sum_i \Gamma_P(12;34) \bar{c}(1) \bar{c}(2) c(4) c(3). \tag{7.7}$$

Again applying the anticommutation properties of c and $\bar{c}$ gives

$$\Gamma_P = -\tfrac{1}{2}\sum_i \Gamma_P(12;34)\bar{c}(1)c(4)\bar{c}(2)c(3). \tag{7.8}$$

This is just an alternate form for the complete particle-hole vertex, so

$$\Gamma(12;34) = -\Gamma_P(14;32). \tag{7.9}$$

This is a second crossing relation which corresponds diagrammatically to crossing lines 2 and 4 in Fig. 6.14. Finally, a similar argument establishes that Γ_P is antisymmetric in its first two and last two labels:

$$\Gamma_P(12;34) = -\Gamma_P(21;34) = -\Gamma_P(12;43). \tag{7.10}$$

The crossing relations and antisymmetry properties of the complete vertices are consequences of the Pauli principle, which should be respected by approximate solutions. The requirement of crossing symmetry is, however, extremely difficult to achieve in practice. The usual approach to perturbation theory [8] for Γ and Γ_P is to define *irreducible particle-hole* and *particle-particle vertices* $\Gamma^{\rm ph}$, $\overline{\Gamma}^{\rm ph}$ and $\Gamma^{\rm pp}$. (The subscript "P" on the irreducible particle-particle vertex may be dropped, since no ambiguity results.) A diagram representing a contribution to Γ_P is said to be particle-particle irreducible if it cannot be split in two by cutting two lines representing a single particle-particle pair. Particle-hole irreducibility is somewhat more complicated since each contribution to Γ may be viewed in two separate channels (horizontal and vertical). Diagrams irreducible in the horizontal channel (i.e., those which cannot be split by a vertical cut through two lines) contribute to $\Gamma^{\rm ph}$, while diagrams irreducible in the vertical channel contribute to $\overline{\Gamma}^{\rm ph}$. The quantities $\Gamma^{\rm ph}$ and $\Gamma^{\rm pp}$ are analogs for particle-hole and particle-particle states of the irreducible self-energy Σ for one-particle states.

The irreducible vertices $\Gamma^{\rm ph}$ and $\overline{\Gamma}^{\rm ph}$ interchange roles when diagrams are alternately examined in the horizontal and vertical channels. Attaching anticommuting c-numbers to keep track of signs, one finds

$$\begin{aligned}
\tfrac{1}{2}\sum_i \Gamma^{\rm ph}(12;34)\bar{c}(1)c(2)\bar{c}(4)c(3) \\
&= -\tfrac{1}{2}\sum_i \Gamma^{\rm ph}(12;34)\bar{c}(4)c(2)\bar{c}(1)c(3) \\
&= -\tfrac{1}{2}\sum_i \overline{\Gamma}^{\rm ph}(42;31)\bar{c}(1)c(2)\bar{c}(4)c(3),
\end{aligned} \tag{7.11}$$

so that

$$\Gamma^{\rm ph}(12;34) = -\overline{\Gamma}^{\rm ph}(42;31). \tag{7.12}$$

The complete vertices Γ and Γ_P may be recovered from the irreducible vertices by using the Bethe–Salpeter equations represented diagrammatically in Fig. 6.15. The intermediate-state lines in this figure represent the

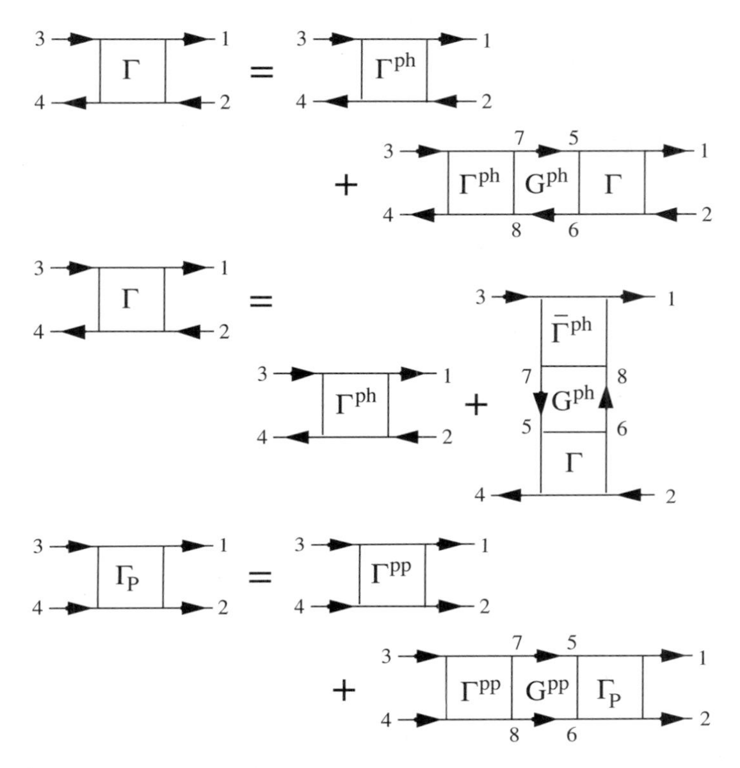

FIGURE 6.15. Bethe–Salpeter equations relating the irreducible vertices Γ^{ph}, $\overline{\Gamma}^{\text{ph}}$ and Γ^{pp} and the two complete vertices Γ and Γ_P. Note these equations are just the analog of Dyson's equation for the one–particle self–energy.

propagators for mutually uncorrelated particle-hole and particle-particle pairs (see Fig. 6.16). The Bethe–Salpeter equations take the form

$$\Gamma(12;34) = \Gamma^{\text{ph}}(12;34) + \Gamma(12;56)G^{\text{ph}}(56;78)\Gamma^{\text{ph}}(78;34) \tag{7.13}$$

$$\Gamma(12;34) = \overline{\Gamma}^{\text{ph}}(12;34) + \Gamma(42;56)G^{\text{ph}}(56;78)\overline{\Gamma}^{\text{ph}}(18;37) \tag{7.14}$$

$$\Gamma_P(12;34) = \Gamma^{\text{pp}}(12;34) + \Gamma_P(12;56)G^{\text{pp}}(56;78)\Gamma^{\text{pp}}(78;34), \tag{7.15}$$

where

$$\begin{aligned} G^{\text{ph}}(12;34) &= \beta\langle c(1)\bar{c}(3)\rangle\langle c(4)\bar{c}(2)\rangle \\ &= \beta G(13)G(42) \end{aligned} \tag{7.16}$$

$$G^{\text{pp}}(12;34) = -\tfrac{1}{2}\beta G(13)G(24). \tag{7.17}$$

A sum is implied for repeated indices.

The factor of $\frac{1}{2}$ in (7.17) arises due to indistinguishability: if it were omitted, double-counting of diagrams would result. As an example, consider the contributions to the complete two-particle vertex in Fig. 6.17. The diagrams

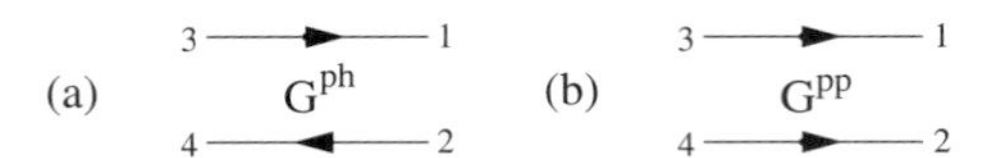

FIGURE 6.16. Propagators for mutually uncorrelated particle-hole and particle-particle pairs. (a) $G^{\mathrm{ph}}(12;34)$. (b) $G^{\mathrm{pp}}(12;34)$.

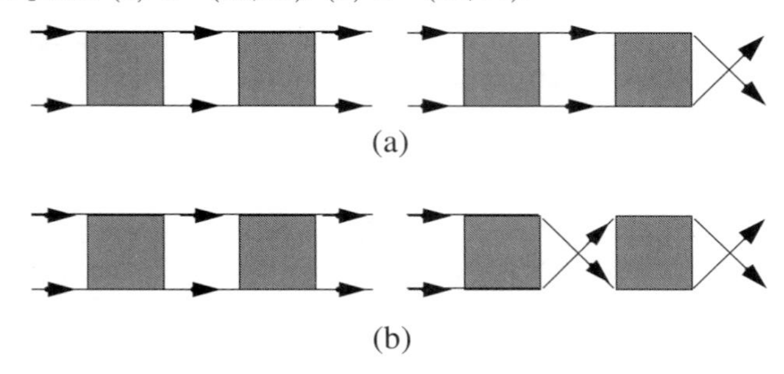

FIGURE 6.17. Contributions to the complete particle-particle vertex. (a) Topologically inequivalent contributions. (b) Topologically equivalent contributions. Double-counting results if these diagrams are included separately.

in Fig. 6.17a are topologically inequivalent and must be counted separately in evaluating the vertex. On the other hand, the contributions in Fig. 6.17b are topologically equivalent and should not be counted separately. Possible double-counting occurs for diagrams in which lines cross in a two-particle intermediate state, since each such diagram may be "twisted" to remove the crossing; the resulting diagram has either no crossed lines or a single set in the final state. Thus a correct prescription for avoiding double-counting in the particle-particle channel is (i) to write down all topologically distinct contributions to the irreducible vertex function (some of which have crossing lines in the initial or final state); and (ii) to associate a factor of $\frac{1}{2}$ with each intermediate-state particle-particle propagator.

For spin-independent interactions (including the zero-range coupling in the Hubbard model), it is convenient to rewrite the complete and irreducible vertices in a form which takes advantage of spin symmetry. The total spin S and total spin projection S^z of a particle-hole or particle-particle pair are conserved in scattering processes. (This should perhaps be restated for the non-instantaneous wave functions which arise in the presence of retarded interactions: all scattering states may be characterized as odd ("singlet") or even ("triplet") under permutation of spin labels.) The particle-hole vertex Γ^{ph} may be decomposed into density ($\Gamma^{\mathrm{ph}}_{\mathrm{d}}$) and magnetic ($\Gamma^{\mathrm{ph}}_{\mathrm{m}}$) components; and the particle-particle vertex Γ^{pp} into singlet ($\Gamma^{\mathrm{pp}}_{\mathrm{s}}$) and triplet ($\Gamma^{\mathrm{pp}}_{\mathrm{t}}$) components. In the absence of external fields, isotropy in spin space guarantees that the vertices for the three magnetic or triplet components ($S^z = 0, \pm 1$) are identical. Note finally that the spin decomposition of the complete vertices Γ and Γ_P does not "diagonalize" the crossing relations: e.g., the complete magnetic vertex in the horizontal channel is related to a combination of magnetic and density vertices in the vertical channel by (7.5).

Expressions for Γ and Γ_P follow immediately once approximate forms are specified for the irreducible vertices $\Gamma^{\rm ph}$ and $\Gamma^{\rm pp}$. (An expression for $\overline{\Gamma}^{\rm ph}$ follows from the crossing relation in (7.12).) Unfortunately low-order choices for $\Gamma^{\rm ph}$ and $\Gamma^{\rm pp}$ lead to violation of the crossing relations for the complete vertices. As an example, consider the Hartree–Fock approximation for the Hubbard modelHubbard model!irreducible vertices. After transforming to the Fourier representation, one finds at this level

$$\begin{aligned} \Gamma^{\rm ph}_{\rm d} &= \text{constant} = U/N \\ \Gamma^{\rm ph}_{\rm m} &= -U/N \\ \Gamma^{\rm pp}_{\rm s} &= 2U/N \\ \Gamma^{\rm pp}_{\rm t} &= 0. \end{aligned} \tag{7.18}$$

(Recall that constant interactions in momentum- and frequency-space result from point couplings in space-time.) The complete $S^z = 1$ magnetic vertex (see Fig. 6.18a) assumes the familiar mean-field form

$$\begin{aligned} \Gamma_{\rm m}(k+Q,k;k'+Q,k') &= -\frac{U}{N}\frac{1}{1-U\chi_{\rm ph}(Q)} \\ \chi_{\rm ph}(Q) &= -\frac{\beta}{N}\sum_k G(k+Q)G(k), \end{aligned} \tag{7.19}$$

with $k \equiv (\mathbf{k}, i\omega_n)$. Applying the crossing operation from (7.5) generates a contribution to the $(\uparrow\uparrow\rightarrow\downarrow\downarrow)$ particle-hole vertex (see Fig. 6.18b):

$$\Gamma^{\rm ph}_{\downarrow\downarrow,\uparrow\uparrow}(k',k;k'+Q,k+Q) = \frac{U}{N}\frac{1}{1-U\chi_{\rm ph}(Q)}, \tag{7.20}$$

i.e.,

$$\Gamma^{\rm ph}_{\downarrow\downarrow,\uparrow\uparrow}(k+Q,k;k'+Q,k') = \frac{U}{N}\frac{1}{1-U\chi_{\rm ph}(k'-k)}. \tag{7.21}$$

This expression is irreducible in the horizontal channel; i.e., it is a contribution to $\Gamma^{\rm ph}$. In contrast, the Hartree–Fock approximation for the irreducible $(\uparrow\uparrow\rightarrow\downarrow\downarrow)$ vertex is just U/N. Thus, the crossing operation generates non-Hartree–Fock expressions for $\Gamma^{\rm ph}$ (and Γ), explicitly demonstrating the violation of (7.5). A similar argument demonstrates the violation of (7.9) at this level (see Fig. 6.18c).

8 Parquet Equations

The Hartree–Fock example in Section 7 illustrates the nature of the problem which must be solved to restore crossing symmetry: the crossing oper-

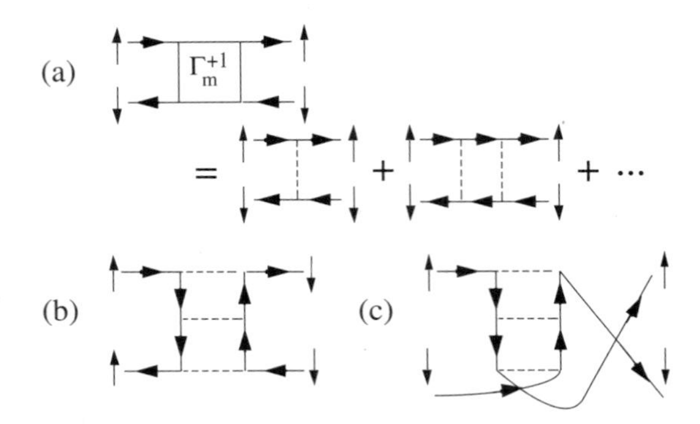

FIGURE 6.18. Illustration of the failure of crossing symmetry at Hartree–Fock level. (a) Complete $S^z = 1$ magnetic vertex $\Gamma_{\rm m}^{+1}$ for the Hubbard model in Hartree–Fock approximation. (b) Contribution to the $\Gamma^{\rm ph}_{\downarrow\downarrow,\uparrow\uparrow}$ vertex which results from crossing. (c) Contribution to the $\Gamma^{\rm pp}_{\uparrow\downarrow,\uparrow\downarrow}$ vertex which results from crossing.

ation transforms (reducible) ladder summations in any of the three scattering channels into contributions to the irreducible vertices in the other two channels. In order to motivate a solution, it is useful to write down a set of *exact* relations between $\Gamma^{\rm ph}$, $\overline{\Gamma}^{\rm ph}$ and $\Gamma^{\rm pp}$. (For brevity the vertices are not decomposed in (S, S^z) channels below; the necessary transformations are discussed in Section 9.) The so-called *parquet equations* for the irreducible vertices take the form shown diagrammatically in Fig. 6.19. The vertices $\Lambda^{\rm irr}$ and $\Lambda_P^{\rm irr}$ are irreducible in all three channels and have the crossing symmetry of the complete vertices Γ and Γ_P:

$$\Lambda^{\rm irr}(12;34) = -\Lambda^{\rm irr}(42;31) = -\Lambda_P^{\rm irr}(14;32). \tag{8.1}$$

A good deal of care is necessary to prevent sign errors in writing the equations out: based on the label-ordering conventions in (7.3) and (7.7), the parquet equations are as follows:

$$\begin{aligned}\Gamma^{\rm ph}(12;34) = \Lambda^{\rm irr}(12;34) &+ \Gamma(42;56)G^{\rm ph}(56;78)\overline{\Gamma}^{\rm ph}(18;37)\\ &+ \Gamma_P(41;56)G^{\rm pp}(56;78)\Gamma^{\rm pp}(78;32) \quad (8.2)\\ \overline{\Gamma}^{\rm ph}(12;34) = \Lambda^{\rm irr}(12;34) &+ \Gamma(12;56)G^{\rm ph}(56;78)\Gamma^{\rm ph}(78;34)\\ &- \Gamma_P(14;56)G^{\rm pp}(56;78)\Gamma^{\rm pp}(78;32) \quad (8.3)\\ \Gamma^{\rm pp}(12;34) = \Lambda_P^{\rm irr}(12;34) &- \Gamma(24;56)G^{\rm ph}(56;78)\overline{\Gamma}^{\rm ph}(18;37)\\ &+ \Gamma(14;56)G^{\rm ph}(56;78)\overline{\Gamma}^{\rm ph}(28;37). \quad (8.4)\end{aligned}$$

Summation is implied for all repeated indices. The right-hand side of (8.4) may alternately be written in terms of $\overline{\Gamma}^{\rm ph}$ and $\Gamma^{\rm ph}$ or in terms of $\Gamma^{\rm ph}$ only.

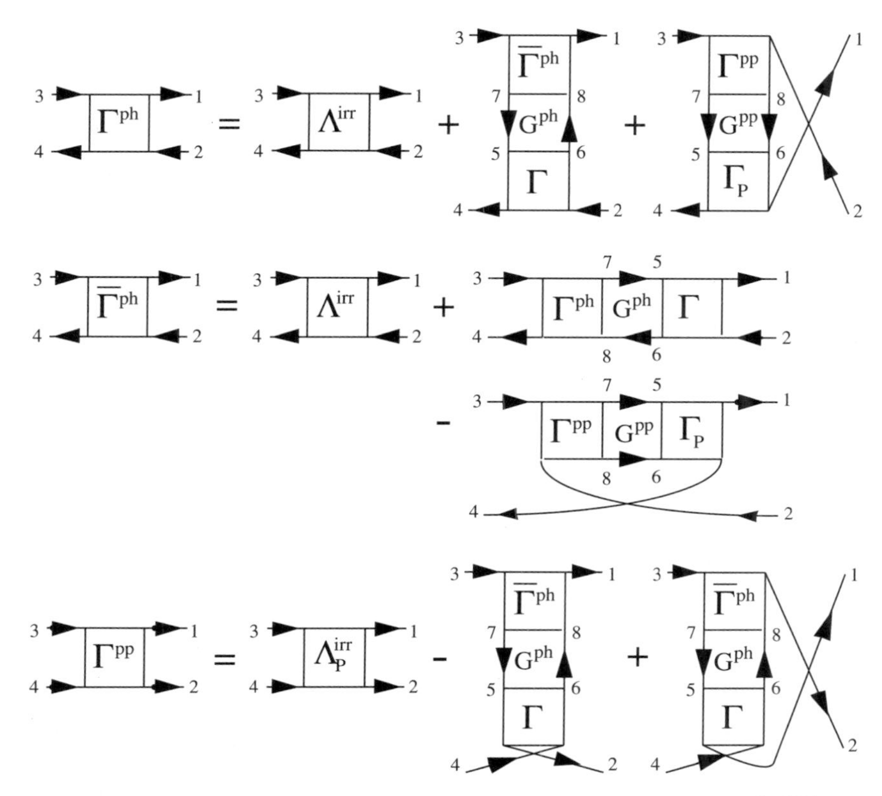

FIGURE 6.19. Parquet equations relating the irreducible vertices Γ^{ph}, $\overline{\Gamma}^{\mathrm{ph}}$ and Γ^{pp}. These equations express the fact that a diagram irreducible in one channel is either irreducible in all three channels or reducible in exactly one of the other two.

It is easy to see that (8.2) and (8.3) are equivalent if (7.12) is satisfied: interchanging labels 1 and 4 and multiplying by -1 in (8.3) gives

$$-\overline{\Gamma}^{\mathrm{ph}}(42;31) = -\Lambda^{\mathrm{irr}}(42;31) - \Gamma(42;56)G^{\mathrm{ph}}(56;78)\Gamma^{\mathrm{ph}}(78;31) + \Gamma_P(41;56)G^{\mathrm{pp}}(56;78)\Gamma^{\mathrm{pp}}(78;32); \quad (8.5)$$

this may be immediately reduced to (8.2) by applying (7.12) and (8.1). This means that only two coupled equations need to be written:

$$\Gamma^{\mathrm{ph}}(12;34) = \Lambda^{\mathrm{irr}}(12;34) - \Gamma(42;56)G^{\mathrm{ph}}(56;78)\Gamma^{\mathrm{ph}}(78;31) + \Gamma_P(41;56)G^{\mathrm{pp}}(56;78)\Gamma^{\mathrm{pp}}(78;32) \quad (8.6)$$

$$\Gamma^{\mathrm{pp}}(12;34) = -\Lambda^{\mathrm{irr}}(14;32) + \Gamma(24;56)G^{\mathrm{ph}}(56;78)\Gamma^{\mathrm{ph}}(78;31) - \Gamma(14;56)G^{\mathrm{ph}}(56;78)\Gamma^{\mathrm{ph}}(78;32). \quad (8.7)$$

The complete vertices Γ and Γ_P may be eliminated from the right-hand side of (8.6) and (8.7) using the Bethe–Salpeter equations represented in

Fig. 6.15. It is convenient at this point to view the vertex functions as matrices with compound indices and to introduce a convention for matrix multiplication:

$$(\Gamma_A \Gamma_B)(12;34) = \Gamma_A(12;56)\Gamma_B(56;34). \tag{8.8}$$

As before, summation is implied for repeated indices. The identity matrix is just

$$\mathbf{1}(12;34) = \delta_{13}\delta_{24}. \tag{8.9}$$

With these conventions (8.6) and (8.7) may be rewritten in the form

$$\Gamma^{\text{ph}}(12;34) = \Lambda^{\text{irr}}(12;34) - \Phi(42;31) + \Psi(41;32) \tag{8.10}$$

$$\Gamma^{\text{pp}}(12;34) = -\Lambda^{\text{irr}}(14;32) + \Phi(24;31) - \Phi(14;32), \tag{8.11}$$

where

$$\Phi(12;34) = \left[\Gamma^{\text{ph}}(\mathbf{1} - G^{\text{ph}}\Gamma^{\text{ph}})^{-1}G^{\text{ph}}\Gamma^{\text{ph}}\right](12;34) \tag{8.12}$$

$$\Psi(12;34) = \left[\Gamma^{\text{pp}}(\mathbf{1} - G^{\text{pp}}\Gamma^{\text{pp}})^{-1}G^{\text{pp}}\Gamma^{\text{pp}}\right](12;34). \tag{8.13}$$

In matrix notation the crossing relations in (7.5) and (7.9) take the form

$$\left[\Gamma^{\text{ph}} + \Phi\right](12;34) = -\left[\Gamma^{\text{ph}} + \Phi\right](42;31) \tag{8.14}$$

$$\left[\Gamma^{\text{ph}} + \Phi\right](12;34) = -\left[\Gamma^{\text{pp}} + \Psi\right](14;32). \tag{8.15}$$

It may be verified explicitly that these equations are satisfied by the parquet solutions. For example, substituting for Γ^{ph} on the left-hand side of (8.14) using (8.10) gives

$$\Lambda^{\text{irr}}(12;34) - \Phi(42;31) + \Psi(41;32) + \Phi(12;34). \tag{8.16}$$

Regrouping the first, third and fourth terms and invoking (8.1), then again applying (8.10) or (8.11) to simplify the resulting expression gives the desired relation.

It is now clear how to generate *approximate* expressions for Γ^{ph} and Γ^{pp} which insure crossing symmetry for the resulting complete vertices: it is necessary to solve the parquet equations starting with an approximate, rather than exact, form for Λ^{irr}. (This approximate form must itself satisfy the crossing relations in (8.1).) The simplest contribution to Λ^{irr} is just the bare interaction v (including both direct and exchange terms). The next higher order contribution is the "envelope" graph of $O(v^4)$ represented schematically in Fig. 6.20. It has long been believed [9–11] that for "weakly correlated" systems like nuclear matter and ^{3}He the parquet solution generated by the bare vertex should be extremely accurate; it may be hoped that this accuracy carries over to systems with intermediate and strong coupling.

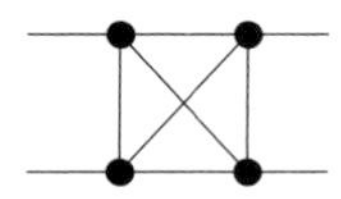

FIGURE 6.20. "Envelope" contribution to the fully irreducible vertices Λ^{irr} and Λ_P^{irr}. Note that the heavy dots, which represent the bare interaction, may be "exploded" into dashed lines representing either direct or exchange processes.

So far the discussion has concentrated entirely on developing approximations for the irreducible two-particle vertices. The SCF problem is considerably complicated by the fact that the one-particle self-energy must be treated simultaneously. In most cases changes in the vertices are largely offset by compensating changes in the self-energy. The exact expression for the self-energy is indicated schematically in Fig. 6.21. Note that in the two-particle scattering term the complete vertex appears only once to avoid overcounting. Diagrams can *still* potentially be double-counted, since two indistinguishable particle lines appear in the intermediate state. For this reason, the contribution Σ_2 written below contains an additional factor of $\frac{1}{2}$. The self-energy equation may be written out explicitly as follows:

$$\Sigma(11') = \Sigma_1(11') + \Sigma_2(11') \tag{8.17}$$

$$\Sigma_1(11') = \delta_{11'} \sum_2 v(1-2)\langle n(2)\rangle - v(1-1')G(11') \tag{8.18}$$

$$\begin{aligned}\Sigma_2(11') &= -\tfrac{1}{2}G(76)\Gamma(17;23)G^{\text{ph}}(23;45)v(45;1'6)\\ &= -\tfrac{1}{2}G(76)[\Gamma G^{\text{ph}}v](17;1'6),\end{aligned} \tag{8.19}$$

where v in the last equation is the bare interaction matrix in the particle-hole channel.

It is useful to expand the scattering term Σ_2 by writing out the Bethe–Salpeter equation for Γ, then applying (8.10) or (8.11) to obtain an expres-

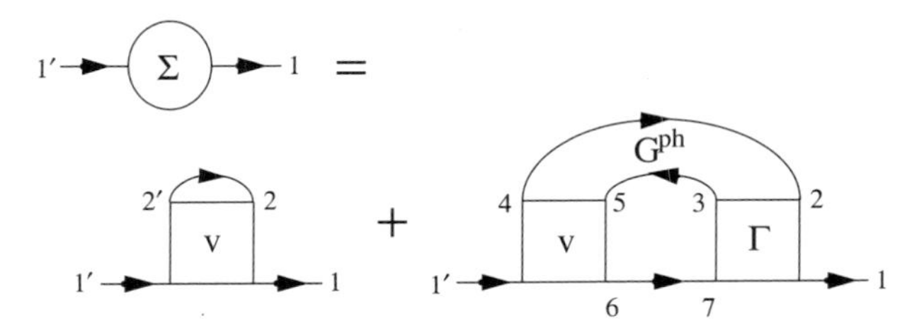

FIGURE 6.21. Diagrammatic representation of the irreducible one-particle self-energy Σ. In the "scattering term," the inclusion of one bare vertex and one complete vertex prevents double-counting. The expression for Γ must follow from solution of the parquet equations (8.10) and (8.11) to insure consistency at both the one- and two-particle levels.

sion symmetric with respect to all three channels. This gives

$$\begin{aligned}\Sigma_2(11') &= -\tfrac{1}{2}G(76)\big[\Gamma^{\mathrm{ph}}(17;23) + \Phi(17;23)\big]G^{\mathrm{ph}}(23;45)v(45;1'6)\\ &= -\tfrac{1}{2}G(76)\big[\Lambda^{\mathrm{irr}}(17;23) + \Phi(17;23) - \Phi(37;21) + \Psi(31;27)\big]\\ &\qquad \times G^{\mathrm{ph}}(23;45)v(45;1'6). \qquad (8.20)\end{aligned}$$

The two particle-hole ladder terms in (8.20) are in fact identical. This can be seen algebraically by noting

$$\begin{aligned}G^{\mathrm{ph}}(23;45)G(76) &= \beta[G(24)G(53)]G(76)\\ &= \beta[G(76)G(53)]G(24)\\ &= G^{\mathrm{ph}}(73;65)G(24) \qquad (8.21)\\ \Phi(37;21) &= \Phi(12;73) \qquad (8.22)\\ v(45;1'6) &= -v(65;1'4). \qquad (8.23)\end{aligned}$$

Equations (8.22) and (8.23) are the equivalents of (7.6) and (7.5) for the particle-hole ladder and the bare interaction. Substituting these expressions in (8.20) and relabeling internal indices establishes the claim.

A further rearrangement in the particle-particle ladder term is convenient: in this case note using (7.16) and (7.17) that

$$G^{\mathrm{ph}}(23;45)G(76) = -2G^{\mathrm{pp}}(72;64)G(53). \qquad (8.24)$$

Furthermore, the equivalents of (7.10) and (7.9) for the particle-particle ladder and the bare interaction in the particle-particle channel give

$$\begin{aligned}\Psi(31;27) &= \Psi(13;72) \qquad (8.25)\\ v(45;1'6) &= -v_P(46;1'5) = v_P(64;1'5). \qquad (8.26)\end{aligned}$$

It follows, again by substituting and relabeling internal indices, that

$$\begin{aligned}&-\tfrac{1}{2}G(76)\Psi(31;27)G^{\mathrm{ph}}(23;45)v(45;1'6)\\ &\qquad = G(67)\Psi(17;23)G^{\mathrm{pp}}(23;45)v_P(45;1'6). \qquad (8.27)\end{aligned}$$

Finally, the term containing Λ^{irr} in (8.20) may be left as it stands or may be rewritten in the particle-particle channel using similar manipulations:

$$-\tfrac{1}{2}G(76)[\Lambda^{\mathrm{irr}}G^{\mathrm{ph}}v](17;1'6) = G(67)[\Lambda_P^{\mathrm{irr}}G^{\mathrm{pp}}v_P](17;1'6). \qquad (8.28)$$

Applying the results from (8.21)–(8.28) to (8.20) and symmetrizing the first term between the particle-hole and particle-particle channels gives

$$\begin{aligned}\Sigma_2(11') &= \tfrac{1}{2}\big\{-\tfrac{1}{2}G(76)[\Lambda^{\mathrm{irr}}G^{\mathrm{ph}}v](17;1'6) + G(67)[\Lambda_P^{\mathrm{irr}}G^{\mathrm{pp}}v_P](17;1'6)\big\}\\ &\quad - G(76)[\Phi G^{\mathrm{ph}}v](17;1'6) + G(67)[\Psi G^{\mathrm{pp}}v_P](17;1'6). \qquad (8.29)\end{aligned}$$

This last form is conveniently prepared for diagonalization with respect to total spin (see Section 9). This form is also convenient because it guarantees that singular contributions due to the presence of low-energy pairs in any of the three channels will survive even when approximations are employed for the irreducible vertices. Note that the multiple-scattering ladder sums in (8.29) begin at $O[v(\Gamma^{\mathrm{ph}})^2]$ and $O[v_P(\Gamma^{\mathrm{pp}})^2]$. The "second-order term" containing Λ^{irr} contains extra symmetry factors of $\frac{1}{2}$ to prevent double-counting.

Equations (8.10), (8.11), and (8.29) are simplified by introducing total spin and spin projection labels in the following section. This leads to the physically appealing picture of electrons interacting through the exchange of four varieties of "bosons": density, magnetic, singlet-pair, and triplet-pair fluctuations. These bosons play a role similar to that of phonons in perturbation theory for the electron-lattice interaction; in contrast to phonons, however, the bosons are composite objects. The boson components are electrons and holes, which themselves strongly interact through the exchange of other bosons. Within this language, the chief difficulty in generating a quantitatively accurate approximation for a theory's correlation functions is insuring that consistent expressions for both electron *and* boson propagators appear throughout. By enforcing the requirement of crossing symmetry for the complete vertices, the parquet equations (8.10), (8.11), and (8.29) automatically build in the nonlinear coupling between dressed electron and boson excitations necessary for full consistency.

9 Spin Diagonalization

It is important in applications of the parquet formalism for spin-independent interactions to exploit the conservation of spin in two-body scattering processes. This allows the separation of vertex functions into components with $S = 0$ (singlet) and $S = 1$ (triplet). Since the scattering vertices are built up from one-particle propagators, a set of transformations between basis sets labeled by (s_1^z, s_2^z) and (S, S^z) is required. We derive these transformations first, then proceed to write down the parquet equations in spin-diagonalized notation.

We begin with the transformation of the particle-hole vertex functions. The equations relating the complete vertex functions Γ in the two spin basis sets are derived below. The equations for irreducible vertex functions Γ^{ph} and for ladders Φ take the same form after the appropriate notational substitutions. In this section we revert to our previous practice of abbreviating space-time variables as

$$1 \equiv (\mathbf{x}_1, \tau_1), \tag{9.1}$$

while writing out spin variables explicitly.

The initial and final states for the $S^z = \pm 1$ magnetic vertex are already spin diagonalized, so that we may write

$$\Gamma_{\mathrm{m}}^{+1}(12;34) = \Gamma_{\mathrm{m}}^{-1}(12;34) = \Gamma_{\uparrow\downarrow,\uparrow\downarrow}(12;34) = \Gamma_{\downarrow\uparrow,\downarrow\uparrow}(12;34). \tag{9.2}$$

The transformation equations for the $S^z = 0$ particle-hole vertices may be derived by examining the normalized density and magnetic operators

$$d(12) = \frac{1}{\sqrt{2}}[\bar{c}_\uparrow(1)c_\uparrow(2) + \bar{c}_\downarrow(1)c_\downarrow(2)] \tag{9.3}$$

$$m_0(12) = \frac{1}{\sqrt{2}}[\bar{c}_\uparrow(1)c_\uparrow(2) - \bar{c}_\downarrow(1)c_\downarrow(2)]. \tag{9.4}$$

The dependence on the space-time variables may be suppressed, giving the relationship between the spin states created by d and m_0 in matrix notation:

$$\begin{bmatrix} |D\rangle \\ |M_0\rangle \end{bmatrix} = \mathbf{S} \begin{bmatrix} |\uparrow\uparrow\rangle \\ |\downarrow\downarrow\rangle \end{bmatrix} \tag{9.5}$$

with

$$\mathbf{S} = \frac{1}{\sqrt{2}} \begin{bmatrix} 1 & 1 \\ 1 & -1 \end{bmatrix}. \tag{9.6}$$

The transformation between the vertex functions in the two basis sets then takes the form

$$\begin{aligned} \begin{bmatrix} \Gamma_{\uparrow\uparrow,\uparrow\uparrow} & \Gamma_{\uparrow\uparrow,\downarrow\downarrow} \\ \Gamma_{\downarrow\downarrow,\uparrow\uparrow} & \Gamma_{\downarrow\downarrow,\downarrow\downarrow} \end{bmatrix} &= \mathbf{S}^{-1} \begin{bmatrix} \Gamma_{\mathrm{d}} & 0 \\ 0 & \Gamma_{\mathrm{m}}^0 \end{bmatrix} \mathbf{S} \\ &= \tfrac{1}{2} \begin{bmatrix} \Gamma_{\mathrm{d}} + \Gamma_{\mathrm{m}}^0 & \Gamma_{\mathrm{d}} - \Gamma_{\mathrm{m}}^0 \\ \Gamma_{\mathrm{d}} - \Gamma_{\mathrm{m}}^0 & \Gamma_{\mathrm{d}} + \Gamma_{\mathrm{m}}^0 \end{bmatrix}. \end{aligned} \tag{9.7}$$

Inverting the transformation gives

$$\begin{aligned} \Gamma_{\mathrm{d}} &= \Gamma_{\uparrow\uparrow,\uparrow\uparrow} + \Gamma_{\uparrow\uparrow,\downarrow\downarrow} \\ \Gamma_{\mathrm{m}}^0 &= \Gamma_{\uparrow\uparrow,\uparrow\uparrow} - \Gamma_{\uparrow\uparrow,\downarrow\downarrow}. \end{aligned} \tag{9.8}$$

Note that since the three components of the magnetic vertex are equal, they may be abbreviated Γ_{m}.

It is useful at this point to illustrate the results above with a concrete example drawn from the Hubbard model. To lowest order the irreducible particle-hole vertex functions in the original spin basis are

$$\begin{aligned} \Gamma_{\uparrow\uparrow,\uparrow\uparrow}^{\mathrm{ph}}(k+Q,k;k'+Q,k') &= 0 \\ \Gamma_{\uparrow\uparrow,\downarrow\downarrow}^{\mathrm{ph}}(k+Q,k;k'+Q,k') &= U/N. \end{aligned} \tag{9.9}$$

The resulting expressions for the irreducible density and magnetic vertices are

$$\begin{aligned} \Gamma_{\mathrm{d}}^{\mathrm{ph}}(k+Q,k;k'+Q,k') &= \quad 0 + (U/N) = U/N \\ \Gamma_{\mathrm{m}}^{\mathrm{ph}}(k+Q,k;k'+Q,k') &= \quad 0 - (U/N) = -U/N. \end{aligned} \tag{9.10}$$

The spin diagonalization of the particle-particle vertex functions can be carried out in like manner. As before we begin by deriving transformation equations for the complete vertex function Γ_P. (The subscript P is suppressed below, since no confusion results.) The $S^z = \pm 1$ components of the triplet vertex are trivial:

$$\Gamma_{\mathrm{t}}^{+1}(12;34) = \Gamma_{\mathrm{t}}^{-1}(12;34) = \Gamma_{\uparrow\uparrow,\uparrow\uparrow}(12;34) = \Gamma_{\downarrow\downarrow,\downarrow\downarrow}(12;34). \tag{9.11}$$

The singlet and $S^z = 0$ triplet vertices are related to the $\uparrow\downarrow$ and $\downarrow\uparrow$ components of Γ_P. As before, the space-time indices of the particle-particle pair may be temporarily suppressed in working out the spin transformation. Since

$$\begin{bmatrix} |S\rangle \\ |T\rangle \end{bmatrix} = \mathbf{S} \begin{bmatrix} |\uparrow\downarrow\rangle \\ |\downarrow\uparrow\rangle \end{bmatrix} \tag{9.12}$$

with

$$\mathbf{S} = \frac{1}{\sqrt{2}} \begin{bmatrix} 1 & -1 \\ 1 & 1 \end{bmatrix}, \tag{9.13}$$

it follows that

$$\begin{aligned} \begin{bmatrix} \Gamma_{\uparrow\downarrow,\uparrow\downarrow} & \Gamma_{\uparrow\downarrow,\downarrow\uparrow} \\ \Gamma_{\downarrow\uparrow,\uparrow\downarrow} & \Gamma_{\downarrow\uparrow,\downarrow\uparrow} \end{bmatrix} &= \mathbf{S}^{-1} \begin{bmatrix} \Gamma_{\mathrm{s}} & 0 \\ 0 & \Gamma_{\mathrm{t}}^0 \end{bmatrix} \mathbf{S} \\ &= \begin{bmatrix} \Gamma_{\mathrm{s}} + \Gamma_{\mathrm{t}}^0 & -\Gamma_{\mathrm{s}} + \Gamma_{\mathrm{t}}^0 \\ -\Gamma_{\mathrm{s}} + \Gamma_{\mathrm{t}}^0 & \Gamma_{\mathrm{s}} + \Gamma_{\mathrm{t}}^0 \end{bmatrix}. \end{aligned} \tag{9.14}$$

Inverting the transformation gives

$$\begin{aligned} \Gamma_{\mathrm{s}} &= \Gamma_{\uparrow\downarrow,\uparrow\downarrow} - \Gamma_{\uparrow\downarrow,\downarrow\uparrow} \\ \Gamma_{\mathrm{t}}^0 &= \Gamma_{\uparrow\downarrow,\uparrow\downarrow} + \Gamma_{\uparrow\downarrow,\downarrow\uparrow}. \end{aligned} \tag{9.15}$$

These equations are in exact analogy to (9.7) and (9.8) for the density and magnetic vertex functions. The three equal components of the triplet vertex may be abbreviated Γ_{t}. It is easy to see that the singlet and triplet vertices have the expected symmetry in space-time, i.e., Γ_{s} is symmetric under interchange of space-time labels in either the initial or final states, while Γ_{t} is antisymmetric. As an example, since

$$\Gamma_{\uparrow\downarrow,\downarrow\uparrow}(12;34) = -\Gamma_{\uparrow\downarrow,\uparrow\downarrow}(12;43) \tag{9.16}$$

by (7.10), it follows that

$$\Gamma_{\mathrm{s}}(12;34) = \Gamma_{\mathrm{s}}(12;43). \tag{9.17}$$

As before, it is useful to illustrate the transformation equations using irreducible vertex functions in the Hubbard model. The lowest-order one has

$$\begin{aligned} \Gamma^{\mathrm{pp}}_{\uparrow\downarrow,\uparrow\downarrow}(k+Q,-k;k'+Q,-k') &= U/N \\ \Gamma^{\mathrm{pp}}_{\uparrow\downarrow,\downarrow\uparrow}(k+Q,-k;k'+Q,-k') &= -U/N. \end{aligned} \tag{9.18}$$

The resulting expression for $\Gamma_{\rm s}^{\rm pp}$ is

$$\Gamma_{\rm s}^{\rm pp}(k+Q,-k;k'+Q,-k') = (U/N) - (-U/N) = 2U/N. \tag{9.19}$$

The expression for $\Gamma_{\rm t}^{\rm pp}$ at this level is just 0.

We can now proceed to write out the parquet equations (8.10) and (8.11) in a spin-diagonalized form. As an example, consider (8.10) for the irreducible density vertex. Applying the result in (9.8) to the irreducible vertex $\Gamma_{\rm d}^{\rm ph}$, one has

$$\begin{aligned}\Gamma_{\rm d}^{\rm ph}(12;34) = \Lambda_{\rm d}^{\rm irr}(12;34) &- \Phi_{\uparrow\uparrow,\uparrow\uparrow}(42;31) - \Phi_{\downarrow\uparrow,\downarrow\uparrow}(42;31)\\ &+ \Psi_{\uparrow\uparrow,\uparrow\uparrow}(41;32) + \Psi_{\uparrow\downarrow,\uparrow\downarrow}(41;32).\end{aligned} \tag{9.20}$$

The ladders can now be re-expressed in spin-diagonal form using the same transformations derived for the complete vertex functions in (9.7) and (9.14). Thus,

$$\begin{aligned}\Gamma_{\rm d}^{\rm ph}(12;34) &= \Lambda_{\rm d}^{\rm irr}(12;34) - \tfrac{1}{2}(\Phi_{\rm d}+\Phi_{\rm m})(42;31) - \Phi_{\rm m}(42;31)\\ &\qquad + \Psi_{\rm t}(41;32) + \tfrac{1}{2}(\Psi_{\rm s}+\Psi_{\rm t})(41;32)\\ &= \Lambda_{\rm d}^{\rm irr}(12;34) - \tfrac{1}{2}\Phi_{\rm d}(42;31) - \tfrac{3}{2}\Phi_{\rm m}(42;31)\\ &\qquad + \tfrac{1}{2}\Psi_{\rm s}(41;32) + \tfrac{3}{2}\Psi_{\rm t}(41;32),\end{aligned} \tag{9.21}$$

where

$$\Phi_{\rm r}(12;34) = \left[\Gamma_{\rm r}^{\rm ph}(\mathbf{1} - G^{\rm ph}\Gamma_{\rm r}^{\rm ph})^{-1}G^{\rm ph}\Gamma_{\rm r}^{\rm ph}\right](12;34) \tag{9.22}$$

for r = d and m; and

$$\Psi_{\rm r}(12;34) = \left[\Gamma_{\rm r}^{\rm pp}(\mathbf{1} - G^{\rm pp}\Gamma_{\rm r}^{\rm pp})^{-1}G^{\rm pp}\Gamma_{\rm r}^{\rm pp}\right](12;34) \tag{9.23}$$

for r = s and t. In like manner the equations for the irreducible magnetic, singlet, and triplet vertex functions may be derived. The results are as follows:

$$\begin{aligned}\Gamma_{\rm m}^{\rm ph}(12;34) = \Lambda_{\rm m}^{\rm irr}(12;34) &- \tfrac{1}{2}\Phi_{\rm d}(42;31) + \tfrac{1}{2}\Phi_{\rm m}(42;31)\\ &- \tfrac{1}{2}\Psi_{\rm s}(41;32) + \tfrac{1}{2}\Psi_{\rm t}(41;32)\end{aligned} \tag{9.24}$$

$$\begin{aligned}\Gamma_{\rm s}^{\rm pp}(12;34) = \Lambda_{\rm s}^{\rm irr}(12;34) &+ \tfrac{1}{2}\Phi_{\rm d}(24;31) - \tfrac{3}{2}\Phi_{\rm m}(24;31)\\ &+ \tfrac{1}{2}\Phi_{\rm d}(14;32) - \tfrac{3}{2}\Phi_{\rm m}(14;32)\end{aligned} \tag{9.25}$$

$$\begin{aligned}\Gamma_{\rm t}^{\rm pp}(12;34) = \Lambda_{\rm t}^{\rm irr}(12;34) &+ \tfrac{1}{2}\Phi_{\rm d}(24;31) + \tfrac{1}{2}\Phi_{\rm m}(24;31)\\ &- \tfrac{1}{2}\Phi_{\rm d}(14;32) - \tfrac{1}{2}\Phi_{\rm m}(14;32).\end{aligned} \tag{9.26}$$

Finally the two-body scattering term in the parquet self-energy equation must be diagonalized with respect to total spin. Reinstating explicit spin labels in (8.29) and applying the transformation equations derived above gives

$$\begin{aligned}\Sigma_2(11') = \tfrac{1}{2}\big\{-\tfrac{1}{2}G(76)\big[\tfrac{1}{2}\Lambda_{\mathrm{d}}^{\mathrm{irr}}G^{\mathrm{ph}}v_{\mathrm{d}} + \tfrac{3}{2}\Lambda_{\mathrm{m}}^{\mathrm{irr}}G^{\mathrm{ph}}v_{\mathrm{m}}\big](17;1'6) \\ + G(67)\big[\tfrac{1}{2}\Lambda_{\mathrm{s}}^{\mathrm{irr}}G^{\mathrm{pp}}v_{\mathrm{s}} + \tfrac{3}{2}\Lambda_{\mathrm{t}}^{\mathrm{irr}}G^{\mathrm{pp}}v_{\mathrm{t}}\big](17;1'6)\big\} \\ - G(76)\big[\tfrac{1}{2}\Phi_{\mathrm{d}}G^{\mathrm{ph}}v_{\mathrm{d}} + \tfrac{3}{2}\Phi_{\mathrm{m}}G^{\mathrm{ph}}v_{\mathrm{m}}\big](17;1'6) \\ + G(67)\big[\tfrac{1}{2}\Psi_{\mathrm{s}}G^{\mathrm{pp}}v_{\mathrm{s}} + \tfrac{3}{2}\Psi_{\mathrm{t}}G^{\mathrm{pp}}v_{\mathrm{t}}\big](17;1'6).\end{aligned} \tag{9.27}$$

This form for the self-energy explicitly displays the four varieties of "boson" mentioned previously. As before, the "second-order" term must be separated out to prevent double-counting.

The simplest, or "basic," parquet approximation is generated by substituting the bare interaction for Λ^{irr}. As an example, for the Hubbard model in the Fourier representation, the values of Λ^{irr} in this basic parquet are

$$\begin{aligned}\Lambda_{\mathrm{d}}^{\mathrm{irr}} &= U/N \\ \Lambda_{\mathrm{m}}^{\mathrm{irr}} &= -U/N \\ \Lambda_{\mathrm{s}}^{\mathrm{irr}} &= 2U/N \\ \Lambda_{\mathrm{t}}^{\mathrm{irr}} &= 0.\end{aligned} \tag{9.28}$$

Finally, we comment on the form of the crossing relations (7.5) and (7.9) in spin-diagonalized form. As mentioned previously, these relations mix the spin channels. As an example, the $S_z = 0$ magnetic vertex may be re-expressed in the following forms:

$$\begin{aligned}\Gamma_{\mathrm{m}}(12;34) = \Gamma_{\mathrm{m}}^{0}(12;34) &= [\Gamma_{\uparrow\uparrow,\uparrow\uparrow} - \Gamma_{\uparrow\uparrow,\downarrow\downarrow}](12;34) \\ &= [-\Gamma_{\uparrow\uparrow,\uparrow\uparrow} + \Gamma_{\downarrow\uparrow,\downarrow\uparrow}](42;31) \\ &= [-\tfrac{1}{2}(\Gamma_{\mathrm{d}} + \Gamma_{\mathrm{m}}^{0}) + \Gamma_{\mathrm{m}}^{-1}](42;31) \\ &= -\tfrac{1}{2}\Gamma_{\mathrm{d}}(42;31) + \tfrac{1}{2}\Gamma_{\mathrm{m}}(42;31)\end{aligned} \tag{9.29}$$

and

$$\begin{aligned}\Gamma_{\mathrm{m}}(12;34) = \Gamma_{\mathrm{m}}^{0}(12;34) &= [\Gamma_{\uparrow\uparrow,\uparrow\uparrow} - \Gamma_{\uparrow\uparrow,\downarrow\downarrow}](12;34) \\ &= [-(\Gamma_P)_{\uparrow\uparrow,\uparrow\uparrow} + (\Gamma_P)_{\uparrow\downarrow,\downarrow\uparrow}](14;32) \\ &= [-\Gamma_{\mathrm{t}}^{+1} + \tfrac{1}{2}(-\Gamma_{\mathrm{s}} + \Gamma_{\mathrm{t}}^{0})](14;32) \\ &= -\tfrac{1}{2}\Gamma_{\mathrm{s}}(14;32) - \tfrac{1}{2}\Gamma_{\mathrm{t}}(14;32).\end{aligned} \tag{9.30}$$

10 Fluctuation Exchange Approximation and Pseudopotential Parquet

The derivation of spin-diagonalized parquet equations in the preceding section may be used to motivate a particularly natural Baym–Kadanoff approximation, which has been studied extensively during the last several years. This is the so-called "FLuctuation EXchange," or FLEX approximation [4, 12–14]. A similar infinite-order approximation for the electron gas dates from the original work of Baym and Kadanoff [1, 2], in which the name "T-matrix approximation" is used. At other points in the literature the name "generalized random phase approximation" has appeared.

The FLEX approximation for the one-particle SCF, or self-energy, results when all factors of $\Lambda^{\rm irr}$, $\Gamma^{\rm ph}$, and $\Gamma^{\rm pp}$ in (9.27) are replaced by the bare interaction v. (Note in comparing results above with those in [4] that different normalizations are employed for the particle-particle propagator and irreducible vertex function in the earlier work.) Thus the non-Hartree–Fock terms in the FLEX SCF take the form

$$\begin{aligned}\Sigma_2(11') = \tfrac{1}{2}\big\{ & -\tfrac{1}{2}G(76)\big[\tfrac{1}{2}v_{\rm d}G^{\rm ph}v_{\rm d} + \tfrac{3}{2}v_{\rm m}G^{\rm ph}v_{\rm m}\big](17;1'6) \\ & + G(67)\big[\tfrac{1}{2}v_{\rm s}G^{\rm pp}v_{\rm s} + \tfrac{3}{2}v_{\rm t}G^{\rm pp}v_{\rm t}\big](17;1'6)\big\} \\ & - G(76)\big[\tfrac{1}{2}v_{\rm d}(1-G^{\rm ph}v_{\rm d})^{-1}(G^{\rm ph}v_{\rm d})^2 \\ & \qquad + \tfrac{3}{2}v_{\rm m}(1-G^{\rm ph}v_{\rm m})^{-1}(G^{\rm ph}v_{\rm m})^2\big](17;1'6) \\ & + G(67)\big[\tfrac{1}{2}v_{\rm s}(1-G^{\rm pp}v_{\rm s})^{-1}(G^{\rm pp}v_{\rm s})^2 \\ & \qquad + \tfrac{3}{2}v_{\rm t}(1-G^{\rm pp}v_{\rm t})^{-1}(G^{\rm pp}v_{\rm t})^2\big](17;1'6). \end{aligned} \tag{10.1}$$

The expressions for irreducible vertex functions within FLEX follow by functional differentiation of the SCF using the method developed in Section 6. Two different types of contribution arise (see Fig. 6.22); for a complete discussion, see [4] and [13]. The two types have been called single-fluctuation-exchange diagrams and Aslamazov–Larkin [15] diagrams. When the vertex functions are spin-diagonalized omitting the Aslamazov–Larkin contributions, the resulting equations reproduce the parquet results (9.21)–(9.26), again with the replacement of $\Lambda^{\rm irr}$, $\Gamma^{\rm ph}$, and $\Gamma^{\rm pp}$ by the bare interaction v on the right-hand sides. For example,

$$\begin{aligned}\Gamma^{\rm ph}_{\rm m}(12;34) = v_{\rm m}(12;34) & + \big[-\tfrac{1}{2}v_{\rm d}(1-G^{\rm ph}v_{\rm d})^{-1}(G^{\rm ph}v_{\rm d}) \\ & \qquad + \tfrac{1}{2}v_{\rm m}(1-G^{\rm ph}v_{\rm m})^{-1}(G^{\rm ph}v_{\rm m})\big](42;31) \\ & + \big[-\tfrac{1}{2}v_{\rm s}(1-G^{\rm pp}v_{\rm s})^{-1}(G^{\rm pp}v_{\rm s}) \\ & \qquad + \tfrac{1}{2}v_{\rm t}(1-G^{\rm pp}v_{\rm t})^{-1}(G^{\rm pp}v_{\rm t})\big](41;32). \end{aligned} \tag{10.2}$$

Some comment is necessary on the role of the Aslamazov–Larkin diagrams, which are omitted in (10.2). These diagrams are essential to guarantee the strict observation of microscopic conservation laws within FLEX (see e.g., [13]). Within the parquet formalism, however, these diagrams represent only the simplest renormalization of two-rung contributions to the particle-hole and particle-particle ladders Φ and Ψ. Since only two rungs are renormalized, these diagrams constitute a tiny subset of the contributions which must be included in a second iteration of the parquet equations away from the FLEX approximation. It is not surprising in this light that

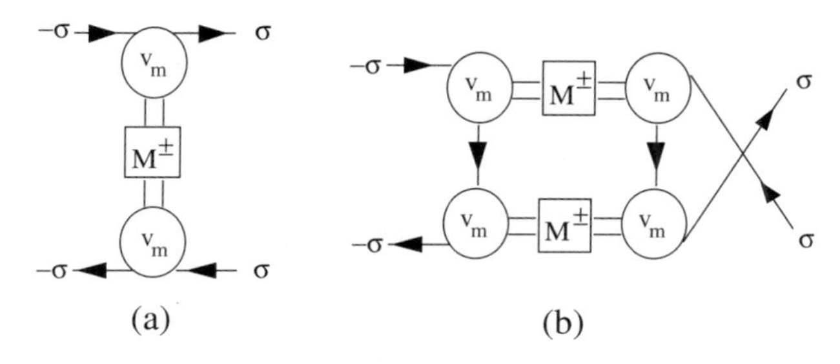

FIGURE 6.22. Example of the two distinct classes of contributions to Γ^{ph} within the FLEX approximation. (a) Single-fluctuation-exchange contribution to $\Gamma^{\mathrm{ph}}_{\sigma\sigma,-\sigma-\sigma}$. The fluctuation is in this case an $S^z = 1$ magnetic fluctuation. (b) Aslamazov–Larkin contribution to $\Gamma^{\mathrm{ph}}_{\sigma\sigma,-\sigma-\sigma}$. In this case a particle-hole (density or magnetic) excitation with $S^z = 0$ decays into two magnetic fluctuations with $S^z = 1$ and -1 (total $S^z = 0$).

the Aslamazov–Larkin diagrams tend to degrade, rather than improve, the overall accuracy of FLEX approximations [13] for two-particle correlation functions.

The intimate connection between the FLEX and parquet expressions for the self-energy and vertex functions justifies the view of FLEX as a parquet with *non*-self-consistent two-particle vertices. While FLEX incorporates all the physical ingredients present in a full parquet, the lack of two-particle self-consistency limits quantitative accuracy. This fact was realized as early [13] as 1990, and attempts have been made to improve on FLEX without solving a full set of space- and time-dependent vertex equations. These attempts have concentrated on the Hubbard model [13] and its zero-dimensional analog, the Anderson impurity model [16, 17]. In both models the bare interaction is instantaneous and zero-range. A physically appealing approach in this case is the so-called "pseudopotential parquet," introduced in [13]. This method amounts to iterating (9.21)–(9.27) with the replacement of the full space- and time-dependent vertex functions Γ^{ph} and Γ^{pp} by instantaneous zero-range pseudopotentials. The lowest order expressions for the vertex functions in (7.18) are replaced by three independent numbers:

$$\begin{aligned} \Gamma^{\mathrm{ph}}_{\mathrm{d}} &\to U_{\mathrm{d}}/N \\ \Gamma^{\mathrm{ph}}_{\mathrm{m}} &\to -U_{\mathrm{m}}/N \\ \Gamma^{\mathrm{pp}}_{\mathrm{s}} &\to 2U_{\mathrm{s}}/N \\ \Gamma^{\mathrm{pp}}_{\mathrm{t}} &\to 0. \end{aligned} \tag{10.3}$$

Note that the triplet pseudopotential necessarily remains zero.

This replacement is an attempt to represent in the simplest possible way the screening or anti-screening of interactions by fluctuations in crossed channels. Of course, if (9.21)–(9.27) are iterated once with the pseudopotentials substituted on the right-hand sides, the resulting approximations for the irreducible vertex functions are no longer instantaneous and zero-range. This is the source of the term "pseudopotential": since each space- and time-dependent matrix is being represented by a single number, only a single target quantity (e.g., a single matrix element or average of matrix elements) may be reproduced exactly. The prescription for calculating revised pseudopotentials after each iteration of (9.21)–(9.27) is as follows: (i) calculate one target in each spin channel using the new space- and time-dependent approximation for $\Gamma^{\rm ph}$ (or $\Gamma^{\rm pp}$)—this may require solving the appropriate Bethe–Salpeter equation; (ii) demand that the calculated targets match the values which result when the irreducible vertices are replaced by revised pseudopotentials. The parquet equations may then be iterated to self-consistency for both the one-particle self-energy *and* the two-particle pseudopotentials. The full space and time dependence specified for the irreducible vertices by the right-hand sides of (9.21)–(9.27) may finally be used to calculate general susceptibilities and correlation functions. (Note that the bare interaction continues to enter the pseudopotential equations wherever v or $\Lambda^{\rm irr}$ is written: the pseudopotentials appear only in the particle-hole and particle-particle ladders Φ and Ψ.)

The targets should be chosen judiciously to represent the behavior of the full matrices as closely as possible. It is reasonable to choose targets which emphasize the frequency and momentum regions in which the complete vertex functions become singular (e.g., $\mathbf{Q} = (\pi, \pi)$, $\Omega = 0$ for the magnetic channel in the nearly-half-filled Hubbard model). In practice susceptibilities have generally been calculated as targets, rather than individual matrix elements of $\Gamma^{\rm ph}$ or Γ. The argument for this choice is that susceptibilities, which involve sums over all values of the relative momentum and frequency, may loosely be viewed as weighted averages of Γ. As an example, the matching condition for the irreducible magnetic vertex in the two-dimensional Hubbard model might be

$$\begin{aligned} \chi_{\rm m}(\Pi) &= \frac{\chi^{\rm ph}(\Pi)}{1 - U_{\rm m}\chi^{\rm ph}(\Pi)} \\ \chi^{\rm ph}(Q) &= -\frac{\beta}{N}\sum_k G(k+Q)G(k) \\ \Pi &= \big((\pi,\pi),0\big). \end{aligned} \tag{10.4}$$

It is straightforward to extend the pseudopotential concept by calculating target quantities for each spin channel and each value of the center-of-mass variable Q. The pseudopotentials become

$$\begin{aligned} \Gamma^{\rm ph}_{\rm d} &\to U_{\rm d}(Q)/N \\ \Gamma^{\rm ph}_{\rm m} &\to -U_{\rm m}(Q)/N \\ \Gamma^{\rm pp}_{\rm s} &\to 2U_{\rm s}(Q)/N \\ \Gamma^{\rm pp}_{\rm t} &\to 0. \end{aligned} \tag{10.5}$$

Note that these vertex functions are still highly local: they describe scattering processes in which the initial-state space and time coordinates of the particle-hole or particle-particle pair coincide (and likewise for the final-state coordinates). In contrast with the simpler choice in (10.3), the initial and final coordinates need no longer be the same. To date numerical calculations have not been carried out using these generalized pseudopotentials.

In several cases of intermediate to strong coupling the pseudopotential parquet approximation in (10.3) has yielded very accurate values for one- and two-particle correlation functions. For a discussion of comparisons with quantum Monte Carlo results for the two-dimensional Hubbard model, see Section 13 and [13]. Our understanding of the reasons for this success is as follows: the parquet (9.21)–(9.27) may always be solved *formally* by an expansion of the irreducible vertex functions Γ^{ph} and Γ^{pp} in powers of the fully irreducible vertex functions Λ^{irr}, which are specified at the outset. Thus, one may write for the basic Hubbard parquet

$$\begin{aligned} \Gamma_{\mathrm{d}}^{\mathrm{ph}} &= U/N + \Delta\Gamma_{\mathrm{d}}^{\mathrm{ph}} \\ \Gamma_{\mathrm{m}}^{\mathrm{ph}} &= -U/N + \Delta\Gamma_{\mathrm{m}}^{\mathrm{ph}} \\ \Gamma_{\mathrm{s}}^{\mathrm{pp}} &= 2U/N + \Delta\Gamma_{\mathrm{s}}^{\mathrm{pp}} \\ \Gamma_{\mathrm{t}}^{\mathrm{pp}} &= 0 + \Delta\Gamma_{\mathrm{t}}^{\mathrm{pp}}. \end{aligned} \tag{10.6}$$

Such an expansion is expected to converge very poorly for values of U of practical interest. The pseudopotential parquet approach amounts to resetting the expansion point for a formal series solution of this type:

$$\begin{aligned} \Gamma_{\mathrm{d}}^{\mathrm{ph}} &= U_{\mathrm{d}}/N + \Delta\Gamma_{\mathrm{d}}^{\mathrm{ph}} \\ \Gamma_{\mathrm{m}}^{\mathrm{ph}} &= -U_{\mathrm{m}}/N + \Delta\Gamma_{\mathrm{m}}^{\mathrm{ph}} \\ \Gamma_{\mathrm{s}}^{\mathrm{pp}} &= 2U_{\mathrm{s}}/N + \Delta\Gamma_{\mathrm{s}}^{\mathrm{pp}} \\ \Gamma_{\mathrm{t}}^{\mathrm{pp}} &= 0 + \Delta\Gamma_{\mathrm{t}}^{\mathrm{pp}}. \end{aligned} \tag{10.7}$$

This is possible in a trivial sense for any choice of the pseudopotentials, just as the one-particle SCF in a Baym–Kadanoff treatment (see (4.11)) may be chosen arbitrarily—by definition, the correction terms restore exact expressions.

A sensible choice of the expansion point makes the correction terms small. In Baym–Kadanoff theories the SCF is chosen exactly to cancel a specified set of skeleton self-energy diagrams when correlation functions are expanded in powers of $S - S^{\mathrm{SCF}}$. It is intended (but rarely verified) that the remaining contributions to the self-energy are small when evaluated perturbatively using G^{SCF}. The pseudopotentials cannot be given such a precise diagrammatic interpretation; however, in principle, the size of corrections may be examined by substituting (10.7) in (9.21)–(9.27) and linearizing the right-hand sides in the $\Delta\Gamma$'s. This gives a set of *linear* matrix equations for the correction terms. It would be of considerable interest to carry out this procedure for the pseudopotential parquet solutions of either the Hubbard or Anderson impurity models. The close agreement of pseudopotential results with available exact results suggests that the correction terms obtained in this way are small for a range of coupling strengths.

11 Analysis of Ordering Instabilities

The formalism developed in previous sections for Baym–Kadanoff and parquet approximations may be used to locate instabilities of an electronic normal state, i.e., symmetry-breaking second-order phase transitions. An instability occurs when the application of an infinitesimal external field is sufficient to produce a finite response, or equivalently when some susceptibility of the system diverges. A divergent susceptibility signals the appearance of a pole in the corresponding complete vertex function Γ_{r}, where r = d, m, s, or t. Rewriting the complete vertex using the Bethe–Salpeter equations (7.13)–(7.15), one finds the condition for an instability is the development of a unit eigenvalue by one of the matrices $G^{\mathrm{ph}}\Gamma_{\mathrm{r}}^{\mathrm{ph}}$ or $G^{\mathrm{pp}}\Gamma_{\mathrm{r}}^{\mathrm{pp}}$. Thus the search for singularities reduces to a linear matrix eigenvalue problem. The matrices of interest are non-Hermitian, so the right- and left-eigenvectors corresponding to a given eigenvalue are different. For cases of interest the eigenvalues are real or occur in complex conjugate pairs [4].

In systems without explicit time dependence, instabilities are always expected at total pair frequency $\Omega = i\Omega = 0$, so calculations may be carried out on the imaginary-frequency axis. (Note that in the particle-particle channel $i\Omega = 0$ corresponds to an electron pair with total energy 2μ, not 0.) As an example, the eigenvalue equation for a $\mathbf{Q} = 0$ singlet instability (i.e., conventional superconductivity) may be written

$$(G^{\mathrm{pp}}\Gamma_{\mathrm{s}}^{\mathrm{pp}})_{kk'}\phi_{k'} = \phi_k, \tag{11.1}$$

where $k = (\mathbf{k}, i\omega)$ is the *relative* momentum-energy of the pair, and

$$\begin{aligned} (G^{\mathrm{pp}})_{kk'} &= -\tfrac{1}{2}\beta\delta_{kk'}G(k)G(-k) \\ (\Gamma_{\mathrm{s}}^{\mathrm{pp}})_{kk'} &= \Gamma_{\mathrm{s}}^{\mathrm{pp}}(k, -k; k', -k'). \end{aligned} \tag{11.2}$$

The pair wave function ϕ generalizes the usual wave function of a two-particle Schrödinger equation to incorporate the effects of retardation. (For further discussion of this point see, e.g., [4].)

In the simplest cases the eigenvalue solution reduces to a Stoner criterion (for itinerant magnetism) or a linearized BCS gap equation (for superconductivity). As an example, in BCS theory the interaction is momentum- and frequency-independent (at least within a limited range), so that

$$(\Gamma_{\mathrm{s}}^{\mathrm{pp}})_{kk'} = \frac{2V}{N} < 0. \tag{11.3}$$

In addition,

$$(G^{\mathrm{pp}})_{kk'} = -\tfrac{1}{2}\beta\delta_{kk'}\frac{T}{i\omega - \varepsilon_{\mathbf{k}}}\frac{T}{-i\omega - \varepsilon_{-\mathbf{k}}} = -\tfrac{1}{2}\delta_{kk'}\frac{T}{\omega^2 + \varepsilon_{\mathbf{k}}^2}. \tag{11.4}$$

The maximum eigenvalue for a point interaction of this type always corresponds to

$$\phi_k = \text{constant}. \tag{11.5}$$

Canceling ϕ_k from both sides of the eigenvalue equation gives

$$|V| \cdot \frac{T}{N}{\sum_{\mathbf{k}\omega}}' \frac{1}{\omega^2 + \varepsilon_{\mathbf{k}}^2} = 1, \tag{11.6}$$

where the prime indicates restriction to a subset of momenta or frequencies. This is precisely the linearized BCS gap equation.

Within FLEX theory the procedure described above has been used to investigate the phase diagrams of a number of models [12,13,18–22] with zero-range bare interactions (see Section 13). As an example, FLEX has provided evidence [12, 13, 18–21] for an unconventional singlet pairing state with $\mathrm{d}_{\mathrm{x}^2-\mathrm{y}^2}$ spatial symmetry in the nearly-half-filled two-dimensional Hubbard model. To date the basic parquet equations have only been solved numerically for the Anderson impurity model, which does not exhibit an ordering instability. The basic parquet solution of the one-dimensional Hubbard model would be of great interest: it is known that the susceptibilities of this model exhibit *simultaneous* power-law divergences in all four scattering channels for $T \to 0$. The basic parquet is the simplest self-consistent technique capable of treating the interplay of singular two-particle propagators in this model.

12 Renormalization Group Solution of SCF Equations

To this point we have been concerned with the derivation of self-consistent equations for the analysis of many-body Fermi systems. In this section we briefly discuss methods for the numerical solution of these equations. We particularly emphasize the use of renormalization group methods [18, 23] for frequency- and momentum-space.

The Baym–Kadanoff and parquet approximations derived in preceding sections all require the solution of nonlinear integral equations in $d+1$ dimensions, with d the spatial dimension. If particle motion is restricted to a discrete set of orbitals on a periodic lattice, the unit-cell index may be conveniently Fourier-transformed, producing a bounded Brillouin zone. (If more than one orbital appears within each cell, an intra-cell index remains.) It is also convenient in most cases to Fourier-transform imaginary-time indices. As discussed previously the resulting Matsubara frequencies are *discrete* and take the values $\omega_n = (2n+1)\pi T$, with T the system temperature. Since the number of allowed frequencies is still infinite, temperature-independent imaginary-axis cutoffs must be established for numerical calculations.

At high temperatures a coarse discretization of the Brillouin zone is sufficient to obtain high numerical accuracy, and the number of Matsubara frequencies which fit within the cutoff range is small. In this case computational storage and time requirements may be minimal. However, as the temperature decreases, an increasingly fine momentum-space mesh is generally required to obtain accurate results, particularly in the region of the Fermi surface, where correlation functions vary rapidly. Furthermore, the number of required Matsubara frequencies within a fixed cutoff is inversely proportional to the system temperature. The requirements of decreasing temperature rapidly begin to tax the storage and speed capabilities of even the largest computers. Increased computational efficiency is clearly essential, particularly for the treatment of models with realistic orbital structures and Coulomb interactions.

In certain cases time requirements for SCF calculations may be reduced drastically from "brute-force" estimates by the use of fast Fourier transforms (FFT's) [14]. Studies of the FLEX approximation for the two-dimensional Hubbard model [14,19–21] have demonstrated the power of this approach, which leads to a reduction from $O(L^2)$ to $O(L \ln L)$ in scaling with the size of the variable domain L. The FFT's utility is limited, however, to approximations in which time requirements are dominated by the computation of convolution integrals. Baym–Kadanoff approximations for more complex models and general parquet approximations are dominated instead by the computation of matrix inverses, for which FFT methods are not applicable. Furthermore the FFT approach leads to no reduction in storage requirements.

An appealing alternative approach to deal with the proliferation of mesh points in general SCF theories is provided by renormalization group (RG) concepts [18, 23]. The central idea, as in all numerical RG applications, is that the need for increasing magnification is not uniform throughout the space of interest (in this case, frequency- and momentum-space). Correlation functions evaluated at high frequencies or at momentum points distant from the Fermi surface eventually become temperature-independent as the temperature is reduced. The temperature below which variations "freeze out" depends on the particular values of ω_n and $\mathbf{k}$ selected. Regions of frequency and momentum may therefore be sequentially removed from calculations. To study the low-temperature limit it is only necessary to tabulate the values of correlation functions at their lowest "active" temperatures, then to carry forward temperature-independent contributions from the eliminated regions to functions in the remaining active regions.

As regions of high frequency and momentum are eliminated, it becomes feasible to increase the magnification scale in those regions which remain without greatly increasing time or storage requirements.

Below we describe an RG prescription for calculating one-particle SCF's in general Baym–Kadanoff theories. (See [23] for a detailed discussion of the RG solution of the FLEX equations for the two-dimensional Hubbard model.) A similar prescription holds for calculating vertex functions [24], but we do not discuss this procedure in detail here.

The central idea in the RG approach is the separation of a variable domain $k = (\mathbf{k}, i\omega_n)$ into "low" and "high" regions. For convenience these regions may be denoted L and H. A general Baym–Kadanoff SCF is determined by an integral equation of the form

$$\Sigma(k) = \int_{k'} F(k, k'), \tag{12.1}$$

with

$$\int_k \equiv \sum_{\mathbf{k}} \sum_{\omega_n}. \tag{12.2}$$

The function $F(k, k')$ depends implicitly on values of Σ throughout the variable space. We ignore this crucial point for the moment and return to it after developing a general technique for solving (12.1).

The integration in (12.1) may be broken into two parts as

$$\begin{aligned} \Sigma(k) &= \int_{k' \in \mathrm{L}} F(k, k') + \int_{k' \in \mathrm{H}} F(k, k') \\ &\equiv \int_{k' \in \mathrm{L}} F(k, k') + \Delta\Sigma(k). \end{aligned} \tag{12.3}$$

Clearly if the function $\Delta\Sigma$ were available from some other source, the labor of solving the integral equation would be greatly reduced: only an integration over L, rather than over the full space, would be required.

For calculations carried out at a sequence of decreasing temperatures, an iterative method for systematically dividing the variable domain and obtaining $\Delta\Sigma$ may indeed be devised. The procedure is as follows: (1) In the first stage, lay down a finite mesh which covers the initial variable domain. For a specified initial temperature T_0 this amounts to choosing a Matsubara cutoff Ω_0 such that

$$\Omega_0 = 2N(T_0)\pi T_0 \tag{12.4}$$

for some integer $N(T_0)$; and dividing the Brillouin zone into a set of uniform blocks centered on a finite set of **k**-points. (Note that the fermion Matsubara frequencies automatically form a discrete set with $2N(T_0)$ elements spaced uniformly between $-[2N(T_0) - 1]\pi T_0$ and $[2N(T_0) - 1]\pi T_0$.)

(2) Solve integral (12.1) for Σ within the discretized variable domain. (3) Divide the variable domain into two parts L and H. The precise condition for the division determines the renormalization group to be employed. The simplest (and most useful) division of the frequency space amounts to the choice of a new cutoff Ω_1, which separates L and H. The most useful division of the momentum space invokes the concept of a Fermi surface S_{F}, with L consisting of wave vectors "close to" S_{F} and H the complement of L. Specific conditions for a frequency-space RG are discussed at length below. (4) For discrete points k in L, calculate the contribution to $\Sigma(k)$ from points k' in H, i.e., explicitly compute

$$\Delta\Sigma(k) = \int_{k' \in \mathrm{H}} F(k, k'). \tag{12.5}$$

(5) Choose a temperature $T_1 < T_0$ for a subsequent calculation of Σ. The new temperature must be consistent with the new frequency cutoff Ω_1 (see below). (6) Lay down a new finite mesh covering the reduced space L. As in step (1), the choice of a temperature T_1 specifies a frequency mesh, which will be finer than the initial mesh since $T_1 < T_0$. The momentum space may also be re-grained, e.g., by subdividing each $\mathbf{k}$-space block in L into smaller blocks, each centered on a new mesh point. (7) Interpolate the function $\Delta\Sigma(k)$ determined in step (4) onto the new L mesh. (8) Solve (12.3) within the reduced domain L using the *predetermined* function $\Delta\Sigma$ from steps (4) and (7). (9) Iterate steps (3)–(8) as desired to study additional temperatures and meshes. Note that the "renormalization correction" $\Delta\Sigma$ obtained at temperature T_i is a sum of terms obtained from the previous i iterations.

As mentioned above, the subtlety in carrying out this procedure is that the kernel $F(k, k')$ itself depends implicitly on the values of Σ throughout the full variable space. This means that F must also be renormalized in a stage-by-stage fashion. The necessary procedure is the analog of vertex renormalization in quantum field theory, just as the calculation of $\Delta\Sigma$ is the analog of mass and wave function renormalization. The precise equations for renormalizing F depend on its functional form. As an example, in the FLEX approximation [18, 23] for the Hubbard and Anderson impurity models (cf. (10.1)) the renormalization of particle-hole exchange diagrams can be accomplished relatively simply by considering the auxiliary function

$$\chi^{\mathrm{ph}}(Q) = -\frac{\beta}{N} \int_k G(k+Q) G(k). \tag{12.6}$$

Just as in the case of Σ, χ^{ph} may be divided into contributions from the L and H regions of the variable domain:

$$\chi^{\mathrm{ph}}(Q) = -\frac{\beta}{N} \int_{k:k \text{ and } k+Q \in L} G(k+Q) G(k) + \Delta\chi^{\mathrm{ph}}(Q) \tag{12.7}$$

with

$$\Delta\chi^{\mathrm{ph}}(Q) = -\frac{\beta}{N}\int_{k:k \text{ or } k+Q\in H} G(k+Q)G(k). \tag{12.8}$$

At each temperature the renormalization correction $\Delta\chi^{\mathrm{ph}}$ must be calculated along with $\Delta\Sigma$, then interpolated onto the mesh to be used in the next RG stage (steps 4 and 7 of the preceding algorithm). We do not consider the renormalization of more general forms for $F(k, k')$ here.

The motivation of the RG procedure is clear: accurate low temperature calculations require the use of fine meshes in frequency and momentum space. The frequency discretization scale is inherently $\Delta\omega = 2\pi T$, while the momentum discretization scale must also tend to zero if one is to resolve possible Fermi surface singularities in correlation functions. Nevertheless, brute-force calculations (and, to a lesser extent, FFT calculations) consume an enormous amount of time in computing functional contributions from regions of high frequency or momentum. One expects that such contributions invariably become temperature-independent below some scale (which depends on the specific frequency or momentum in question). By carrying out a series of calculations on meshes of increasing magnification in domains of decreasing size, the total labor required to solve a problem may be enormously reduced. In some cases the total calculational time to reach a temperature T using the RG approach scales as $|\log T|$, rather than as T^{-2} (brute-force approach) or $T^{-1}|\log T|$ (FFT approach).

In order for the FLEX RG procedure to work, the contribution to $\Delta\Sigma$ from a given domain H must (a) be nearly independent of temperature for all temperatures below the scale at which it is actually calculated; and (b) vary smoothly on the scale of the mesh for which it is calculated. The first condition is essential since once a domain H is eliminated, it appears at no subsequent stage (i.e., lower temperature) in a calculation. The second condition is also important since any useful RG sequence introduces meshes of increasing magnification, for which interpolation is required. By renormalizing sufficiently slowly (i.e., by choosing L sufficiently close to the initial variable domain), both conditions may be satisfied. In FLEX calculations for the two-dimensional Hubbard model [23], accuracy at the 1% level has been obtained using surprisingly large RG steps, with improvements in efficiency of several orders of magnitude at low temperature. Such accuracy is completely acceptable for thermodynamic applications and for a limited range of dynamic applications as well.

In general SCF equations may be solved using (a) a frequency-space RG with fixed momentum discretization, (b) a momentum-space RG with fixed frequency discretization (and fixed temperature), or (c) a general RG which eliminates regions of frequency- and momentum-space concurrently. We describe below the construction of a general frequency-space RG. For a detailed discussion of a two-dimensional momentum-space RG, see [23].

A frequency-space RG sequence may be defined by choosing a set of temperatures T_i and frequency cutoffs Ω_i such that

$$T_0 > T_1 > T_2 > \cdots \tag{12.9}$$

and

$$\Omega_0 \geq \Omega_1 \geq \Omega_2 \geq \cdots . \tag{12.10}$$

These quantities cannot be chosen at random, but must satisfy several conditions discussed below. The total number of positive fermion Matsubara frequencies in the initial stage of the calculation can be denoted $N(T_0)$. This number defines the frequency cutoff Ω_0 through (12.4). This initial cutoff must be chosen sufficiently large that calculated quantities are cutoff-independent.

The cutoffs and temperatures for subsequent RG stages must satisfy the conditions

$$\Omega_i = 2N(T_i)\pi T_i \tag{12.11}$$

and

$$\Omega_i = 2K(T_i)\pi T_{i-1}, \tag{12.12}$$

for some integers $N(T_i)$ and $K(T_i)$ such that

$$1 \leq K(T_i) \leq N(T_{i-1}). \tag{12.13}$$

Both conditions are essential if the RG is to be length-preserving on the imaginary frequency axis. The total length of the initial frequency domain is

$$2\pi T_0 \sum_{\omega_n(T_0)} 1 = 2\pi T_0 \big[2N(T_0)\big] = 2\Omega_0. \tag{12.14}$$

After the first RG division into L and H this total length can be rewritten as

$$\begin{aligned} 2\pi T_1 \sum_{\omega_n(T_1)\in \mathrm{L}} 1 + \Delta\ell &= 2\pi T_1 \big[2N(T_1)\big] + \Delta\ell \\ &= 2\pi T_0 \big[2K(T_1)\big] + \Delta\ell, \end{aligned} \tag{12.15}$$

where

$$\Delta\ell = 2\pi T_0 \sum_{\omega_n(T_0)\in \mathrm{H}} 1 = 2\pi T_0 \big[2N(T_0) - 2K(T_1)\big]. \tag{12.16}$$

The total length of the original frequency interval is preserved. This is insured by requiring that the cutoffs coincide exactly with boson Matsubara frequencies: if this were not true, the length of the original frequency domain could not be reproduced exactly by repeated RG iterations, and more complicated summations would also be distorted.

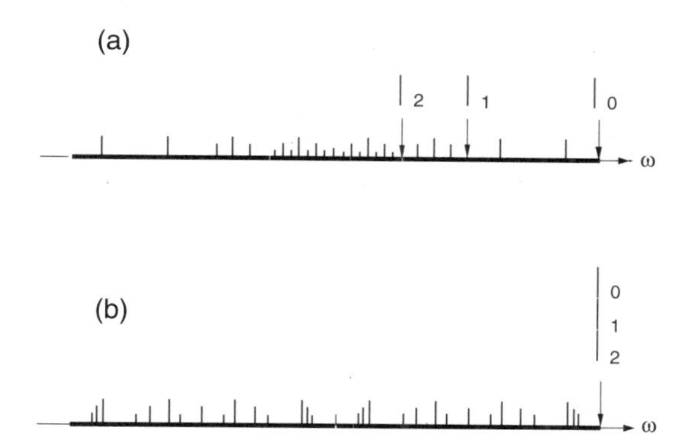

FIGURE 6.23. Imaginary frequency discretization for two frequency-space renormalization groups. (a) Initial three stages of the "factor-of-two" RG. The cutoffs Ω_0, Ω_1, and Ω_2 are represented by arrows. Four positive and four negative fermion frequencies appear for each cutoff. The fermion frequencies are represented by tick marks of varying length, the longest for $i = 0$ and the shortest for $i = 2$. (b) Initial three stages of the "constant-cutoff" RG with $N(T_0) = 4$, $N(T_1) = 5$, and $N(T_2) = 6$. In this case there is effectively no renormalization, since the entire space is retained as L (i.e., H is the null set). As in (a), fermion frequencies are represented by tick marks, the longest for $i = 0$ and the shortest for $i = 2$.

A convenient choice for rapid renormalization is

$$\begin{aligned} K(T_i)/N(T_i) &= T_i/T_{i-1} = \tfrac{1}{2} \\ K(T_i)/N(T_{i-1}) &= \Omega_i/\Omega_{i-1} = \tfrac{1}{2} \end{aligned} \qquad (12.17)$$

for $i = 1, 2, \ldots$ This amounts to simultaneous reduction of the temperature scale and the frequency cutoff by a factor of 2 at each RG stage (see Fig. 6.23a). The total number of fermion frequencies employed in each stage is then the same as the number employed in the first stage, i.e., $2N(T_0)$. This means that if the number of iterations to reach convergence remains constant from stage to stage, each stage takes exactly the same calculational time. The total calculational time to pass from a temperature T_0 to a temperature $T_N = 2^{-N}T_0$ is then just $N + 1$ times the calculational time for temperature T_0. This logarithmic scaling with T_N should be compared with the performance of brute-force (time $\propto T_N^{-2}$) or FFT (time $\propto T_N^{-1}|\log T_N|$) algorithms. (In fact the number of iterations to reach convergence at a single stage slowly increases as the temperature decreases. Since this is true for each type of algorithm, it does not alter the basic comparison of efficiencies.)

While rapid renormalization is the key to calculational efficiency, it may in some cases be necessary to scan slowly through a specified temperature range. For example, a slow scan may be preferable in the vicinity of a phase transition, where one or more susceptibilities exhibit power-law divergences. The simplest way to accomplish a slow scan is to keep the cutoff *fixed* while slowly decreasing the temperature, i.e., to set

$$K(T_i) = N(T_{i-1}) \tag{12.18}$$

with $N(T_i)$ a slowly increasing integer function of i. (For a pictorial illustration of this process see Fig. 6.23b.) The temperatures T_i satisfy the relation

$$T_i = \frac{K(T_i)}{N(T_i)} T_{i-1}. \tag{12.19}$$

After scanning the temperature range of interest more rapid renormalization can again be introduced.

An important technical component in a successful RG is accurate interpolation between mesh scales. We do not discuss interpolation techniques in detail here, but refer the reader to the original studies.

RG methods have to date been used to solve a number of different SCF problems. Frequency-space RG's have been applied in FLEX studies of the normal [23] and superconducting [18] states of the two-dimensional Hubbard model; basic parquet studies of the Anderson impurity model [24]; and FLEX studies of the normal and superconducting states of a two-dimensional CuO_2 model with extended-range interactions [25]. Momentum-space RG's are conceptually as straightforward, but computationally more involved. To date a momentum-space RG has only been applied in FLEX studies of the normal state [23] of the two-dimensional Hubbard model.

13 Some Numerical Examples

It is perhaps useful at this point to present a few simple examples of SCF calculations which illustrate the techniques developed in the preceding sections. We begin with results [13] for the two-dimensional Hubbard model obtained using FLEX and pseudopotential parquet theory. Energies are measured in units of the near-neighbor hopping parameter t. Results are shown for on-site Coulomb energy $U/t = 4$ and average electron number per site $\langle n \rangle = 0.875$. An 8×8 discretization of the Brillouin zone is employed.

The one-particle propagator $G(\mathbf{k}, \tau)$ is plotted in Fig. 6.24 for $\mathbf{k} = (0, 0)$, $(\pi/2, 0)$, and $(\pi, 0)$. Results obtained using FLEX (Eq. 10.1), pseudopotential parquet (Eq. 10.3), and direct numerical simulation (quantum Monte Carlo) are compared. The result for a noninteracting system with the same electron density is also shown. It is clear that the FLEX result incorporates the right qualitative effects, but fails quantitatively. The FLEX propagator is smaller in magnitude than the exact propagator because FLEX overestimates the effect of fluctuations in the one-particle SCF: in particular, the non-self-consistent expression for the irreducible magnetic vertex is too attractive. This defect is remedied by the pseudopotential parquet solution,

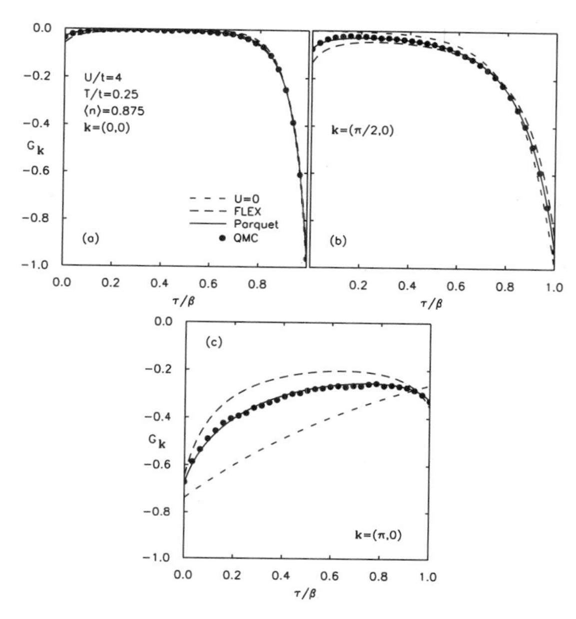

FIGURE 6.24. Comparison of FLEX, pseudopotential parquet, and QMC propagators $G(\mathbf{k}, \tau)$ for $\langle n \rangle = 0.875$ in the two-dimensional Hubbard model. All results are for an 8×8 mesh. (a) $\mathbf{k} = (0, 0)$. (b) $\mathbf{k} = (\pi/2, 0)$. (c) $\mathbf{k} = (\pi, 0)$.

and the agreement becomes quantitative. The temperature-dependent values of the three pseudopotentials U_{d}, U_{m}, and U_{s} are shown in Fig. 6.25.

The calculation of two-particle properties within the pseudopotential parquet is illustrated in Fig. 6.26, which compares time-dependent magnetic correlation functions $\Delta\chi_{\mathrm{m}}(\mathbf{Q}, \tau)$ for $\mathbf{Q} = (0, 0)$, $(\pi, 0)$, and (π, π). (The zero-frequency components of the correlation functions have been removed; for a discussion of this point, see [4]). It is noteworthy that $\Delta\chi_{\mathrm{m}}(0, \tau)$ is not identically zero within the pseudopotential parquet solution: this demonstrates a violation of the global conservation law for S^z. As noted previously, such violations are to be expected in general parquet solutions. In this case the violation is tolerably small, particularly in view of the excellent agreement of the correlation functions for non-zero $\mathbf{Q}$.

As an illustration of the power of the instability analysis discussed in Section 11, the FLEX phase diagram for the two-dimensional Hubbard model is shown in Fig. 6.27. The two ordered phases are an antiferromagnet, which should be interpreted as an "almost-ordered" state with exponentially long correlation lengths, in view of the dimensionality; and a singlet superconductor with discrete $\mathrm{d}_{\mathrm{x}^2-\mathrm{y}^2}$ spatial symmetry. This phase diagram has much in common with the present picture of superconductivity in the high-temperature cuprate superconductors, though the identification remains controversial.

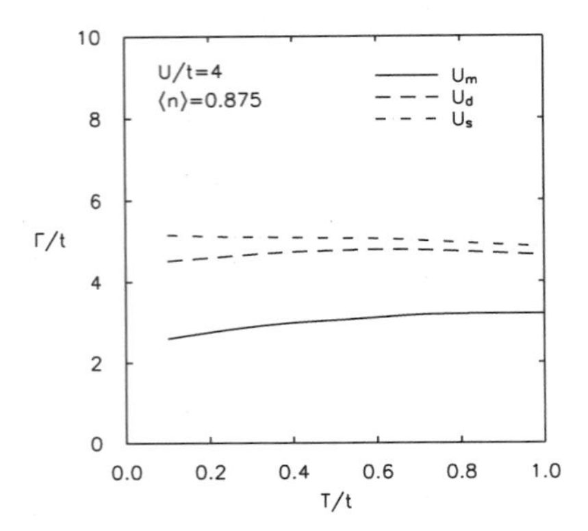

FIGURE 6.25. Temperature dependence of the Hubbard parquet pseudopotentials $U_{\rm d}$, $U_{\rm m}$, and $U_{\rm s}$ for $U/t = 4$ and $\langle n \rangle = 0.875$.

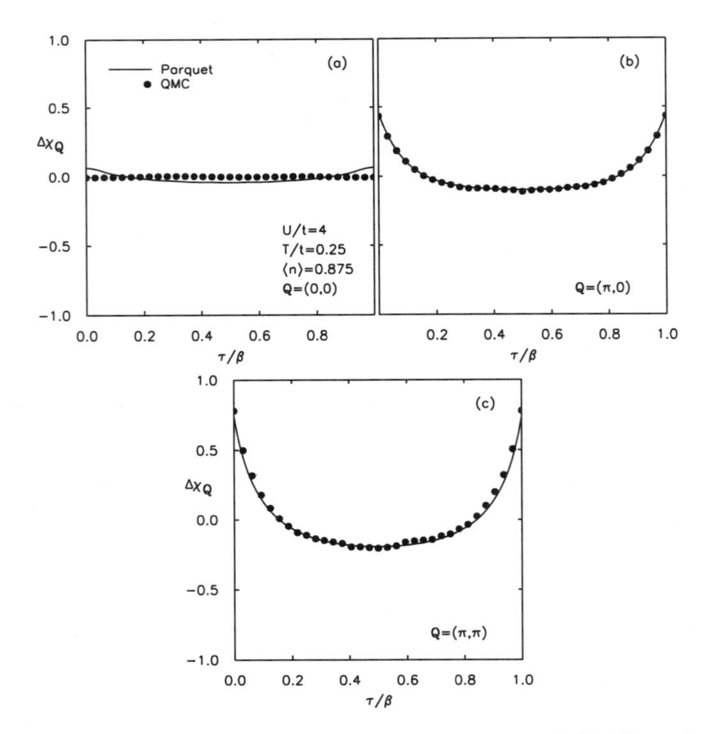

FIGURE 6.26. Comparison of pseudopotential parquet and QMC results for the finite-frequency part of the magnetic correlation function $\Delta\chi_{\rm m}(\mathbf{Q}, \tau)$. The parameters are as in Figs. 6.24 and 6.25. (a) $\mathbf{Q} = (0, 0)$. (b) $\mathbf{Q} = (\pi, 0)$. (c) $\mathbf{Q} = (\pi, \pi)$.

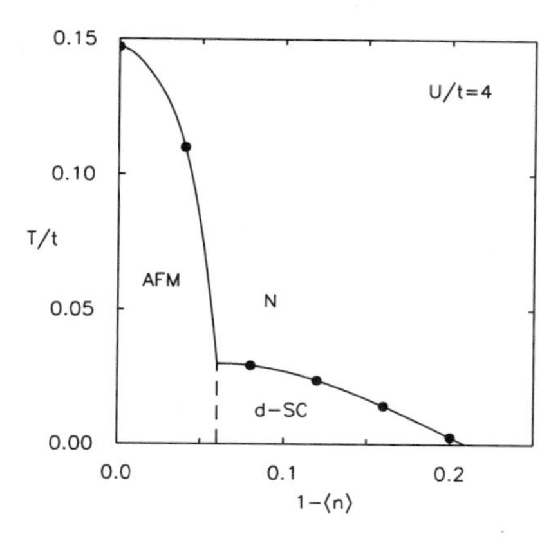

FIGURE 6.27. FLEX phase diagram for the partially filled two-dimensional Hubbard model. Only the boundary between the normal and ordered states is calculated: additional transitions (e.g., a first-order transition from a commensurate to incommensurate antiferromagnetic state) may occur at lower temperatures. The dashed line separating the antiferromagnetic and superconducting states is only a guide to the eye. All transitions are to be interpreted in the mean-field sense.

Finally, as an illustration of results from the basic parquet, we show in Fig. 6.28 a calculation of the magnetic susceptibility of the Anderson impurity model [17]. The result for $T\chi_{\mathrm{m}}$ from the basic parquet is compared with the exact result of a Bethe Ansatz calculation. The parameters are Γ, the half-width of the impurity resonance; U, the impurity Coulomb energy; and ε_{d}, the position of the impurity level relative to the chemical potential of the host metal. For comparison with higher-dimensional models, 2Γ is a rough equivalent of the electron bandwidth. The agreement of the results for $U/\pi\Gamma = 1.3$ is striking for the following reason: the quantitative accuracy of parquet theory is expected to degrade as the system dimensionality decreases and fluctuation effects increase—zero-dimensional systems are potentially the most difficult to treat. The importance of fluctuations can be appreciated by noting that the Hartree–Fock solution of the Anderson impurity model has a magnetically ordered phase with a high transition temperature. In contrast the magnetic susceptibility in the exact solution remains finite down to zero temperature (but becomes exponentially large as U increases).

The failure of the basic parquet to reproduce the exact susceptibility for $U/\pi\Gamma = 2.6$ illustrates the ultimate limitations of this approximation. Note again that if 2Γ is viewed as the analog of the electron bandwidth W, this parameter set corresponds to $U/W = 4$. The results in Fig. 6.28 lend great encouragement for the application of the basic parquet in higher-dimensional systems with realistic Coulomb interactions.

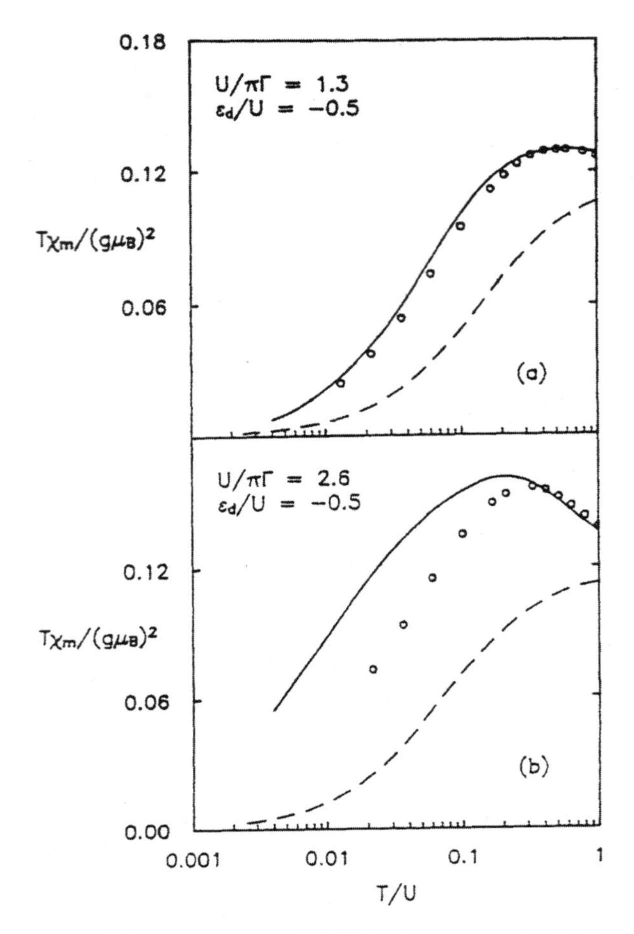

FIGURE 6.28. $T\chi_{\rm m}(T)$ as a function of T/U in the symmetric Anderson impurity model. Solid line: Bethe Ansatz; dashed line: non-magnetic mean-field; circles: basic parquet. (a) $U/\pi\Gamma = 1.3$. (b) $U/\pi\Gamma = 2.6$.

14 Conclusion

Self-consistent field methods beyond Hartree–Fock theory have a long formal history but have only been applied in large-scale numerical calculations in recent years. Advances in computer power and in the formulation of solution algorithms should allow continued progress. As an example of an application which builds on the formalism discussed here, the FLEX approximation has recently been used for the solution of a two-dimensional model with three bands and extended-range Coulomb interactions [25]. The RG method introduced in Section 12 allows the solution of this problem on a large workstation, rather than a supercomputer. In addition, FLEX studies have recently been performed on a model with both zero-range Coulomb and phonon-mediated interactions [26]. Phonons introduce the complication of frequency dependence in interaction vertices, a feature which necessarily also appears in parquet treatments.

A central challenge remains the solution of the basic parquet approximation in a spatially extended system. Results from the pseudopotential analysis of the two-dimensional Hubbard model and from the basic parquet analysis of the Anderson impurity model suggest that quantitative accuracy can potentially be obtained for a wide range of parameter values.

Acknowledgments: This work was partially supported by the National Science Foundation (grant DMR–9520636). Some of the numerical calculations discussed in Section 13 were performed using the resources of the San Diego Supercomputer Center.

15 References

[1] G. Baym and L.P. Kadanoff, Phys. Rev. **124**, 287 (1961).

[2] G. Baym, Phys. Rev. **127**, 1391 (1962).

[3] C. de Dominicis and P.C. Martin, J. Math. Phys. **5**, 14, 31 (1964).

[4] N.E. Bickers and D.J. Scalapino, Ann. Phys. (N.Y.) **193**, 206 (1989).

[5] A more detailed discussion of the topics in Section 3 is available elsewhere. For a pedagogical treatment see, e.g., N.E. Bickers, *The Large Degeneracy Expansion in Dilute Magnetic Alloys*, Ph.D. thesis, Cornell University, 1986.

[6] See, e.g., M. Sigrist and K. Ueda, Rev. Mod. Phys. **63**, 239 (1991).

[7] Parts of our discussion below are an expansion of an earlier set of published lecture notes (N.E. Bickers, Int. J. Mod. Phys. **5**, 253 (1991)).

[8] See, e.g., P. Nozières, *Theory of Interacting Fermi Systems* (W.A. Benjamin, New York, 1964).

[9] S. Babu and G.E. Brown, Ann. Phys. (N.Y.) **78**, 1 (1973).

[10] M. Pfitzner and P. Wölfle, Phys. Rev. B**35**, 4699 (1987).

[11] K.F. Quader, K.S. Bedell, and G.E. Brown, Phys. Rev. B**36**, 156 (1987), and references therein.

[12] N.E. Bickers, D.J. Scalapino, and S.R. White, Phys. Rev. Lett. **62**, 961 (1989).

[13] N.E. Bickers and S.R. White, Phys. Rev. B**43**, 8044 (1991).

[14] J.W. Serene and D.W. Hess, Phys. Rev. B**44**, 3391 (1991).

[15] L.G. Aslamazov and A.I. Larkin, Fiz. Tverd. Tela. **10**, 1104 (1968). [English translation: Soviet Phys.—Solid State **10**, 875 (1968).

[16] C.-X. Chen, J. Luo, and N.E. Bickers, J. Appl. Phys. **69**, 5469 (1991).

[17] C.-X. Chen and N.E. Bickers, Solid State Commun. **82**, 311 (1992).

[18] C.-H. Pao and N.E. Bickers, Phys. Rev. Lett. **72**, 1870 (1994); Phys. Rev. B**51**, 16310 (1995).

[19] P. Monthoux and D.J. Scalapino, Phys. Rev. Lett. **72**, 1874 (1994).

[20] St. Lenck, J.P. Carbotte, and R.C. Dynes, Phys. Rev. B**50**, 10149 (1994).

[21] T. Dahm and L. Tewordt, Phys. Rev. Lett. **74**, 793 (1995).

[22] J. Luo and N.E. Bickers, Phys. Rev. B**47**, 12153 (1993); Phys. Rev. B**48**, 15983 (1993).

[23] C.-H. Pao and N.E. Bickers, Phys. Rev. B**49**, 1586 (1994).

[24] C.-X. Chen and N.E. Bickers, unpublished.

[25] G. Esirgen and N.E. Bickers, Phys. Rev. B**55**, 2122 (1997); Phys. Rev. B**57**, 5376 (1998).

[26] C.-H. Pao and H.-B. Schüttler, preprint, 1998.

7

Fermi and Non-Fermi Liquid Behavior of Quantum Impurity Models: A Diagrammatic Pseudo-Particle Approach

J. Kroha and P. Wölfle

Dedicated to Erwin Müller-Hartmann on the occasion of his 60th birthday

ABSTRACT We review a systematic many-body method capable of describing Fermi liquid and non-Fermi liquid behavior of quantum impurity models at low temperatures on the same footing. The crossover to the high temperature local moment regime is covered as well. The approach makes use of a pseudo-particle representation of the impurity Hilbert space in the limit of infinite Coulomb repulsion U as well as for finite U. Approximations are derived from a generating Luttinger–Ward functional, in terms of renormalized perturbation theory in the hybridization V. Within a "conserving T-matrix approximation" (CTMA), including all two-particle vertex functions, an infinite series of leading infrared singular skeleton diagrams is included. The local constraint is strictly enforced. Applied to the $\mathrm{SU}(N) \times \mathrm{SU}(M)$ multichannel Anderson model the method allows to recover the Fermi liquid behavior of the single channel case, as well as the non-Fermi liquid behavior in the multi-channel case. The results are compared with the "noncrossing approximation" (NCA) and with data obtained by the numerical renormalization group and the Bethe ansatz. The generalization of the method to the case of finite on-site repulsion U is presented in a systematic way and solved on the level of a generalized NCA, fully symmetric with respect to all virtual excitations of the model.

1 Introduction

Over the past two decades the problem of correlated electrons on a lattice has emerged as a central theme of condensed matter theory. Since the mid-1970s numerous metallic systems have been found in which the conduction electron system is believed to be dominated by a strong on-site Coulomb interaction. Examples are the heavy fermion compounds [1] and the cuprate superconductors [2], but systems like the two-dimensional electron gas in quantum Hall devices [3] are governed by similar physics. Typical model

Hamiltonians are the Hubbard model and the t-J model for the single-band situation, and the Anderson or Kondo lattice models in the multi-band case. In the event that the local Coulomb repulsion in these models exceeds the bandwidth, conventional many-body theory, i.e., perturbation theory (including infinite resummations), in the Coulomb repulsion does not work anymore. The obvious alternative, perturbation expansion in terms of the kinetic energy, meets with severe difficulty, mainly due to the infinite degeneracy of the ground state of a lattice model in the limit of zero hopping (a collection of independent atoms) and the noncanonical commutation relations of the field operators of the eigenstates in the atomic limit. It is therefore desirable to develop new methods which are capable of dealing with these problems.

To make clear what is the principal obstacle for applying the well-established many-body techniques to this class of problems let us consider the Hubbard model with large on-site repulsion U, where each lattice site can either be empty (state $|0\rangle$), singly occupied ($|\uparrow\rangle$, $|\downarrow\rangle$ for an electron with spin projection $\uparrow$, $\downarrow$) or doubly occupied ($|2\rangle$). The dynamics of an electron will be very different according to whether it resides on a singly or doubly occupied site. For less than half band filling and large repulsive U the doubly occupied states will be very costly in energy and will contribute to the low energy physics only via virtual processes. This is the typical feature of strongly correlated electrons: the dynamics is constrained to a subspace of the total Hilbert space. The necessary projection onto this subspace is difficult to perform within conventional many-body theory and requires the development of new theoretical tools. This article deals with the method of auxiliary particles introduced to effect the projection in Hilbert space, while keeping most of the desirable features of renormalized perturbation theory.

The idea of auxiliary particles is simple: for each of the four states $|0\rangle$, $|\uparrow\rangle$, $|\downarrow\rangle$, $|2\rangle$ at a given lattice site (considering one orbital per site), a particle is introduced, which is created out of a vacuum state without any lattice site at all. The Fermi character of the electron requires that two of the auxiliary particles be fermions, e.g., the ones representing $|\uparrow\rangle$ and $|\downarrow\rangle$, and the remaining two be bosons. To ensure that a lattice site is in one of the four physical states, the number of auxiliary particles is constrained to be exactly one. Compared to alternative ways of effecting the projection in Hilbert space, the auxiliary particle method has the advantage of allowing one to use the machinery of quantum field theory, provided the constraint can be dealt with in a satisfactory way. As will be shown below, the constraint is closely related to a fictitious gauge symmetry in the Fock space of pseudo-particles. The formulation in terms of renormalized perturbation theory in the kinetic energy presented in this paper allows to conserve the gauge symmetry in any chosen conserving approximation. Conserving approximations, on the other hand, are generated by functional derivation of a Luttinger–Ward functional. The Luttinger–Ward functional is chosen taking into account information on important physical processes as well as using expansion in a small parameter, e.g., the hybridization energy in units of the conduction electron bandwidth. Auxiliary particle representations were pioneered by Abrikosov [4], who first introduced a fermionic representation for local spins, and later by Barnes [5], who was the first to define and use the pseudo-particle representation of a local impurity level we will be working with below.

Somewhat later these representations were applied to the N-fold degenerate Anderson-impurity models in conjunction with a mean-field approximation (MFA) of the slave boson fields [6, 7]. As will be discussed below, the MFA, inspite of its initial success, suffers from severe problems: (1) it leads invariably to the Fermi liquid ground state, even for multi-channel models, in contradiction to the exact non-Fermi liquid behavior in those cases; (2) even for the single-channel case the MFA features a spurious first order transition separating the low-temperature Kondo screened state from the high temperature local moment regime. On the technical side, the MFA violates the gauge symmetry mentioned above. The latter problem may be corrected by including the leading fluctuations around the saddle point in $1/N$ expansion [7], whereas the former problems remain. We shall therefore avoid using the mean-field approximation, which solves the problem of the spurious phase transition.

2 Single- and Multi-Channel Quantum Impurity Models

Quantum impurity models such as the Kondo and Anderson impurity models were discussed first in the context of magnetic impurities in metals [8]. In these models an N-fold spin degenerate local impurity level (here called d-level) hybridizes with the conduction band. In the presence of a sufficiently strong Coulomb repulsion U between any two electrons in the local impurity level and provided the level lies sufficiently far below the Fermi energy the local level will be singly occupied and a magnetic moment forms. The antiferromagnetic exchange interaction between the local moment and the local conduction electron spin density leads to the Kondo effect. The physics of the Kondo effect is not restricted to magnetic ions in metals. It depends only on the local level being degenerate (any two-level or multi-level system qualifies) and the local Coulomb interaction U being sufficiently strong. The Kondo effect can therefore occur as well for quadrupole moments, the two-level systems in amorphous metals [9] or for the energy level system of quantum dots [10]. We will consider the prototype Anderson impurity Hamiltonian [11]

$$H = H_c + H_d + H_{\mathrm{hyb}}, \tag{2.1}$$

with the conduction electron Hamiltonian

$$H_c = \sum_{\vec{k},\sigma} \varepsilon_k c_{\vec{k}\sigma}^{+} c_{\vec{k}\sigma} \quad ; \quad \sigma = 1, \ldots, N, \tag{2.2}$$

the local d-electron Hamiltonian

$$H_d = E_d \sum_{\sigma} d_\sigma^+ d_\sigma + U \sum_{\sigma<\sigma'} n_{d\sigma} n_{d\sigma'}, \tag{2.3}$$

and the hybridization term

$$H_{\mathrm{hyb}} = V \sum_{\vec{k}\sigma} (c_{\vec{k}\sigma}^+ d_\sigma + \mathrm{H.c.}). \tag{2.4}$$

In lowest-order perturbation theory in V the d-level at energy E_d acquires a width $\Gamma = \pi V^2 \mathcal{N}(0)$, with $\mathcal{N}(0)$ the conduction electron density of states at the Fermi level $\omega = 0$. In the Kondo limit of the model, defined by $E_d < 0$, $2E_d + U > 0$, $\Gamma/|E_d| \ll 1$ and $\Gamma/(2E_d + U) \ll 1$, the low-energy sector of the Anderson model may be mapped onto the Kondo model [12] featuring a local antiferromagnetic exchange interaction J of the local spin of the single electron in the d-level with the local conduction electron spin, where $J = V^2[1/|E_d| + 1/(2E_d + U)]$. These models have been studied extensively by means of Wilson's renormalization group [13], by the Bethe ansatz method [14, 15], and by means of a phenomenological Fermi liquid theory [16]. In this way the following physical picture has emerged [8]: The model contains a dynamically generated low-temperature scale, the Kondo temperature, which, for $U \to \infty$, is expressed in terms of the parameters of the Anderson Hamiltonian as

$$T_K = D(N\Gamma/D)^{(M/N)} \mathrm{e}^{-\pi E_d/N\Gamma}, \tag{2.5}$$

with $\mathcal{N}(0)$ and $D = 1/\mathcal{N}(0)$ the density of states at the Fermi energy and the high-energy band cutoff, respectively. N, M are the degeneracy of the local level and the number of conduction electron channels (see below). In the intermediate temperature regime, $T \gtrsim T_K$, resonant spin flip scattering of electrons at the Fermi surface off the local degenerate level leads to logarithmic corrections to the magnetic susceptibility $\chi(T)$, the linear specific heat coefficient $\gamma(T)$, and the resistivity $\rho(T)$ and to a breakdown of perturbation theory at $T \simeq T_K$. In the single-channel case ($M = 1$), below T_K a collective many-body spin singlet state develops in which the impurity spin is screened by the conduction electron spins as lower and lower energy scales are successively approached, leaving the system with a pure potential scattering center. The spin singlet formation is sketched in Fig. 7.1a and corresponds to a vanishing entropy at $T = 0$, $S(0) = 0$. It also leads to saturated behavior of physical quantities below T_K, like

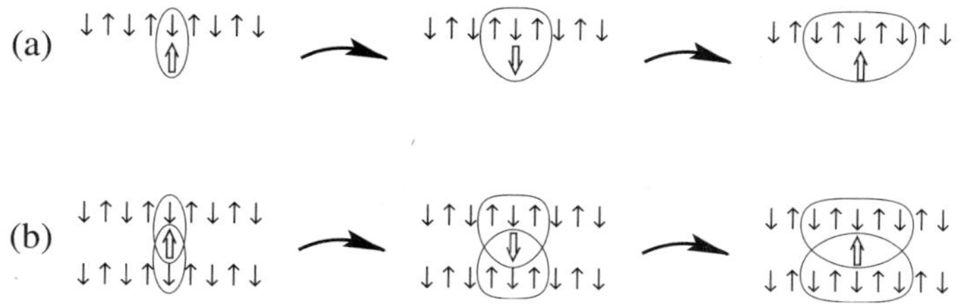

FIGURE 7.1. Sketch of the renormalization group for (a) the single-channel Kondo model (local moment compensation) and (b) the two-channel Kondo model (local moment over-compensation). Small arrows denote conduction electron spins $\frac{1}{2}$, a heavy arrow a localized spin $\frac{1}{2}$. The curved arrows indicate successive renormalization steps.

$\chi(T) = \text{const}$, $\gamma(T) = c(T)/T = \text{const}$, and $\Delta\rho(T) = \rho(T) - \rho(0) \propto T^2$, i.e., to Fermi liquid behavior.

Multichannel quantum impurity models are characterized by the completely symmetric coupling of M degenerate conduction bands to the impurity. Let us assume for definiteness that the ground state of the impurity is a magnetic singlet, with M-fold orbital degeneracy, $|\Gamma\mu\rangle$, where $\mu = 1, \ldots, M$, and the first excited state, obtained by adding one electron to the impurity, is magnetic and N-fold degenerate, but is an orbital singlet, $|\Gamma'\sigma\rangle$, $\sigma = 1, \ldots, N$. An $SU(N) \times SU(M)$ symmetric Anderson model may then be defined in the limit $U \to \infty$ as [9]

$$H = \sum_{\mu} H_{c\mu} = +E_d \sum_{\sigma} |\Gamma'\sigma\rangle\langle\Gamma'\sigma| + H_{\text{hyb}}, \tag{2.6}$$

where

$$H_{c\mu} = \sum_{\vec{k}\sigma} \varepsilon_{\vec{k}} c^{\dagger}_{\vec{k}\mu\sigma} c_{\vec{k}\mu\sigma} \tag{2.7}$$

describes the conduction electron continuum in orbital channel μ and

$$H_{\text{hyb}} = V \sum_{\vec{k},\sigma,\mu} (c^{+}_{\vec{k}\mu\sigma} |\Gamma\mu\rangle\langle\Gamma'\sigma| + \text{h.c.}). \tag{2.8}$$

The effect of the Coulomb interaction is to restrict the average occupation to $n_d = \sum_{\sigma} |\Gamma'\sigma\rangle\langle\Gamma'\sigma| \leq 1$.

As in the single channel case, in the multi-channel case, too, the conduction electrons of each channel *separately* screen the impurity moment by multiple spin scattering at temperatures below the Kondo scale T_K. However, in this case, the local moment is over-compensated, since the impurity spin can never form a singlet state with both conduction electron channels at the same time in this way, as can be seen in Fig. 7.1b. As a consequence of this frustration, there is not a unique ground state, leading to a finite residual entropy [17, 18] at $T = 0$ of $S(0) = k_B \ln \sqrt{2}$ in the two-channel model. In particular, the precondition of FL theory of a 1:1 correspondence between interacting and noninteracting states is violated. As a consequence, characteristic singular temperature dependence [17, 19] of physical quantities persists for $T \lesssim T_K$ down to $T = 0 : \chi(T) \propto -\ln(T/T_K)$, $c(T)/T \propto -\ln(T/T_K)$ and $\rho(T) - \rho(0) \propto -\sqrt{T/T_K}$.

3 Pseudo-Particle Representation

As discussed above, the dynamics of an electron occupying a local level of a quantum impurity will depend in an essential way on whether the level is singly occupied or multiply occupied, provided the Coulomb interaction U between two electrons in the same level is large. It is therefore useful to divide the Hilbert space into sectors labeled by the occupation number of the quantum impurity. For each Fock state $|\alpha\rangle$ of the impurity one may define a creation operator a_α^+, which, when operating on a vacuum state $|\mathrm{vac}\rangle$, creates the state $|\alpha\rangle : |\alpha\rangle = a_\alpha^+|\mathrm{vac}\rangle$. For a usual impurity level with spin degeneracy $N = 2$ one thus defines two bosons b and a, and two fermions f_σ, $\sigma = \uparrow, \downarrow$, creating the empty and doubly occupied states $|0\rangle = b^+|\mathrm{vac}\rangle$, $|2\rangle = a^+|\mathrm{vac}\rangle$, and the singly occupied states $|\sigma\rangle = f_\sigma^+|\mathrm{vac}\rangle$. The creation of an electron in the empty level, effected by a fermion creation operator d_σ^+, according to $|\sigma\rangle = d_\sigma^+|0\rangle$ or $|2\rangle = d_\uparrow^+|\downarrow\rangle = -d_\downarrow^+|\uparrow\rangle$, is described in terms of the pseudo-particle operators as

$$d_\sigma^+ = f_\sigma^+ b + \eta_\sigma a^+ f_{-\sigma} \tag{3.1}$$

where the factor $\eta_\sigma = \pm 1$ for $\sigma = \uparrow, \downarrow$ accounts for the antisymmetry of the doubly occupied state. By the introduction of the pseudo-particles the Fock space has been vastly extended. The physical sector of this artificial Fock space is defined by the requirement that the impurity should be occupied by exactly one pseudo-particle at any time (corresponding to a definite state), as expressed by the constraint

$$\sum_\sigma f_\sigma^+ f_\sigma + b^+ b + a^+ a = 1. \tag{3.2}$$

In the following we will largely consider the case of infinite Coulomb repulsion U, implying that double (multiple) occupancy of the impurity is not possible. The operators a^+, a will therefore not appear in this case. In the multi-channel SU(N) $\times$ SU(M) Anderson model discussed in the last section, we will, however, need to introduce M boson operators b_μ^+, $\mu = 1, \ldots, M$ creating the M ground states $|\Gamma\mu\rangle$, and N fermion operators f_σ^+, $\sigma = 1, \ldots, N$. In terms of these operators the Hamiltonian of the SU(N) $\times$ SU(M) Anderson model takes the form

$$H = \sum_{\vec{k},\sigma,\mu} \varepsilon_{\vec{k}} c_{\vec{k}\mu\sigma}^\dagger c_{\vec{k}\mu\sigma} + E_d \sum_\sigma f_\sigma^\dagger f_\sigma + V \sum_{\vec{k},\sigma,\mu} (c_{\vec{k}\mu\sigma}^\dagger b_{\bar{\mu}}^\dagger f_\sigma + h.c.). \tag{3.3}$$

In order for H to be SU(M) invariant, the slave boson multiplet $b_{\bar{\mu}}$ transforms according to the conjugate representation of the SU(M). In addition, the operator constraint

$$Q \equiv \sum_{\sigma} f_{\sigma}^{\dagger} f_{\sigma} + \sum_{\mu} b_{\bar{\mu}}^{\dagger} b_{\bar{\mu}} = 1 \tag{3.4}$$

has to be satisfied at all times. One might interpret the constraint as a statement of charge quantization, with the integer Q the conserved, quantized charge. Similar to quantum field theories with conserved charges, the charge conservation is intimately related to the existence of a local gauge symmetry. Indeed, the system defined by the Hamiltonian (3.3) is invariant under simultaneous local U(1) gauge transformations $f_{\sigma} \to f_{\sigma} \mathrm{e}^{i\phi(\tau)}$, $b_{\bar{\mu}} \to b_{\bar{\mu}} \mathrm{e}^{i\phi(\tau)}$, with $\phi(\tau)$ an arbitrary time-dependent phase.

3.1 Exact Projection onto the Physical Hilbert Space

While the gauge symmetry guarantees the conservation of the quantized charge Q, it does not single out any particular Q, such as $Q = 1$. In order to effect the projection onto the sector of Fock space with $Q = 1$, one may use a procedure first proposed by Abrikosov [4]: Consider first the grand-canonical ensemble with respect to Q, defined by the statistical operator

$$\hat{\rho}_G = \frac{1}{Z_G} \mathrm{e}^{-\beta(H+\lambda Q)}, \tag{3.5}$$

where $Z_G = \mathrm{tr}[\exp\{-\beta(H+\lambda Q)\}]$ is the grand-canonical partition function with respect to Q, $-\lambda$ is the associated chemical potential, and the trace extends over the complete Fock space, including summation over Q. The expectation value of an observable $\hat{A}$ in the grand-canonical ensemble is given by

$$\langle \hat{A} \rangle_G = \mathrm{tr}[\hat{\rho}_G \hat{A}]. \tag{3.6}$$

The physical expectation value of $\hat{A}$, $\langle \hat{A} \rangle$, is to be evaluated in the canonical ensemble where $Q = 1$. It can be calculated from the grand-canonical ensemble by differentiating with respect to the fugacity $\zeta = \mathrm{e}^{-\beta\lambda}$ and taking λ to infinity [6],

$$\langle \hat{A} \rangle = \lim_{\lambda\to\infty} \frac{\frac{\partial}{\partial\zeta} \mathrm{tr}[\hat{A} e^{-\beta(H+\lambda Q)}]}{\frac{\partial}{\partial\zeta} \mathrm{tr}[e^{-\beta(H+\lambda Q)}]} = \lim_{\lambda\to\infty} \frac{\langle Q\hat{A} \rangle_G}{\langle Q \rangle_G}. \tag{3.7}$$

3.1.1 Projecting Operators Acting on the Impurity States

We list two important results, which follow straightforwardly from (3.7): First, the canonical partition function in the subspace $Q = 1$ is

$$\begin{aligned} Z_C &= \lim_{\lambda\to\infty} \mathrm{tr}\left[Q e^{-\beta(H+\lambda(Q-1))}\right] \\ &= \lim_{\lambda\to\infty} \left(e^{\beta\lambda} \langle Q \rangle_G(\lambda)\right) Z_{Q=0}, \end{aligned} \tag{3.8}$$

where the subscripts G and C denote the grand-canonical and the canonical ($Q = 1$) expectation values, respectively. Second, the canonical $Q = 1$ expectation value of any operator $\hat{A}$ which has a zero expectation value in the $Q = 0$ subspace, $\hat{A}|Q = 0\rangle = 0$, is given by

$$\langle \hat{A} \rangle_C = \lim_{\lambda \to \infty} \frac{\langle \hat{A} \rangle_G(\lambda)}{\langle Q \rangle_G(\lambda)}. \tag{3.9}$$

Note that $\hat{A}|Q = 0\rangle = 0$ holds true for any physically observable operator acting on the impurity. Examples are the physical electron operator $d^\dagger_{\mu\sigma} = f^\dagger_\sigma b_{\bar{\mu}}$ or the local spin operator $\vec{S} = \sum_{\sigma\sigma'} \frac{1}{2} f^\dagger_\sigma \vec{\tau}_{\sigma\sigma'} f_{\sigma'}$. In this case the operator Q appearing in the numerator of (3.7) is not necessary to project away the $Q = 0$ sector. In particular, the constrained d-electron Green's function is given in terms of the grand-canonical one ($G_d(\omega, T, \lambda)$) as

$$G_d(\omega) = \lim_{\lambda \to \infty} \frac{G_d(\omega, T, \lambda)}{\langle Q \rangle_G(\lambda)}. \tag{3.10}$$

In the enlarged Hilbert space ($Q = 0, 1, 2, \dots$) $G_d(\omega, T, \lambda)$ may be expressed in terms of the grand-canonical pseudo-fermion and slave boson Green's functions using Wick's theorem. These auxiliary particle Green's functions, which constitute the basic building blocks of the theory, are defined in imaginary time representation as

$$\mathcal{G}_{f\sigma}(\tau_1 - \tau_2) = -\langle \widehat{T}\{f_\sigma(\tau_1) f^\dagger_\sigma(\tau_2)\}\rangle_G \tag{3.11}$$

$$\mathcal{G}_{b\bar{\mu}}(\tau_1 - \tau_2) = -\langle \widehat{T}\{b_{\bar{\mu}}(\tau_1) b^\dagger_{\bar{\mu}}(\tau_2)\}\rangle_G, \tag{3.12}$$

where $\widehat{T}$ is the time-ordering operator. The Fourier transforms of $\mathcal{G}_{f,b}$ may be expressed in terms of the exact self-energies $\Sigma_{f,b}$ as

$$\mathcal{G}_{f,b}(i\omega_n) = \left\{[\mathcal{G}^0_{f,b}(i\omega_n)]^{-1} - \Sigma_{f,b}(i\omega_n)\right\}^{-1}, \tag{3.13}$$

where

$$\mathcal{G}^0_{f\sigma}(i\omega_n) = (i\omega_n - E_d - \lambda)^{-1} \tag{3.14}$$

$$\mathcal{G}^0_{b\bar{\mu}}(i\omega_n) = (i\omega_n - \lambda)^{-1}. \tag{3.15}$$

Since, as a consequence of the projection procedure $\lambda \to \infty$, the energy eigenvalues of $H + \lambda Q$ scale to infinity as λQ, it is useful to shift the zero of the auxiliary particle frequency scale by λ (in the $Q = 1$ sector) and to define the "projected" Green's functions as

$$G_{f,b}(\omega) = \lim_{\lambda \to \infty} \mathcal{G}_{f,b}(\omega + \lambda). \tag{3.16}$$

Note that this does not affect the energy scale of physical quantities (like the local d electron Green's function), which is the *difference* between the pseudo-fermion and the slave-boson energy.

3.1.2 Canonical Expectation Values of Conduction Electron Operators

The canonical (i.e., projected onto the $Q = 1$ subspace), local conduction electron Green function is given as

$$G_{c\mu\sigma}(i\omega_n) = \left\{[G^0_{c\mu\sigma}(i\omega_n)]^{-1} - \Sigma_{c\mu\sigma}(i\omega_n)\right\}^{-1} \tag{3.17}$$

with

$$G^0_{c\mu\sigma}(i\omega_n) = \sum_{\vec{k}} G^0_{c\mu\sigma}(\vec{k}, i\omega_n) = \sum_{\vec{k}} (i\omega_n + \mu_c - \varepsilon_{\vec{k}})^{-1}, \tag{3.18}$$

where μ_c is the chemical potential of the conduction electrons. The canonical, local c-electron self-energy, $\Sigma_{c\mu\sigma}(i\omega_n)$, cannot be obtained from the grand-canonical one by simply taking the limit $\lambda \to \infty$, since the c-electron density has a nonvanishing expectation value in the $Q = 0$ subspace. However, it follows immediately from the Anderson Hamiltonian, (2.6)–(2.8), that the exact, canonical conduction electron t-matrix $t_{\sigma\mu}(i\omega)$, defined by $G_c = G^0_c[1 + tG^0_c]$, is proportional to the full, projected d-electron Green's function, $t_{\mu\sigma}(i\omega) = |V|^2 G_{d\mu\sigma}$. Thus, we have as an exact relation,

$$G_{c\mu\sigma}(i\omega_n) = G^0_{c\mu\sigma}(i\omega_n)\left[1 + |V|^2 G_{d\mu\sigma}(i\omega_n) G^0_{c\mu\sigma}(i\omega_n)\right], \tag{3.19}$$

and by comparison with (3.17) we obtain the local conduction electron self-energy respecting the constrained dynamics in the impurity orbital,

$$\Sigma_{c\mu\sigma}(i\omega_n) = \frac{|V|^2 G_{d\mu\sigma}(i\omega_n)}{1 + |V|^2 G^0_{c\mu\sigma}(i\omega_n) G_{d\mu\sigma}(i\omega_n)}. \tag{3.20}$$

Using phenomenological Fermi liquid theory [16] and also by means of perturbation theory to infinite order in the on-site repulsion U [20] it has been shown for the Fermi liquid case $M = 1$ of the symmetric Anderson model ($2E_d = -U$) that the exact d-electron propagator $G_{d\sigma}(\omega)$ and the d-electron self-energy $\Sigma_{d\sigma}(\omega) \equiv \omega - G_{d\sigma}(\omega)^{-1}$ obey the following local Fermi liquid relations in the limit $\omega \to 0 - i0$, $T \to 0$:

Luttinger theorem

$$\int d\omega\, f(\omega) \frac{\partial \Sigma_{d\sigma}(\omega)}{\partial \omega} G_{d\sigma}(\omega) = 0 \tag{3.21}$$

Friedel–Langreth

$$\frac{1}{\pi} \operatorname{Im} G_{d\sigma}(\omega) = \frac{1}{\Gamma} \sin^2\left(\frac{\pi n_d}{N}\right) - c\left[\left(\frac{\omega}{T_K}\right)^2 + \left(\frac{\pi T}{T_K}\right)^2\right] \tag{3.22}$$

$$\operatorname{Im} \Sigma_{d\sigma}(\omega) = \frac{\Gamma}{\sin^2(\pi n_d/N)} + c\left(\frac{\Gamma}{\sin^2(\pi n_d/N)}\right)^2 \left[\left(\frac{\omega}{T_K}\right)^2 + \left(\frac{\pi T}{T_K}\right)^2\right], \tag{3.23}$$

where c is a constant of $O(1)$. Combining (3.20) and (3.22), it follows that (for $M = 1$, away from particle hole symmetry) $\Sigma_{c\sigma}$ exhibits (in an exact theory) local Fermi liquid behavior as well, $\mathrm{Im}\,\Sigma_{c\sigma}(\omega - i0, T = 0) = a + b(\omega/T_K)^2$ for $\omega \to 0$. Note that this quantity is different from the grand-canonical conduction electron self-energy and has a finite imaginary part at the Fermi level.

The momentum-dependent conduction electron Green's function in the presence of a single impurity is given in terms of the canonical d-electron propagator as

$$G_{c\mu\sigma}(\vec{k},\vec{k}';i\omega_n)G^0_{c\mu\sigma}(\vec{k},i\omega_n)\left[\delta_{\vec{k},\vec{k}'} + |V|^2 G_d(i\omega_n)G^0_{c\mu\sigma}(\vec{k}',i\omega_n)\right]. \quad (3.24)$$

The latter expression is the starting point for treating a random system of many Anderson impurities [21].

3.2 Analytical Properties and Infrared Behavior

The Green's functions $G_{f,b,c}$ have the following spectral representations

$$G_{f,b,c}(i\omega_n) = \int_{-\infty}^{\infty} d\omega' \, \frac{A_{f,b,c}(\omega')}{i\omega_n - \omega'} \quad (3.25)$$

with the normalization of the spectral functions $A_{f,b,c}$

$$\int_{-\infty}^{\infty} d\omega \, A_{f,b,c}(\omega) = 1. \quad (3.26)$$

Taking the limit $\lambda \to \infty$ has important consequences on the analytical structure of the auxiliary particle Green's functions:

(1) It follows directly from the definitions (3.11) and (3.16), using (3.5) and (3.6), that the traces appearing in the canonical functions $G_{f,b}$ are taken purely over the $Q = 0$ sector of Fock space.[1] Thus, the backward-in-time ($\tau_1 < \tau_2$) or hole-like contribution to the auxiliary propagators in (3.11) vanishes after projection, and we have

$$G_{f\sigma}(\tau_1 - \tau_2) = -\Theta(\tau_1 - \tau_2) \lim_{\lambda\to\infty} \langle f_\sigma(\tau_1) f^\dagger_\sigma(\tau_2)\rangle_G \quad (3.27)$$

$$G_{b\bar{\mu}}(\tau_1 - \tau_2) = -\Theta(\tau_1 - \tau_2) \lim_{\lambda\to\infty} \langle b_{\bar{\mu}}(\tau_1) b^\dagger_{\bar{\mu}}(\tau_2)\rangle_G . \quad (3.28)$$

[1] This means that the auxiliary particle propagators are *not* calculated in the canonical ($Q = 1$) ensemble. The projection onto the $Q = 1$ sector of Fock space is achieved only when they are combined to calculate expectation values of physically observable operators like $G_{d\sigma}$, $\langle \vec{S}\rangle$ etc. The latter can be seen explicitly, e.g., from (5.4), 2nd equality.

Consequently, their spectral functions $A_{f,b}$ have the Lehmann representation

$$A_{f\sigma}(\omega) = \sum_{m,n\geq 0} \mathrm{e}^{-\beta E_m^0} |\langle 1,n|f_\sigma^\dagger|0,m\rangle|^2 \delta\big(\omega - (E_n^1 - E_m^0)\big) \tag{3.29}$$

$$A_{b\bar{\mu}}(\omega) = \sum_{m,n\geq 0} \mathrm{e}^{-\beta E_m^0} |\langle 1,n|b_{\bar{\mu}}^\dagger|0,m\rangle|^2 \delta\big(\omega - (E_n^1 - E_m^0)\big), \tag{3.30}$$

where E_n^Q are the energy eigenvalues ($E_0^0 \leq E_n^Q$ is the ground-state energy) and $|Q,n\rangle$ the many-body eigenstates of H in the sector Q of Fock space. At zero temperature, A_f reduces to $A_{f\sigma}(\omega) = \sum_{n\geq 0} |\langle 1,n|f_\sigma^\dagger|0,0\rangle|^2 \times \delta\big(\omega - (E_n^1 - E_0^0)\big)$ and similar for A_b. It is seen that the $A_{f,b}$ have threshold behavior at $\omega = E_0 \equiv E_0^1 - E_0^0$, with $A_{f,b}(\omega) \equiv 0$ for $\omega < E_0$, $T = 0$. The vanishing imaginary part at frequencies $\omega < 0$ may be shown to be a general property of all quantities involving slave particle operators, e.g., also of auxiliary particle self-energies and vertex functions.

(2) As will be seen in Section 5.2 (Eq. 5.4), physical expectation values not only involve the particle-like auxiliary propagators (Eq. 3.27, (3.28)) but also hole-like contributions. It is, therefore, useful to define the "anti-fermion" and "anti-boson" propagators (in imaginary time representation)

$$G_{f\sigma}^-(\tau_1 - \tau_2) = -\Theta(\tau_2 - \tau_1) \lim_{\lambda\to\infty} \langle f_\sigma^\dagger(\tau_2) f_\sigma(\tau_1)\rangle_G \tag{3.31}$$

$$G_{b\bar{\mu}}^-(\tau_1 - \tau_2) = -\Theta(\tau_2 - \tau_1) \lim_{\lambda\to\infty} \langle b_{\bar{\mu}}^\dagger(\tau_2) b_{\bar{\mu}}(\tau_1)\rangle_G \,, \tag{3.32}$$

whose spectral functions have the Lehmann representations

$$A_{f\sigma}^-(\omega) = \sum_{m,n\geq 0} \mathrm{e}^{-\beta E_m^1} |\langle 0,n|f_\sigma|1,m\rangle|^2 \delta\big(\omega - (E_n^0 - E_m^1)\big) \tag{3.33}$$

$$A_{b\bar{\mu}}^-(\omega) = \sum_{m,n\geq 0} \mathrm{e}^{-\beta E_m^1} |\langle 0,n|b_{\bar{\mu}}|1,m\rangle|^2 \delta\big(\omega - (E_n^0 - E_m^1)\big). \tag{3.34}$$

E_0^1 is the ground-state energy in the $Q = 1$ sector. The expressions (3.29), (3.30) and (3.33), (3.34) immediately imply a relation between $A_{f,b}$ and $A_{f,b}^-$,

$$A_{f,b}^-(\omega) = \mathrm{e}^{-\beta\omega} A_{f,b}(\omega). \tag{3.35}$$

(3) The property of only forward-in-time propagation (Eq. 3.27, (3.28)) means that the auxiliary particle propagators $G_{f,b}$ are formally identical to the core hole propagators of the well-known X-ray threshold problem [22–24]. Thus, the knowledge of the infrared behavior of the latter may be directly applied to the former. In particular, the spectral functions are found (see below) to diverge at the threshold E_0 in a power law fashion (infrared singularity)

$$A_{f,b}(\omega) \sim |\omega - E_0|^{-\alpha_{f,b}} \theta(\omega - E_0) \tag{3.36}$$

due to a diverging number of particle-hole excitation processes in the conduction electron sea as $\omega \to E_0$. For the single channel case ($M = 1$), i.e., the usual Kondo or mixed valence problem, the exponents α_f and α_b can be found analytically from the following chain of arguments: Anticipating that in this case, the impurity spin is completely screened by the conduction electrons at temperature $T = 0$, leaving a pure-potential scattering center, the ground state $|1, 0\rangle$ is a Slater determinant of one-particle scattering states, characterized by scattering phase shifts η_σ in the s-wave channel (assuming for simplicity a momentum independent hybridization matrix element V). To calculate the fermion spectral function $A_{f\sigma}(\omega)$ at $T = 0$ from (3.29), one needs to evaluate $\langle 1, n|f_\sigma^\dagger|0, 0\rangle$, which is just the overlap of two slater determinants, an eigenstate of the fully interacting Kondo system, $|1, n\rangle$, on the one hand, and the ground state of the conduction electron system in the absence of the impurity combined with the decoupled impurity level occupied by an electron with spin σ, $f_\sigma^\dagger|0, 0\rangle$, on the other hand. As shown by Anderson [22], the overlap of the two ground-state slater determinants, $\langle 1, 0|f_\sigma^\dagger|0, 0\rangle$, tends to zero in the thermodynamic limit (orthogonality catastrophe). Analogous relations hold for the boson spectral function $A_b(\omega)$. As a result, the exponential relaxation into the interacting ground state for long times is inhibited, leading to the infrared power law divergence of the spectral functions, (3.36).

The X-ray threshold exponents can be expressed in terms of the scattering phase shifts at the Fermi level by the exact relation [25]

$$\alpha_{f,b} = 1 - \sum_{\sigma'} \left(\frac{\eta_{f,b\sigma'}}{\pi} \right)^2 . \tag{3.37}$$

Here the $\eta_{f\sigma'}$ ($\eta_{b\sigma'}$) are the scattering phase shifts of the single-particle wave functions in channel σ' of the fully interacting ground state $|1, 0\rangle$, relative to the wave functions of the free state $f_\sigma^\dagger|0, 0\rangle$ ($b^\dagger|0, 0\rangle$). Via the Friedel sum rule, the scattering phase shifts are, in turn, related to the change $\Delta n_{c\sigma'}$ of the average number of conduction electrons per scattering channel σ' due to the presence of the impurity: $\eta_{f,b\sigma'} = \pi\Delta n_{c\sigma'}$. Obviously, $\Delta n_{c\sigma'}$ is equal and opposite in sign to the difference of the average impurity occupation numbers of the states $|1, 0\rangle$ and $f_\sigma^\dagger|0, 0\rangle$ ($b^\dagger|0, 0\rangle$). Thus, in the pseudo-fermion propagator $G_{f\sigma}$ we have the phase shifts,

$$\eta_{f\sigma'} = -\pi \left(\frac{n_d}{N} - \delta_{\sigma\sigma'} \right) \tag{3.38}$$

and in the slave boson propagator G_b,

$$\eta_b = -\pi \frac{n_d}{N}, \tag{3.39}$$

where n_d denotes the total occupation number of the impurity level in the interacting ground state. (The term $\delta_{\sigma\sigma'}$ in $\eta_{f\sigma'}$ appears because $f_\sigma^\dagger|0, 0\rangle$

has impurity occupation number 1 in the channel σ.) For example, in the Kondo limit $n_{\to}1$ and for a spin $\frac{1}{2}$ impurity ($N=2$), this leads to resonance scattering, $\eta_{f,b\sigma'}=\pi/2$. As a result, one finds [26] for the threshold exponents

$$\alpha_f = \frac{2n_d - n_d^2}{N} \tag{3.40}$$

$$\alpha_b = 1 - \frac{n_d^2}{N}. \tag{3.41}$$

These results have been found independently from Wilson's numerical renormalization group approach [27, 28] and using the Bethe ansatz solution and boundary conformal field theory [29]. It is interesting to note that

(i) the exponents depend on the level occupancy n_d (in the Kondo limit $n_d \to 1$, $\alpha_f = 1/N$ and $\alpha_b = 1-1/N$, whereas in the opposite, empty orbital, limit $n_d \to 0$, $\alpha_f \to 0$ and $\alpha_b \to 1$)

(ii) the sum of the exponents $\alpha_f + \alpha_b = 1 + 2\big(n_d(1-n_d)\big)(N) \geq 1$.

We stress that the above derivation of the infrared exponents $\alpha_{f,b}$ holds true only if the impurity complex acts as a pure potential scattering center at $T=0$. This is equivalent to the statement that the conduction electrons behave locally, i.e., at the impurity site, like a Fermi liquid. Conversely, in the multi-channel (non-FL) case, $N \geq 2$, $M \geq N$, the exponents have been found from a conformal field theory solution [19] of the problem in the Kondo limit to be

$$\alpha_f = M/(M+N)$$
$$\alpha_b = N/(M+N),$$

which differ from the FL values. Thus, one may infer from the values of $\alpha_{f,b}$ as a function of n_d whether the system is in a local Fermi liquid state.

4 Mean Field Approach and $1/N$ Expansion at $U \to \infty$

For physical situations of interest, the $s-d$ hybridization of the Anderson model (3.3) is much smaller than the conduction bandwidth, $\mathcal{N}(0)V \ll 1$, where $\mathcal{N}(0) = 1/D$ is the local conduction electron density of states at the Fermi level. This suggests a perturbation expansion in $\mathcal{N}(0)V$. A straightforward expansion in terms of bare Green's functions is not adequate, as it would not allow us to capture the physics of the Kondo screened state, or else the infrared divergencies of the auxiliary particle spectral functions discussed in the last section. In the framework of the slave boson representation, two types of nonperturbative approaches have been developed. The first one is mean field theory for both the slave boson amplitude $\langle b \rangle$ and the constraint ($\langle Q \rangle = 1$ rather than $Q = 1$). The second one is resummation of the perturbation theory to infinite order.

4.1 Slave Boson Mean Field Theory

Slave boson mean field theory is based on the assumption that the slave bosons condense at low temperatures such that $\langle b_{\bar{\mu}} \rangle \neq 0$. Replacing the operator $b_{\bar{\mu}}$ in $H + \lambda Q$ by $\langle b_{\bar{\mu}} \rangle$ (see [30]), where λ is a Lagrange multiplier to be adjusted such that $\langle Q \rangle = 1$, one arrives at a resonance level model for the pseudo-fermions. The position of the resonance, $E_d + \lambda$, is found to be given by the Kondo temperature T_K, and is thus close to the Fermi energy. The resonance generates the low-energy scale T_K, and leads to local Fermi liquid behavior. While this is qualitatively correct in the single-channel case, it is in blatant disagreement with the exactly known behavior in the multi-channel case. The mean field theory can be shown to be exact for $M = 1$ in the limit $N \to \infty$ for a model in which the constraint is softened to be $Q = N/2$. However, for finite N the breaking of the local gauge symmetry, which would be implied by the condensation of the slave boson field, is forbidden by Elitzur's theorem [31]. It is known that for finite N the fluctuations in the phase of the complex expectation value $\langle b_{\bar{\mu}} \rangle$ are divergent and lead to the suppression of $\langle b_{\bar{\mu}} \rangle$ to zero (see also [32–34]). This is true in the cartesian gauge, whereas in the radial gauge the phase fluctuations may be shown to cancel at least in lowest order. It has not been possible to connect the mean field solution, an apparently reasonable description at low temperatures and for $M = 1$, to the high temperature behavior ($T \gg T_K$), dominated by logarithmic temperature dependence, in a continuous way [30]. Therefore, it seems that the slave boson mean field solution does not offer a good starting point even for only a qualitatively correct description of quantum impurity models.

4.2 1/N Expansion versus Self-Consistent Formulation

The critical judgment of mean field theory is corroborated by the results of a straightforward $1/N$-expansion in the single channel case, keeping the exact constraint, and not allowing for a finite bose field expectation value [35]. Within this scheme the exact behavior of the thermodynamic quantities (known from the Bethe ansatz solution) at low temperatures as well as high temperatures is recovered to the considered order in $1/N$. Also, the exact auxiliary particle exponents $\alpha_{f,b}$ are reproduced in order $1/N$, using a plausible exponentiation scheme [36].

In addition, dynamical quantities like the d-electron spectral function and transport coefficients can be calculated exactly to a desired order in $1/N$ within this approach. However, as clear-cut and economical this method may be, it does have serious limitations. Firstly, the $1/N$ expansion is not uniformly convergent as a function of temperature [37]. Rather, the expansions at low temperature and at high temperature have to be done around two different saddle points (the limits $N \to \infty$). It is not known how to match these expansions in the crossover region $T \sim T_K$ in a systematic way. Secondly, the experimentally most relevant case of $N = 2$ or somewhat larger is not accessible in $1/N$ expansion. Thirdly, non-Fermi liquid behavior, being necessarily nonperturbative in $1/N$, cannot be dealt with in a controlled way on the basis of a $1/N$-expansion. To access these latter two regimes, a new approach nonperturbative in $1/N$ is necessary.

5 Conserving Approximations: Gauge-Invariant Self-Consistent Perturbation Theory in the Hybridization

We conjecture that this new approach is a gauge-invariant many-body theory of pseudo-fermions and slave bosons. As long as gauge-symmetry-violating objects such as Bose field expectation values or fermion pair correlation functions do not appear in the theory, gauge invariance of physical quantities can be guaranteed in suitably chosen approximations by the proper match of pseudo-fermion and slave boson properties, without introducing an additional gauge field. This requires the use of conserving approximations [38], derived from a Luttinger–Ward functional Φ.

5.1 Generating Functional

Φ consists of all vacuum skeleton diagrams built out of fully renormalized Green's functions $G_{b,f,c}$ and the bare vertex V. The self-energies $\Sigma_{b,f,c}$ are obtained by taking the functional derivative of Φ with respect to the corresponding Green's function (cutting the Green's function line in each diagram in all possible ways),

$$\Sigma_{b,f,c} = \delta\Phi/\delta G_{b,f,c}. \tag{5.1}$$

Irreducible vertex functions, figuring as integral kernels in two-particle Bethe–Salpeter equations, are generated by second-order derivatives of Φ.

The choice of diagrams for Φ defines a given approximation. It should be dictated by the dominant physical processes and by expansion in a small parameter, if available. Even in the presence of a small parameter (in this case $\mathcal{N}_0 V$), straightforward low order renormalized perturbation theory may not give an even qualitatively correct result, if singular vertex functions appear. It is therefore necessary to check whether vertex functions become singular at the level where approximate single particle Green's

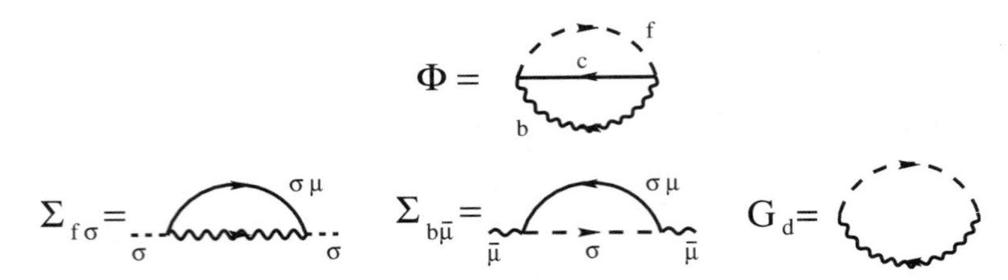

FIGURE 7.2. Diagrammatic representation of the generating functional Φ of the NCA. Also shown are the pseudo-particle self-energies and the local electron Green's function derived from Φ (Eqs. 3.11–3.13). Throughout this chapter, dashed, wavy and solid lines represent fermion, boson, and conduction electron lines, respectively. In the diagram for $\Sigma_{f\sigma}$ the spin labels are shown explicitly to demonstrate that there are no coherent spin fluctuations taken into account.

functions are used to calculate the integral kernels of the respective vertex functions. Should this be the case, the vertex functions have to be included into the approximation in a gauge-invariant way. As we shall see, in the present case it is required to take all two-particle vertex functions into consideration. Vertex functions involving three or more interacting particles will be omitted in the hope that the corresponding phase space is small so that they will contribute less even if they are singular. This is the first fundamental approximation. The second one is that we will approximate the irreducible kernels in the Bethe–Salpeter equations for the vertex functions by the lowest-order diagram in $\mathcal{N}_0 V$.

5.2 Noncrossing Approximation (NCA)

As noted before, in the present context, we may take the hybridization V to be a small quantity (dimensionless parameter $\mathcal{N}_0 V$). This suggests starting with the lowest-order (in V) diagram of Φ, which is second order (see Fig. 7.2). The self-energies generated from this obey after analytic continuation to real frequencies ($i\omega \to \omega - i0$) and projection the following equations of self-consistent second order-perturbation theory:

$$\Sigma_{f\sigma}^{(\mathrm{NCA})}(\omega - i0) = \Gamma \sum_{\mu} \int \frac{d\varepsilon}{\pi} [1 - f(\varepsilon)] A^0_{c\mu\sigma}(\varepsilon) G_{b\bar{\mu}}(\omega - \varepsilon - i0) \tag{5.2}$$

$$\Sigma_{b\bar{\mu}}^{(\mathrm{NCA})}(\omega - i0) = \Gamma \sum_{\sigma} \int \frac{d\varepsilon}{\pi} f(\varepsilon) A^0_{c\mu\sigma}(\varepsilon) G_{f\sigma}(\omega + \varepsilon - i0) \tag{5.3}$$

$$\begin{aligned} G_{d\mu\sigma}^{(\mathrm{NCA})}(\omega - i0) &= \int d\varepsilon\, \mathrm{e}^{-\beta\varepsilon} [G_{f\sigma}(\omega + \varepsilon - i0) A_{b\bar{\mu}}(\varepsilon) \\ &\qquad - A_{f\sigma}(\varepsilon) G_{b\bar{\mu}}(\varepsilon - \omega + i0)] \\ &= \int d\varepsilon\, [G_{f\sigma}(\omega + \varepsilon - i0) A^-_{b\bar{\mu}}(\varepsilon) \\ &\qquad - A^-_{f\sigma}(\varepsilon) G_{b\bar{\mu}}(\varepsilon - \omega + i0)], \end{aligned} \tag{5.4}$$

where $A^0_{c\mu\sigma} = 1\pi \operatorname{Im} G^0_{c\mu\sigma}/\mathcal{N}(0)$ is the (free) conduction electron density of states per spin and channel, normalized to the density of states at the Fermi level $\mathcal{N}(0)$, and $f(\varepsilon) = 1/(\exp(\beta\varepsilon)+1)$ denotes the Fermi distribution function. Together with the expressions (3.13), (3.14) for the Green's functions, (5.2)–(5.4) form a set of self-consistent equations for $\Sigma_{b,f,c}$, composed of all diagrams without any crossing propagator lines and are, thus, known as the "noncrossing approximation," in short NCA [39, 40].

At zero temperature and for low frequencies, (5.2) and (5.3) may be converted into a set of linear differential equations for G_f and G_b [41], which allow us to find the infrared exponents as $\alpha_f = M/M+N$; $\alpha_b = N/M+N$, independent of n_d. For the single channel case these exponents do not agree with the exact exponents derived in Section 3.2. This indicates that the NCA is not capable of recovering the local Fermi liquid behavior for $M = 1$. A numerical evaluation of the d-electron Green's function, which is given by the local self-energy Σ_c divided by V^2 and hence is given by the boson-fermion bubble within NCA (Fig. 7.2), shows indeed a spurious singularity at the Fermi energy [42]. The NCA performs somewhat better in the multi-channel case, where the exponents α_f and α_b yield the correct non-Fermi liquid exponents of physical quantities as known from the Bethe ansatz solution [15] and conformal field theory [19]. However, the specific heat and the residual entropy are not given correctly in NCA. Also, the limiting low-temperature scaling laws for the thermodynamic quantities are attained only at temperatures substantially below T_K, in disagreement with the exact Bethe ansatz solution.

5.3 *Evaluation of the Self-Consistency Equations at Low Temperatures*

In order to enter the asymptotic power law regime of the auxiliary spectral functions, the self-consistent equations must be evaluated for temperatures several orders of magnitude below T_K, the low temperature scale of the model. The equations are solved numerically by iteration. In the following we describe the two main procedures to make the diagrammatic auxiliary particle technique suitable for the lowest temperatures.

The grand-canonical expectation value of the auxiliary particle number appearing in (3.10) is given in terms of the grand-canonical (unprojected) auxiliary particle spectral functions $\mathcal{A}_{f,b}(\omega, \lambda)$ by

$$\langle Q\rangle_G(\lambda) = \int d\omega \left[f(\omega) \sum_{\sigma} \mathcal{A}_{f\sigma}(\omega, \lambda) + b(\omega) \sum_{\mu} \mathcal{A}_{b\bar{\mu}}(\omega, \lambda) \right], \tag{5.5}$$

where $f(\omega)$, $b(\omega)$ are the Fermi and Bose distribution functions, respectively. Substituting this into the expression (3.8) for the canonical partition function we obtain after carrying out the transformation $\omega \to \omega + \lambda$, and taking the limit $\lambda \to \infty$

$$e^{-\beta F_{\rm imp}(T)} \equiv \frac{Z_C}{Z_{Q=0}} = \lim_{\lambda\to\infty} e^{\beta\lambda} \langle Q\rangle_G(\lambda)$$
$$= \int d\omega\, e^{-\beta\omega} \Big[\sum_\sigma A_{f\sigma}(\omega) + \sum_\mu A_{b\bar\mu}(\omega) \Big]. \quad (5.6)$$

By definition $F_{\rm imp} = -1/\beta \ln(Z_C/Z_{Q=0})$ is the impurity contribution to the free energy.

The numerical evaluation of expectation values like $\langle Q\rangle_G(\lambda \to \infty)$ (Eq. 5.6) or $G_{d\mu\sigma}(\omega, \lambda \to \infty)$ (Eq. 5.4) is nontrivial, (1) because at $T = 0$ the auxiliary spectral functions $A_{f,b}(\omega, T)$ are divergent at the threshold frequency E_0, where the exact position of E_0 is a priori not known, and (2) because the Boltzmann factors $e^{-\beta\omega}$ diverge exponentially for $\omega < 0$. Therefore, we apply the following transformations:

(1) Before performing the projection $\omega \to \omega + \lambda$, $\lambda \to \infty$ we redefine the frequency scale of all auxiliary particle functions $A_{f,b}$ according to $\omega \to \omega + \lambda_0$, where λ_0 is a finite parameter. In each iteration, λ_0 is then determined such that

$$\int d\omega\, e^{-\beta\omega} \Big[\sum_\sigma A_{f\sigma}(\omega) + \sum_\mu A_{b\bar\mu}(\omega) \Big] = 1, \quad (5.7)$$

where $A_{f,b}(\omega) = \lim_{\lambda\to\infty} A_{f,b}(\omega + \lambda_0, \lambda)$ is now an auxiliary spectral function with the new reference energy. It can be seen by comparison with (5.6) that $\lambda_0(T) = F_{\rm imp}(T) = F_{Q=1}(T) - F_{Q=0}(T)$, i.e., λ_0 is the chemical potential for the auxiliary particle number Q, or equivalently the impurity contribution to the free energy. The difference of the free energies becomes equal to the threshold energy $E_0 = E^{\rm GS}_{Q=1} - E^{\rm GS}_{Q=0}$ at $T = 0$. More importantly, however, the above way of determining a "threshold" is less ad hoc than, for example, defining it by a maximum in some function appearing in the NCA equations. It can also be seen from (5.7) that this procedure defines the frequency scale of the auxiliary particles such that the $T = 0$ threshold divergence of the spectral functions is at the *fixed* frequency $\omega = 0$. This substantially increases the precision as well as the speed of numerical evaluations. (3.10) for the projected d-electron Green's function becomes

$$G_d(\omega) = \lim_{\lambda\to\infty} e^{\beta\lambda} G_d(\omega, T, \lambda). \quad (5.8)$$

(2) The divergence of the Boltzmann factors implies that the self-consistent solutions for $A_{f,b}(\omega)$ vanish exponentially $\sim e^{\beta\omega}$ for negative frequencies, confirming their threshold behavior. It is convenient not to formulate the self-consistent equations in terms of $A_{f,b}$ as in earlier evaluations [43],

but to define new functions $\tilde{A}_{f,b}(\omega)$ and $\mathrm{Im}\,\tilde{\Sigma}_{f,b}(\omega)$ such that

$$A_{f,b}(\omega) = f(-\omega)\tilde{A}_{f,b}(\omega) \tag{5.9}$$

$$\mathrm{Im}\,\Sigma_{f,b}(\omega) = f(-\omega)\,\mathrm{Im}\,\tilde{\Sigma}_{f,b}(\omega). \tag{5.10}$$

After fixing the chemical potential λ_0 and performing the projection onto the physical subspace, the canonical partition function (Eq. 3.8) behaves as $\lim_{\lambda\to\infty} e^{\beta(\lambda-\lambda_0)} Z_C(T) = 1$, and from (3.35) we have $A^{-}_{f,b}(\omega) = f(\omega)\tilde{A}_{f,b}(\omega)$. In this way all exponential divergencies are absorbed by one single function for each particle species. As an example, the NCA equations in terms of these functions are free of divergencies of the statistical factors and read

$$\mathrm{Im}\,\tilde{\Sigma}_{f\sigma}(\omega - i0) = \Gamma\sum_{\mu}\int d\varepsilon\,\frac{f(-\varepsilon)(1-f(\omega-\varepsilon))}{1-f(\omega)} \times A^{0}_{c\mu\sigma}(\varepsilon)\tilde{A}_{b\bar{\mu}}(\omega-\varepsilon) \tag{5.11}$$

$$\mathrm{Im}\,\tilde{\Sigma}_{b\bar{\mu}}(\omega - i0) = \Gamma\sum_{\sigma}\int d\varepsilon\,\frac{f(\varepsilon)(1-f(\omega+\varepsilon))}{1-f(\omega)} \times A^{0}_{c\mu\sigma}(\varepsilon)\tilde{A}_{f\sigma}(\omega+\varepsilon) \tag{5.12}$$

$$\langle Q\rangle(\lambda_0, \lambda\to\infty) = \int d\omega\, f(\omega)\left[\sum_{\sigma}\tilde{A}_{f\sigma}(\omega) + \sum_{\mu}\tilde{A}_{b\bar{\mu}}(\omega)\right] = 1 \tag{5.13}$$

$$\mathrm{Im}\,G_{d\sigma}(\omega - i0) = \int d\varepsilon\,[f(\varepsilon+\omega)f(-\varepsilon) + f(-\varepsilon-\omega)f(\varepsilon)] \times \tilde{A}_{f\sigma}(\varepsilon+\omega)\tilde{A}_{b}(\varepsilon). \tag{5.14}$$

The real parts of the self-energies Σ_f, Σ_b are determined from $\mathrm{Im}\,\Sigma_f$, $\mathrm{Im}\,\Sigma_b$ through a Kramers–Kroenig relation, and the auxiliary functions $\tilde{A}_{f\sigma}(\omega) = 1/\pi\,\mathrm{Im}\,\tilde{\Sigma}_{f\sigma}(\omega - i0)/[(\omega + \lambda_0 - i0 - E_d - \mathrm{Re}\,\Sigma_{f\sigma}(\omega - i0))^2 + \mathrm{Im}\,\Sigma_{f\sigma}(\omega - i0)^2]$, $\tilde{A}_{b\bar{\mu}}(\omega) = \frac{1}{\pi}\,\mathrm{Im}\,\tilde{\Sigma}_{b\bar{\mu}}(\omega - i0)/[(\omega + \lambda_0 - i0 - \mathrm{Re}\,\Sigma_{b\bar{\mu}}(\omega - i0))^2 + \mathrm{Im}\,\Sigma_{b\bar{\mu}}(\omega - i0)^2]$, thus closing the above set of equations.

The method described above allows us to solve the NCA equations effectively for temperatures down to typically $T = 10^{-4}T_K$. It may be shown that the procedures described above can also be applied to self-consistently compute vertex corrections beyond the NCA (see Section 6), thus avoiding any divergent statistical factors in the selfconsistency equations.

6 Conserving T-Matrix Approximation (CTMA) at $U \to \infty$

6.1 Dominant Contributions at Low Energy

In order to eliminate the shortcomings of the NCA mentioned above, the guiding principle should be to find contributions to the vertex functions which renormalize the auxiliary particle threshold exponents to their correct values, since this is a necessary condition for the description of FL

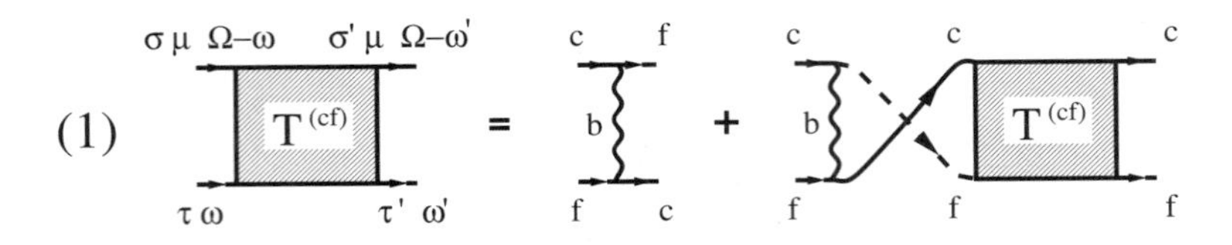

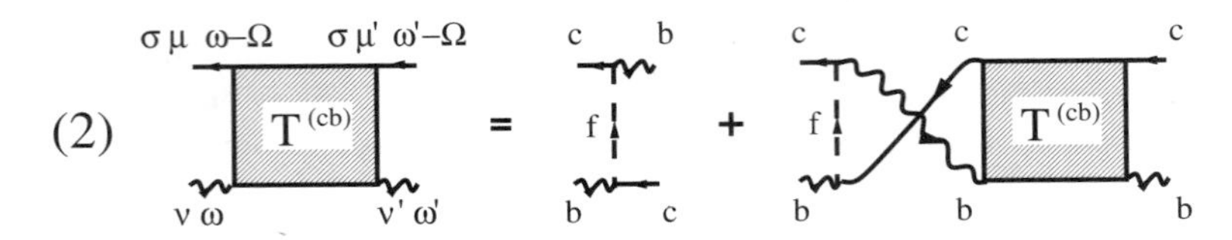

FIGURE 7.3. Diagrammatic representation of the Bethe–Salpeter equation for (1) the conduction electron-pseudo-fermion T-matrix $T^{(cf)}$ (Eq. 6.1), and (2) the conduction electron-slave boson T-matrix $T^{(cb)}$ (Eq. 6.2). $T^{(cb)}$ is obtained from $T^{(cf)}$ by interchanging $f \leftrightarrow b$ and $c \leftrightarrow c^{\dagger}$.

and non-FL behavior, as discussed in Section 3.2. Furthermore, it is instructive to realize that in NCA any coherent spin flip and charge transfer processes are neglected, as can be seen explicitly from (5.2) and (5.3) or from Fig. 7.2. These processes are known to be responsible for the quantum coherent collective behavior of the Anderson impurity complex below T_K. The existence of collective excitations in general is reflected in a singular behavior of the corresponding two-particle vertex functions. In view of the tendency of Kondo systems to form a collective spin singlet state, we focus our attention on the two-particle vertex functions, in particular, in the spin singlet channel of the pseudo-fermion-conduction electron vertex function and in the slave boson-conduction electron vertex function. It may be shown by power counting arguments (compare Appendix Appendix A) that there are no corrections to the NCA exponents in any finite order of perturbation theory [44]. Thus, we are led to search for singularities in the aforementioned vertex functions arising from an infinite resummation of terms.

From the preceding discussion it is natural to perform a partial resummation of those terms which, at each order in the hybridization V, contain the maximum number of spin flip or charge fluctuation processes, respectively. This amounts to calculating the conduction electron-pseudo-fermion vertex function in the "ladder" approximation defined in Fig. 7.3, where the irreducible vertex is given by $V^2 G_b$. In analogy to similar resummations for an interacting one-component Fermi system, we call the total $c-f$ vertex function T-matrix $T^{(cf)}$. The Bethe–Salpeter equation for $T^{(cf)}$ reads

(Fig. 7.3(1)),

$$
\begin{aligned}
T^{(cf)\mu}_{\sigma\tau,\sigma'\tau'}&(i\omega_n, i\omega'_n, i\Omega_n) \\
&= +V^2 G_{b\bar{\mu}}(i\omega_n + i\omega'_n - i\Omega_n)\delta_{\sigma\tau'}\delta_{\tau\sigma'} \\
&\quad - V^2 T \sum_{\omega''_n} G_{b\bar{\mu}}(i\omega_n + i\omega''_n - i\Omega_n) G_{f\sigma}(i\omega''_n) G^0_{c\mu\tau}(i\Omega_n - i\omega''_n) \\
&\qquad\qquad\qquad \times T^{(cf)\mu}_{\tau\sigma,\sigma'\tau'}(i\omega''_n, i\omega'_n, i\Omega_n). \qquad (6.1)
\end{aligned}
$$

A similar integral equation holds for the charge fluctuation T-matrix $T^{(cb)}$ (Fig. 7.3(2)),

$$
\begin{aligned}
T^{(cb)\sigma}_{\mu\nu,\mu'\nu'}&(i\omega_n, i\omega'_n, i\Omega_n) \\
&= +V^2 G_{f\sigma}(+i\omega_n + i\omega'_n - i\Omega_n)\delta_{\mu\nu'}\delta_{\nu\mu'} \\
&\quad - V^2 T \sum_{\omega''_n} G_{f\sigma}(i\omega_n + i\omega''_n - i\Omega_n) G_{b\bar{\mu}}(i\omega''_n) G^0_{c\nu\sigma}(-i\omega''_n - i\Omega_n) \\
&\qquad\qquad\qquad \times T^{(cb)\sigma}_{\nu\mu,\mu'\nu'}(i\omega''_n, i\omega'_n, i\Omega_n). \qquad (6.2)
\end{aligned}
$$

In the above Bethe–Salpeter equations σ, τ, σ', τ' represent spin and μ, ν, μ', ν' channel indices. We note that these are the only two-particle vertex functions after projection. The principal approximation adopted here is the form of the irreducible kernel, which we approximate by the lowest-order diagram.

Inserting NCA Green's functions for the intermediate state propagators of (6.1) and solving it numerically, we find at low temperatures and in the Kondo regime ($n_d \gtrsim 0.7$) a pole of $T^{(cf)}$ in the singlet channel (see Appendix A) as a function of the center-of-mass (COM) frequency Ω, at a frequency which scales with the Kondo temperature, $\Omega = \Omega_{cf} \simeq -T_K$. This is shown in Fig. 7.4. The threshold behavior of the imaginary part of $T^{(cf)}$ as a function of Ω with vanishing spectral weight at negative frequencies and temperature $T = 0$ is clearly seen. In addition, a very sharp structure appears, whose broadening is found to vanish as the temperature tends to zero, indicative of a pole in $T^{(cf)}$ at the *real* frequency Ω_{cf}, i.e., the tendency to form a collective singlet state between the conduction electrons and the localized spin. Similarly, the corresponding T-matrix $T^{(cb)}$ in the conduction electron-slave boson channel, evaluated within the analogous approximation, develops a pole at negative values of Ω in the empty orbital regime ($n_d \lesssim 0.3$). In the mixed valence regime ($n_d \simeq 0.5$) the poles in both $T^{(cf)}$ and $T^{(cb)}$ coexist. The appearance of poles in the two-particle vertex functions $T^{(cf)}$ and $T^{(cb)}$, which signals the formation of collective states, may be expected to influence the behavior of the system in a major way.

6.2 Self-Consistent Formulation: CTMA

On the level of approximation considered so far, the description is not yet consistent: In the limit of zero temperature the spectral weight of $T^{(cf)}$ and

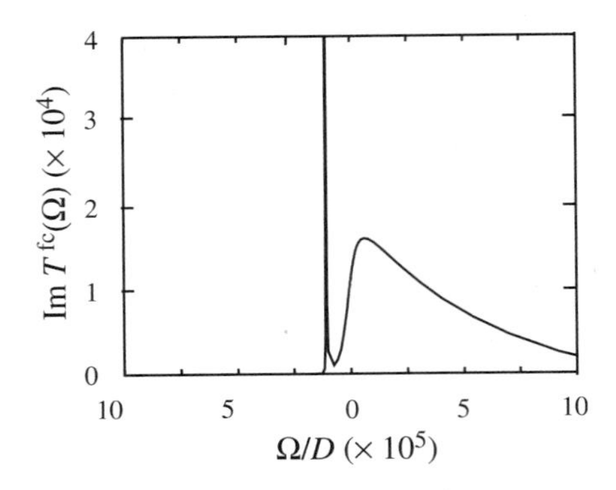

FIGURE 7.4. Imaginary part of the conduction electron-pseudo-fermion T-matrix $T^{(cf)}$ as a function of the COM frequency Ω for the single-channel case $M = 1$, $N = 2$, evaluated by inserting NCA solutions for the intermediate state propagators ($E_d = -0.67D$, $\Gamma = 0.15D$, $T = 4 \cdot 10^{-3} T_K$). The contribution from the pole positioned at a negative frequency $\Omega = \Omega_{cf} \simeq -T_K$ (compare text) is clearly seen.

$T^{(cb)}$ at negative frequencies Ω should be strictly zero (threshold property). Nonvanishing spectral weight at $\Omega < 0$ like a pole contribution for negative Ω in $T^{(cf)}$ or $T^{(cb)}$ would lead to a diverging contribution to the self-energy, which is unphysical. However, recall that a minimum requirement on the approximation used is the preservation of gauge symmetry. This requirement is not met when the integral kernel of the T-matrix equation is approximated by the NCA result. Rather, the approximation should be generated from a Luttinger–Ward functional. The corresponding generating functional is shown in Fig. 7.5. It is defined as the infinite series of all vacuum skeleton diagrams which consist of a single ring of auxiliary particle propagators, where each conduction electron line spans at most two hybridization vertices. As shown in Appendix A by means of a cancellation theorem, the CTMA includes, at any given loop order, all infrared singular contributions to leading and subleading order in the frequency ω. The first diagram of the infinite series of CTMA terms corresponds to NCA (Fig. 7.2). The diagram containing two boson lines is excluded, since it is not a skeleton. Although the spirit of the present theory is different from a large N expansion, it should be noted that the sum of the Φ diagrams containing up to four boson lines includes all terms of a $1/N$ expansion up to $O(1/N^2)$ [45]. By functional differentiation with respect to the conduction electron Green's function and the pseudo-fermion or the slave boson propagator, respectively, the shown Φ functional generates the ladder approximations $T^{(cf)}$, $T^{(cb)}$ for the total conduction electron-pseudo-fermion vertex function and for the total conduction electron-slave boson vertex function (Fig. 7.3). The auxiliary particle self-energies are obtained in the conserving scheme as the functional derivatives of Φ with respect to G_f or

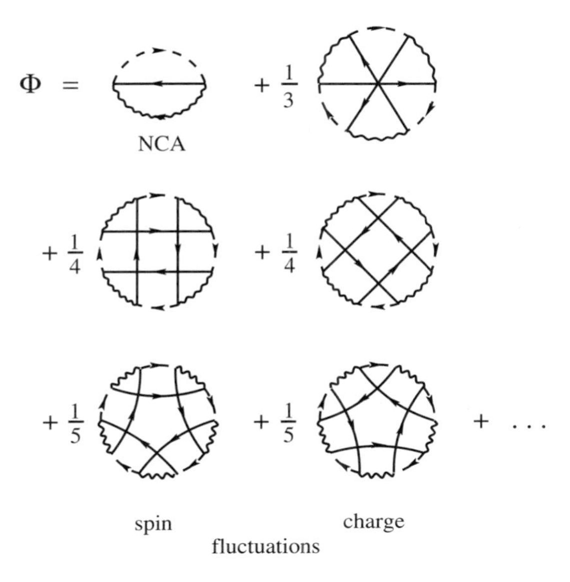

FIGURE 7.5. Diagrammatic representation of the Luttinger–Ward functional generating the conserving T-matrix approximation (CTMA). The terms with the conduction electron lines running clockwise (labeled "spin fluctuations") generate the T-matrix $T^{(cf)}$, while the terms with the conduction electron lines running counter-clockwise (labeled "charge fluctuations") generate the T-matrix $T^{(cb)}$.

G_b, respectively, (Eq. 5.1). This defines a set of self-consistency equations which we term conserving T-matrix approximation (CTMA), where the self-energies are given as nonlinear and nonlocal (in time) functionals of the Green's functions, while the Green's functions are in turn expressed in terms of the self-energies. The solution of these equations requires that the T-matrices have vanishing spectral weight at negative COM frequencies Ω. Indeed, the numerical evaluation shows that the poles of $T^{(cf)}$ and $T^{(cb)}$ are shifted to $\Omega = 0$ by self-consistency, where they merge with the continuous spectral weight present for $\Omega > 0$, thus renormalizing the threshold exponents of the auxiliary spectral functions.

In the following we give explicitly the self-consistent equations which determine the auxiliary particle self-energies within CTMA. For that purpose, it is useful to define conduction electron-fermion and conduction electron-boson vertex functions $T^{(cf)(\pm)}$, $T^{(cb)(\pm)}$ without $(+)$ or with $(-)$ an alternating sign between terms with even and odd number of rungs (compare Fig. 7.3). In the Matsubara representation, these vertex functions, to be labeled "even" $(+)$ and "odd" $(-)$ below, are given by the following Bethe–

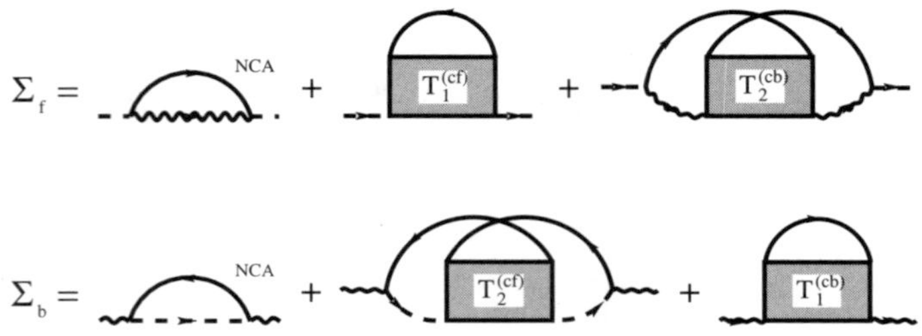

FIGURE 7.6. Diagrammatic representation of the CTMA expressions for pseudo-particle self-energies Σ_f and Σ_b. The first term drawn on the righthand side of Σ_f and Σ_b, respectively, is the NCA diagram. The diagrammatic parts $T_{1,2}^{(cf)}$, $T_{1,2}^{(cb)}$ are explained in the text (Eqs. 6.16–6.19).

Salpeter equations:

$$\begin{aligned}
&T^{(cf)(\pm)\mu}_{\sigma,\tau}(i\omega_n, i\omega'_n, i\Omega_n)(i\omega_n, i\omega_n, i\Omega_n) \\
&\quad = U^{(cf)\mu}_{\sigma,\tau} \pm V^2 \frac{1}{\beta}\sum_{\omega''_n} G_{b\bar{\mu}}(i\omega_n + i\omega''_n - i\Omega_n) G_{f\sigma}(i\omega''_n) \\
&\qquad\qquad \times G^0_{c\mu\tau}(i\Omega_n - i\omega''_n) T^{(cf)(\pm)\mu}_{\tau,\sigma}(i\omega''_n, i\omega'_n, i\Omega_n) \qquad (6.3)
\end{aligned}$$

$$\begin{aligned}
&U^{(cf)\mu}_{\sigma,\tau}(i\omega_n, i\omega'_n, i\Omega_n) \\
&\quad = -V^4 \frac{1}{\beta}\sum_{\omega''_n} G_{b\bar{\mu}}(i\omega_n + i\omega''_n - i\Omega_n) G_{f\sigma}(i\omega''_n) \\
&\qquad\qquad \times G^0_{c\mu\tau}(i\Omega_n - i\omega''_n) G_{b\bar{\mu}}(i\omega'_n + i\omega''_n - i\Omega_n) \qquad (6.4)
\end{aligned}$$

and

$$\begin{aligned}
T^{(cb)(\pm)\sigma}_{\mu,\nu}(i\omega_n, i\omega'_n, i\Omega_n) &= U^{(cb)\sigma}_{\mu,\nu}(i\omega_n, i\omega'_n, i\Omega_n) \\
&\pm V^2 \frac{1}{\beta}\sum_{\omega''_n} G_{f\sigma}(i\omega_n + i\omega''_n - i\Omega_n) G_{b\bar{\mu}}(i\omega''_n) \\
&\qquad \times G^0_{c\nu\sigma}(i\omega''_n - i\Omega_n) T^{(cb)(\pm)\sigma}_{\nu,\mu}(i\omega''_n, i\omega'_n, i\Omega_n) \qquad (6.5)
\end{aligned}$$

$$\begin{aligned}
&U^{(cb)\sigma}_{\mu,\nu}(i\omega_n, i\omega'_n, i\Omega_n) \\
&\quad = -V^4 \frac{1}{\beta}\sum_{\omega''_n} G_{f\sigma}(i\omega_n + i\omega''_n - i\Omega_n) \\
&\qquad \times G_{b\bar{\mu}}(i\omega''_n) G^0_{c\nu\sigma}(i\omega''_n - i\Omega_n) G_{f\sigma}(i\omega'_n + i\omega''_n - i\Omega_n). \qquad (6.6)
\end{aligned}$$

Note that, in addition to the alternating sign, these vertex functions differ from the T-matrices defined in (6.1) and (6.2) in that they contain only terms with two or more rungs, since the inhomogeneous parts $U^{(cf)}$ and $U^{(cb)}$ represent terms with two bosonic or fermionic rungs, respectively. The terms with a single rung correspond to the NCA diagrams and are evaluated separately (see below).

The spin degrees of freedom of $T^{(cf)(\pm)}$ are uniquely determined by the spin indices σ, τ of the ingoing conduction electron and pseudo-fermion lines (Fig. 7.3). It is instructive to note that in the spin $S=\frac{1}{2}$ case ($N=2$) the singlet and triplet vertex functions (which correspond to the two-particle Green's functions in the singlet channel, $\phi^s \sim \sum_\sigma \langle T\{(c_\sigma f_{-\sigma} - c_{-\sigma} f_\sigma)(c_\sigma^\dagger f_{-\sigma}^\dagger - c_{-\sigma}^\dagger f_\sigma^\dagger)\}\rangle$, and in the triplet channel with magnetic quantum number $m=0,\pm 1$, $\phi^t_{m=0} \sim \sum_\sigma \langle T\{(c_\sigma f_{-\sigma} + c_{-\sigma} f_\sigma)(c_\sigma^\dagger f_{-\sigma}^\dagger + c_{-\sigma}^\dagger f_\sigma^\dagger)\}\rangle$, $\phi^t_{m=\pm 1} \sim \langle T\{c_{\pm 1/2} f_{\pm 1/2} c^\dagger_{\pm 1/2} f^\dagger_{\pm 1/2}\}\rangle$, respectively) may be identified in the following way,

$$T^{(cf)s} = \sum_\sigma T^{(cf)(-)}_{\sigma,-\sigma} \tag{6.7}$$

$$T^{(cf)t}_{m=0} = \sum_\sigma T^{(cf)(+)}_{\sigma,-\sigma} \tag{6.8}$$

$$T^{(cf)t}_{m=\pm 1} = T^{(cf)(+)}_{\pm 1/2, \pm 1/2}. \tag{6.9}$$

Analogous relations hold for the conduction electron-boson vertex function in terms of the channel degrees of freedom μ, ν. The total CTMA pseudo-particle self-energies, as derived by functional differentiation from the generating functional Φ, Fig. 7.5, are shown in Fig. 7.6 and consist of three terms each,

$$\Sigma_{f\sigma}(i\omega_n) = \Sigma^{(\mathrm{NCA})}_{f\sigma}(i\omega_n) + \Sigma^{(cf)}_{f\sigma}(i\omega_n) + \Sigma^{(cb)}_{f\sigma}(i\omega_n) \tag{6.10}$$

$$\Sigma_{b\bar{\mu}}(i\omega_n) = \Sigma^{(\mathrm{NCA})}_{b\bar{\mu}}(i\omega_n) + \Sigma^{(cf)}_{b\bar{\mu}}(i\omega_n) + \Sigma^{(cb)}_{b\bar{\mu}}(i\omega_n). \tag{6.11}$$

The first term of Σ_f and Σ_b represents the NCA self-energies, (5.2) and (5.3). The second and third terms arise from the spin and the charge fluctuations, respectively, and are given for pseudo-fermions by

$$\Sigma^{(cf)}_{f\sigma}(i\omega_n) = M\frac{1}{\beta}\sum_{\Omega_n} G^0_c(i\Omega_n - i\omega_n) T^{(cf)}_1(i\omega_n, i\omega_n, i\Omega_n) \tag{6.12}$$

$$\begin{aligned}\Sigma^{(cb)}_{f\sigma}(i\omega_n) = -MV^2\frac{1}{\beta^2}\sum_{\omega_n'\omega_n''} & G^0_c(i\omega_n - i\omega_n') G_b(i\omega_n') \\ & \times T^{(cb)}_2(i\omega_n', i\omega_n'', i\omega_n' + i\omega_n'' - i\omega_n) \\ & \times G^0_c(i\omega_n - i\omega_n'') G_b(i\omega_n'')\end{aligned} \tag{6.13}$$

and for slave bosons by

$$\begin{aligned}\Sigma^{(cf)}_{b\bar{\mu}}(i\omega_n) = -NV^2\frac{1}{\beta^2}\sum_{\omega_n'\omega_n''} & G^0_c(i\omega_n' - i\omega_n) G_f(i\omega_n') \\ & \times T^{(cf)}_2(i\omega_n', i\omega_n'', i\omega_n' + i\omega_n'' - i\omega_n) \\ & \times G^0_c(i\omega_n'' - i\omega_n) G_f(i\omega_n'')\end{aligned} \tag{6.14}$$

$$\Sigma^{(cb)}_{b\bar{\mu}}(i\omega_n) = N\frac{1}{\beta}\sum_{\Omega_n} G^0_c(i\omega_n - i\Omega_n) T^{(cb)}_1(i\omega_n, i\omega_n, i\Omega_n), \tag{6.15}$$

where the vertex functions appearing in these expressions are defined as

$$T_1^{(cf)} = \frac{N+1}{2} T^{(cf)(+)} + \frac{N-1}{2} T^{(cf)(-)} - NU^{(cf)} \tag{6.16}$$

$$T_2^{(cf)} = \frac{N+1}{2} T^{(cf)(+)} - \frac{N-1}{2} T^{(cf)(-)} - U^{(cf)} \tag{6.17}$$

$$T_1^{(cb)} = \frac{M-1}{2} T^{(cb)(+)} + \frac{M+1}{2} T^{(cb)(-)} - MU^{(cb)} \tag{6.18}$$

$$T_2^{(cb)} = \frac{M-1}{2} T^{(cb)(+)} - \frac{M+1}{2} T^{(cb)(-)} - U^{(cb)} . \tag{6.19}$$

These combinations of the even and odd vertex functions ensure the proper spin and channel summations in the self-energies. For the sake of clarity, the spin and channel indices as well as the frequency variables are not shown explicitly. In (6.16) and (6.18) the terms with two rungs, $N\,U^{(cf)}$, $M\,U^{(cb)}$, have been subtracted, since they would generate nonskeleton self-energy diagrams. Likewise, in (6.17) and (6.19) the two-rung terms have been subtracted in order to avoid a double counting of terms in the self-energies.

We now turn to the analytic continuation to real frequencies of the expressions derived above. Transforming the Matsubara summations into contour integrals shows that integrations along branch cuts of auxiliary particle Green's functions carry an additional factor $\exp(-\beta\lambda)$ as compared to integrations along branch cuts of physical Green's functions, which vanishes upon projection onto the the physical Fock space, $\lambda \to \infty$. Thus, as a general rule, only integrations along branch cuts of the c-electron propagators contribute to the auxiliary particle self-energies. Therefore, by performing the analytic continuation, $i\omega_n \to \omega - i0 \equiv \omega$ in all frequency variables, we obtain the advanced pseudo-fermion self-energy,

$$\Sigma_{f\sigma}^{(cf)}(\omega) = M \int \frac{d\varepsilon}{\pi} f(\varepsilon - \omega) A_c^0(\varepsilon - \omega) \pi \mathcal{N}(0) T_1^{(cf)}(\omega, \omega, \varepsilon) \tag{6.20}$$

$$\begin{aligned} \Sigma_{f\sigma}^{(cb)}(\omega) = -M\Gamma \int \frac{d\varepsilon}{\pi} \int \frac{d\varepsilon'}{\pi} & f(\varepsilon - \omega) f(\varepsilon' - \omega) A_c^0(\omega - \varepsilon) G_b(\varepsilon) \\ & \times \pi \mathcal{N}(0) T_2^{(cb)}(\varepsilon, \varepsilon', \varepsilon + \varepsilon' - \omega) \\ & \times A_c^0(\omega - \varepsilon') G_b(\varepsilon') \end{aligned} \tag{6.21}$$

and the advanced slave boson self-energy,

$$\begin{aligned} \Sigma_{b\bar{\mu}}^{(cf)}(\omega) = -N\Gamma \int \frac{d\varepsilon}{\pi} \int \frac{d\varepsilon'}{\pi} & f(\varepsilon - \omega) f(\varepsilon' - \omega) A_c^0(\varepsilon - \omega) G_f(\varepsilon) \\ & \times \pi \mathcal{N}(0) T_2^{(cf)}(\varepsilon, \varepsilon', \varepsilon + \varepsilon' - \omega) \\ & \times A_c^0(\varepsilon' - \omega) G_f(\varepsilon') \end{aligned} \tag{6.22}$$

$$\Sigma_{b\bar{\mu}}^{(cb)}(\omega) = -N \int \frac{d\varepsilon}{\varphi} f(\varepsilon - \omega) A_c^0(\omega - \varepsilon) \pi \mathcal{N}(0) T_1^{(cb)}(\omega, \omega, \varepsilon), \tag{6.23}$$

where the vertex functions are given by (6.16), (6.17) and (6.18), (6.19) with

$$T_{\sigma,\tau}^{(cf)(\pm)\mu}(\omega,\omega',\Omega) = U_{\sigma,\tau}^{(cf)(\pm)\mu}(\omega,\omega',\Omega) \pm (-\Gamma)\int \frac{d\varepsilon}{\pi} f(\varepsilon-\Omega) G_{b\bar{\mu}}(\omega+\varepsilon-\Omega) G_{f\sigma}(\varepsilon) \times A^0_{c\mu\tau}(\Omega-\varepsilon) T_{\tau,\sigma}^{(cf)(\pm)\mu}(\varepsilon,\omega',\Omega) \quad (6.24)$$

$$\pi\mathcal{N}(0) U_{\sigma,\tau}^{(cf)(\pm)\mu}(\omega,\omega',\Omega) = +\Gamma^2 \int \frac{d\varepsilon}{\pi} f(\varepsilon-\Omega) G_{b\bar{\mu}}(\omega+\varepsilon-\Omega) G_{f\sigma}(\varepsilon) \times A^0_{c\mu\tau}(\Omega-\varepsilon) G_{b\bar{\mu}}(\omega'+\varepsilon-\Omega) \quad (6.25)$$

and

$$T_{\mu,\nu}^{(cb)(\pm)\sigma}(\omega,\omega',\Omega) = U_{\mu,\nu}^{(cb)(\pm)\sigma}(\omega,\omega',\Omega) \pm (+\Gamma)\int \frac{d\varepsilon}{\pi} f(\varepsilon-\Omega) G_{f\sigma}(\omega+\varepsilon-\Omega) G_{b\bar{\mu}}(\varepsilon) \times A^0_{c\nu\sigma}(\varepsilon-\Omega) T_{\nu,\mu}^{(cb)(\pm)\sigma}(\varepsilon,\omega',\Omega) \quad (6.26)$$

$$\pi\mathcal{N}(0) U_{\mu,\nu}^{(cb)(\pm)\sigma}(\omega,\omega',\Omega) = -\Gamma^2 \int \frac{d\varepsilon}{\pi} f(\varepsilon-\Omega) G_{f\sigma}(\omega+\varepsilon-\Omega) G_{b\bar{\mu}}(\varepsilon) \times A^0_{c\nu\sigma}(\varepsilon-\Omega) G_{f\sigma}(\omega'+\varepsilon-\Omega) \quad (6.27)$$

In the above expressions, as in the NCA equations (5.2)–(5.4), we have used the dimensionless conduction electron spectral density, $A_c^0(\omega) = (1/\pi)\,\mathrm{Im}\,G^0_{c\mu\sigma}(\omega-i0)/\mathcal{N}(0)$, and we have suppressed obvious spin and channel indices. All frequency variables are to be understood as the limit $\omega \equiv \omega - i0$.

Equations (6.20)–(6.23), supplemented by the vertex functions (Eqs. 6.16–6.19, 6.24–6.27) form, together with the NCA contributions (Eqs. 5.2, 5.3) and the definitions of the auxiliary particle Green functions (Eqs. 3.13, 3.14), the closed set of self-consistent CTMA equations [46]. It is seen that in these equations only those branches of the T-matrix vertex functions appear which are advanced with respect to all three frequency variables, although in general the T-matrix consists of 2^3 independent analytical branches. This simplification is a consequence of the exact projection onto the physical sector of Fock space. Inspection of the analytically continued CTMA equations also shows that the slave boson self-energy is obtained from the pseudo-fermion self-energy, including the proper signs, by simply replacing $G_f \leftrightarrow G_b$ and inverting the frequency argument of A_c^0 in all expressions.

Equations (6.20)–(6.27) may be rewritten in terms of the spectral functions without threshold, $\tilde{A}_{f,b}$, in a straightforward way, as explained in Section 5.3, thus avoiding divergent statistical factors in the d-electron Green's function.

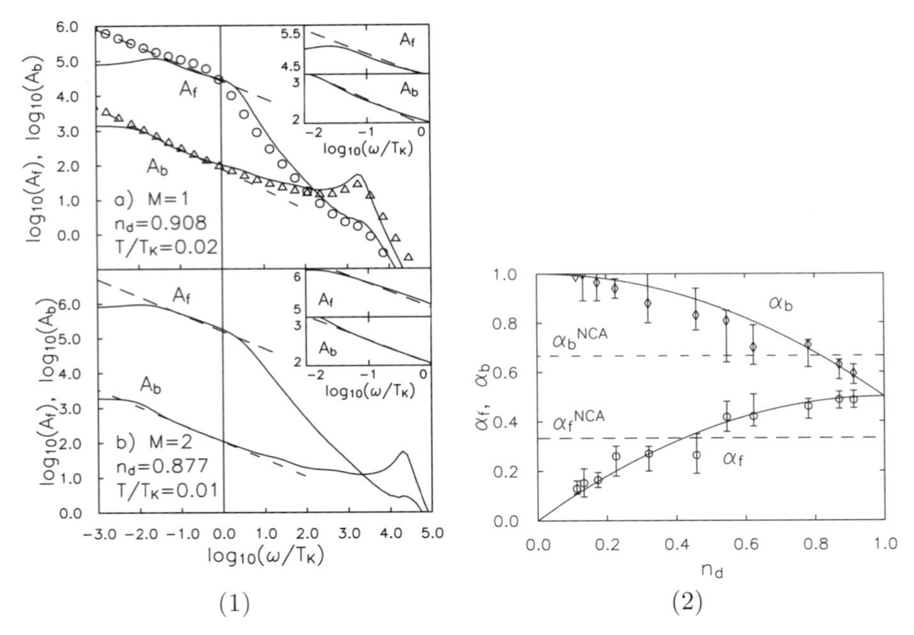

FIGURE 7.7. (1) Pseudo-fermion and slave boson spectral functions A_f and A_b in the Kondo regime ($N = 2$; $E_d = -0.05$, $\Gamma = 0.01$ in units of the half-bandwidth D), for (top) the single-channel ($M = 1$) and (bottom) the multi-channel ($M = 2$) case. In the single-channel case, the symbols represent the results of NRG for the same parameter set, $T = 0$. The slopes of the dashed lines indicate the exact threshold exponents as derived in Section 3.2 for $M = 1$ and as given by conformal field theory for $M = 2$. The insets show magnified power law regions. (2) CTMA results (symbols with error bars) for the threshold exponents α_f and α_b of A_f and A_b, ($N = 2$, $M = 1$). Solid lines: exact values (Section 3.2), dashed lines: NCA results (Section 5.2).

The self-consistent solutions are obtained by first solving the linear Bethe–Salpeter equations for the T-matrices by matrix inversion, computing the auxiliary particle self-energies from $T^{(cf)}$ and $T^{(cb)}$, and then constructing the fermion and boson Green's functions from the respective self-energies. This process is iterated until convergence is reached. We have obtained reliable solutions down to temperatures of the order of at least $10^{-2}T_K$ both for the single-channel and for the two-channel Anderson model. Note that $T_K \to 0$ in the Kondo limit; in the mixed valence and empty impurity regimes, significantly lower temperatures may be reached, compared to the low temperature scale of the model.

6.3 Results for the Auxiliary Particle Spectral Functions

As shown in Fig. 7.7(1a), the auxiliary particle spectral functions obtained from CTMA [46] are in good agreement with the results of a numerical renormalization group (NRG) calculation [27] (zero temperature results), given the uncertainties in the NRG at higher frequencies. Typical behavior in the Kondo regime is obtained: a broad peak of width $\sim \Gamma$ in A_b at $\omega \simeq |E_d|$, representing the hybridizing d-level and a structure in A_f at $\omega \simeq T_K$. Both functions display power law behavior at frequencies below T_K, which at finite T is cut off at the scale $\omega \simeq T$. The exponents extracted from the frequency range $T < \omega < T_K$ of our finite T results compare well with the exact result also shown (see insets of Fig. 7.7(1)a)). A similar analysis has been performed for a number of parameter sets spanning the complete range of d-level occupation numbers n_d. The extracted power law exponents are shown in Fig. 7.7(2), together with error bars estimated from the finite frequency ranges over which the fit was made. The comparatively large error bars in the mixed valence regime arise because here spin flip and charge fluctuation processes, described by the poles in $T^{(cf)}$ and $T^{(cb)}$, respectively, are of equal importance, impeding the convergence of the numerical procedure. In this light, the agreement with the exact results (solid curves) is very good, the exact value lying within the error bars or very close in each case.

In the multi-channel case ($N \geq 2$, $M \geq N$) NCA has been shown [44] to reproduce asymptotically the correct threshold exponents, $\alpha_f = M/(M + N)$, $\alpha_b = N/(M + N)$, in the Kondo limit. Calculating the T-matrices using NCA Green's functions (as discussed in the single-channel case) we find again a pole in the singlet channel of $T^{(cf)}$. However, in this case the CTMA does not renormalize the NCA exponents in the Kondo limit of the two-channel model; i.e., the threshold exponents obtained from the CTMA solutions are very close to the exact ones, $\alpha_f = \frac{1}{2}$, $\alpha_b = \frac{1}{2}$, as shown in Fig. 7.7(1b).

We note in passing that we have also solved the spinless case ($N = 1$, $M = 1$) using the auxiliary particle method [47]. The Anderson impurity model (Eqs. 2.6–2.8), then reduces to a noninteracting resonant level model. In this sense, the spinless case may be seen as the most extreme case of a quantum impurity model with a Fermi liquid ground state. It does, however, remain nontrivial in the slave particle representation (Eq. 3.3) and, therefore, constitutes a test case for the description of Fermi liquid behavior within the auxiliary particle method. As discussed in Section 3.2, the auxiliary particle spectral functions A_f, A_b in general display an infrared threshold with power law behavior which is induced by an orthogonality catastrophy analogous to that of the X-ray problem. In the spinless case, A_f and A_b have been shown to correspond precisely to the photoemission and X-ray absorption spectral densities of the X-ray problem, respectively [47]. The results of the CTMA for the $N = 1$, $M = 1$ case are shown in Fig. 7.8. Good quantitative agreement of the X-ray exponents extracted from the numerical evaluation of the CTMA equations is found in the regions $n_d \lesssim 1$ and $n_d \gtrsim 0$ which may be identified with the regions of a strong (scattering

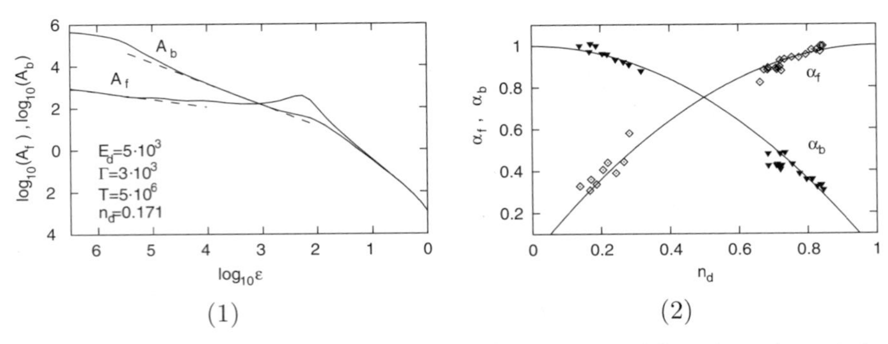

FIGURE 7.8. (1) Pseudo-fermion and slave boson spectral functions A_f and A_b for the X-ray problem ($N = 1, M = 1$; $E_d = 0.005$, $\Gamma = 0.003$ in units of the half-bandwidth D) [48]. The slopes of the dashed lines indicate the exact threshold exponents as derived in Section 3.2. (2) CTMA results (symbols) for the threshold exponents α_f and α_b of the spectral functions A_f and A_b, ($N = 1$, $M = 1$). Solid lines: exact values (Section 3.2).

phase shift $\eta_\sigma \simeq \pi/2$) and a weak potential (scattering phase shift $\eta_\sigma \simeq 0$) scatterer.

The agreement of the CTMA exponents with their exact values in the Kondo, mixed valence, and empty impurity regimes of the Spin $\frac{1}{2}$ single-channel model, in the spinless model, and in the Kondo regime of the two-channel model may be taken as evidence that the T-matrix approximation correctly describes both the Fermi liquid ($N = 1, 2$, $M = 1$) and the non-Fermi liquid ($N = 2$, $M = 1, 2$) regimes of the SU(N) $\times$ SU(M) Anderson model. Therefore, we expect the CTMA to describe correctly physically observable quantities of the SU(N) $\times$ SU(M) Anderson impurity model as well.

6.4 *Results for Physical Quantities: Spin Susceptibility*

We have calculated the static spin susceptibility χ of the Anderson model in the Kondo regime by solving the CTMA equations in a finite magnetic field H coupled to the impurity spin and taking the derivative of the magnetization $M = \frac{1}{2} g\mu_B \langle n_{f\uparrow} - n_{f\downarrow}\rangle$ with respect to H. The resulting $\chi(T) = (\partial M/\partial H)_T$ is shown in Fig. 7.9 both for the single-channel case ($N = 2$, $M = 1$) and for the two-channel Anderson model ($N = 2$, $M = 2$). It is seen that in the single-channel case CTMA correctly reproduces the constant Pauli susceptibility (Fermi liquid behavior) below T_K, while NCA gives an incorrect, nonanalytic temperature dependence of $\chi(T) - \chi(0) \propto -T^{1/3}$ at low T. In the two-channel case CTMA describes the non-Fermi liquid behavior, i.e., the exact [15] logarithmic temperature dependence of the susceptibility below the Kondo scale T_K. In contrast, the NCA solution recovers the logarithmic behavior only far below T_K.

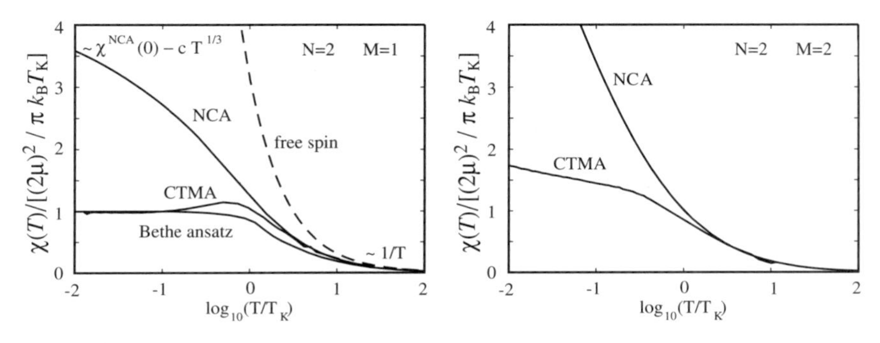

FIGURE 7.9. Static susceptibility of the single-channel ($N = 2$, $M = 1$) and the two-channel ($N = 2$, $M = 2$) Anderson impurity model in the Kondo regime ($E_d = -0.8D$, $\Gamma = 0.1D$, Landé factor $g = 2$). In the single-channel case, the CTMA and NCA results are compared with the Bethe ansatz result for the Kondo model [14], where the CTMA as well as the Bethe ansatz curves are scaled by their respective Kondo temperatures [14]. The CTMA susceptibility is universal in that it depends only on the ratio T/T_K for various values of the parameters of the Anderson model (not shown).

Other physical quantities have been calculated for the Anderson model at low temperatures as well and will be published in forthcoming work [48].

7 Anderson Model at Finite U: Generalized NCA and CTMA

The single-channel Anderson model at finite U may be represented in terms of the pseudo-fermion operators f_σ and slave boson operators a, b (discussed in the beginning of Section 3) as

$$H = H_c + E_d(\Sigma_\sigma f_\sigma^+ f_\sigma + 2a^+ a) + U a^+ a \\ + V \Sigma_{\vec{k}\sigma} [c_{\vec{k}\sigma}^+ (b^+ f_\sigma + \eta_\sigma f_{-\sigma}^+ a) + h.c.] \quad (7.1)$$

subject to the constraint

$$\sum_\sigma f_\sigma^+ f_\sigma + a^+ a + b^+ b = 1. \quad (7.2)$$

There are now two bosons, a "light" boson b (as in the case $U \to \infty$ considered before) and a "heavy" boson a. As seen from (7.1), the strong two-particle interaction U is transformed into a potential term by this representation. The corresponding diagrams are made of propagator lines for the light boson (wiggly line, as before) and for the heavy boson (zig-zag line) as well as a new vertex, where incoming conduction electron lines and pseudo-fermion lines merge into an outgoing heavy boson line.

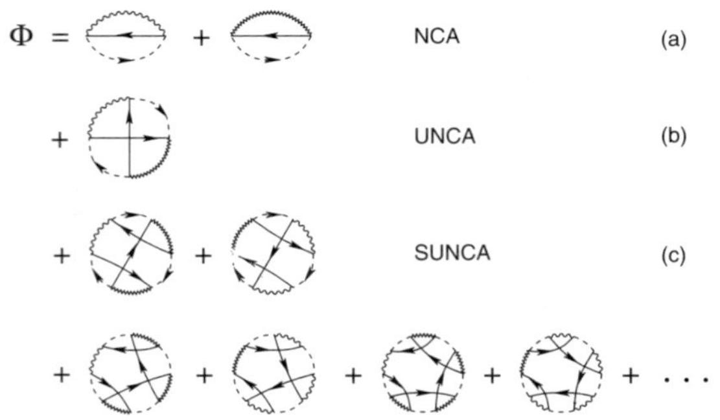

Φ = + NCA (a)

\+ UNCA (b)

\+ + SUNCA (c)

\+ + + + + . . .

FIGURE 7.10. Diagrammatic representation of the generating functional for the Anderson impurity model at finite U. Solid, dashed, wiggly, and zig-zag lines correspond, respectively, to conduction electron c, pseudo-fermion f, "light" boson b, and "heavy" boson a propagators. (a) NCA including light (empty impurity) and heavy (doubly occupied impurity) boson lines. (b) UNCA containing, in addition to an infinite number of (bare) hybridization processes into the empty impurity state (nonskeleton diagrams), a single hybridization into the doubly occupied impurity state. (c) Symmetrized, finite-U NCA (SUNCA): The infinite series of all diagrams of (a)–(c) generates an approximation where the hybridization processes into the empty and doubly occupied state are treated in a symmetric way; i.e., for each contribution with a bare light boson line there is a corresponding contribution with the bare light boson line replaced by a bare heavy boson line, and vice versa.

7.1 Generating Functional

It seems straightforward to define a generalization of NCA for finite U by adding to the second-order skeleton diagram for the generating functional Φ a second one where the light boson is replaced by the heavy boson (see Fig. 7.10a). This approximation was considered sometime ago [49,50], where it was found to fail badly: Not even the Kondo energy scale is recovered in the so defined approximation. The reason for this failure is obvious: In the Kondo regime ($n_d \sim 1$) the local spin is coupled to the conduction electron spin density at the impurity through the antiferromagnetic exchange coupling $J = V^2(-1/E_d + 1/E_d + U)$. The two terms on the r.h.s. of this relation arise from virtual transitions into the empty and doubly occupied local level, which contribute equally in the symmetric case $|E_d| = E_d + U$. The symmetric occurrence of these two virtual processes in all intermediate states is not included in the simple extension of NCA proposed above. The self-energy insertions in each of the two diagrams always contain only one of the processes, leading to an effective J which is only one-half of the correct value. Correspondingly, the Kondo temperature $T_K \sim \exp(-1/\mathcal{N}(0)J)$ comes out to scale as the square of the correct value, which is orders of magnitude too small.

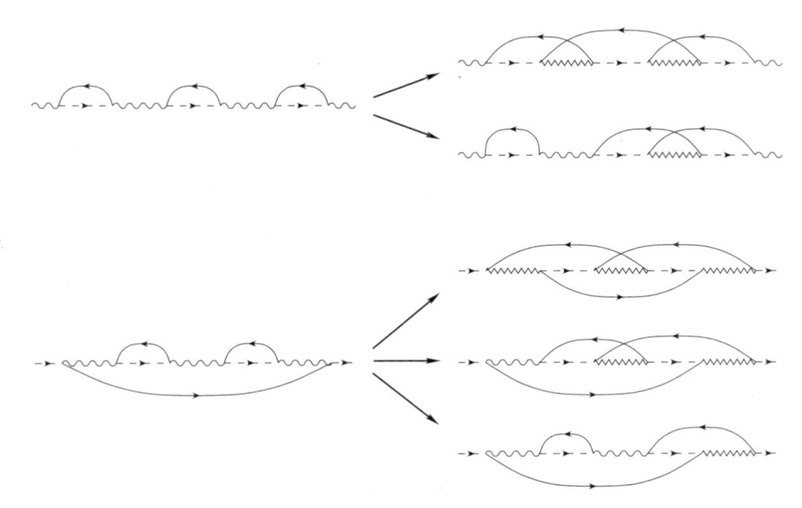

FIGURE 7.11. Two examples of how diagrams involving hybridization into the doubly occupied impurity state are generated from the bare noncrossing diagrams of the infinite U case by replacing light with heavy boson lines (see text).

To correct this deficiency it is necessary to include additional diagrams, restoring the symmetry between the two virtual processes, excitation into the empty or into the doubly occupied impurity state. Due to the exact projection onto the physical Hilbert space, each diagram contributing to Φ contains exactly one auxiliary particle ring (see, e.g., Fig. 7.5). As a first step one may add the next order diagram to Φ (see Fig. 7.10b). As will be seen below, this approximation, later referred to as UNCA, helps to recover a large part of the correct behavior of T_K. A completely symmetric treatment, however, requires the summation of infinite classes of diagrams. These diagrams are generated by replacing a light boson line with a heavy boson line in each of the bare (nonskeleton) diagrams of the elementary NCA diagrams of Fig. 7.11. Each replacement leads to a crossing of conduction electron lines spanning one fermion and at most two boson lines. As a result, the class of diagrams of Φ shown in Fig. 7.10c are obtained. These diagrams look similar to the CTMA diagrams shown in Fig. 7.5, but contain one light boson line and an arbitrary number of heavy boson lines, or vice versa. Diagrams with, for example, two light boson lines and an arbitrary number of heavy boson lines (and conduction electron lines spanning at most one fermion line) are reducible and do not appear. We will call the approximation defined by the generating functional given by the sum of the diagrams of Figs. 7.10(a–c) "symmetrized, finite-U noncrossing approximation" (SUNCA) [51, 52]. It is interesting to note that the symmetric treatment of light and heavy bosons leads to diagrams for Φ of a structure similar to the ones defining the CTMA (Fig. 7.5). The above approximation corresponding to the CTMA at $U \to \infty$, termed "symmetrized, finite-U Conserving T-Matrix Approximation (SUCTMA) is thus defined in a natural way by summing up all skeleton diagrams with an arbitrary number of light or heavy boson lines, dressed by conduction electron lines spanning only one fermion line. The rationale for keeping only these crossings is that (as explained in the appendix), diagrams containing conduction electron lines spanning two or more fermion lines may be grouped into sets where the most singular contributions cancel. Thus, the SUCTMA is defined by adding to the diagrams of the SUNCA the CTMA diagrams (Fig. 7.5) with (only) light boson lines or (only) heavy boson lines. The SUCTMA equations have not yet been evaluated.

7.2 Results of SUNCA

The pseudo-particle self-energies Σ_a, Σ_b and Σ_f as well as the local conduction electron self-energy Σ_c are obtained as functional derivatives with respect to the corresponding Green functions. They can be expressed in terms of the two vertex functions defined in Fig. 7.12 as solutions of Bethe–Salpeter equations. It is seen that the vertex functions are obtained by ladder summations involving a heavy boson or a light boson line as the rung of the ladder. These three-point vertex functions depend on two frequencies and are much less difficult to calculate than the four-point T-matrices of the CTMA. The full expressions of the self-energies are given in Fig. 7.13. The local d-electron Green's function after projection is again proportional to Σ_c:

$$G_d(\omega) = \frac{1}{V^2}\Sigma_c(\omega). \tag{7.3}$$

The self-consistent set of equations defined by Fig. 7.13 and the expression for the pseudo-particle Green function (Eq. 3.13), supplemented by the expression for the heavy boson Green's function

$$\mathcal{G}_a(i\omega_n) = \{i\omega_n - (2E_d + U) - \Sigma_a(i\omega_n)\}^{-1}, \tag{7.4}$$

have been solved numerically [51,52]. As expected, the pseudo-particle spectral functions are found to display power law divergencies at the infrared threshold. The exact power law exponents, derived from the Friedel sum

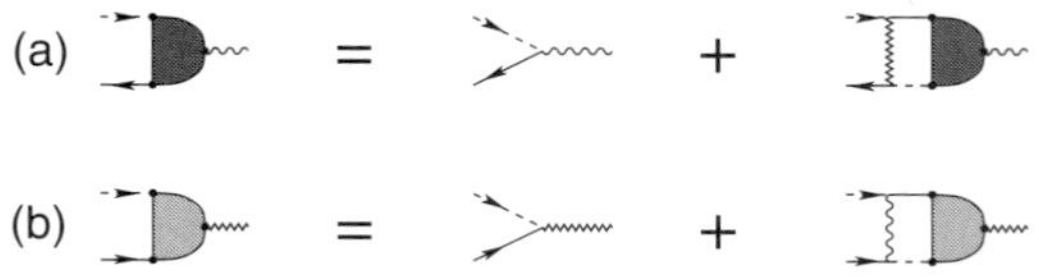

(a) = +

(b) = +

FIGURE 7.12. Diagrammatic representation of the Bethe–Salpeter equations for (a) the renormalized light boson (empty impurity) and (b) the renormalized heavy boson (doubly occupied impurity) vertex, as generated by the SUNCA Luttinger–Ward functional (Fig. 7.11).

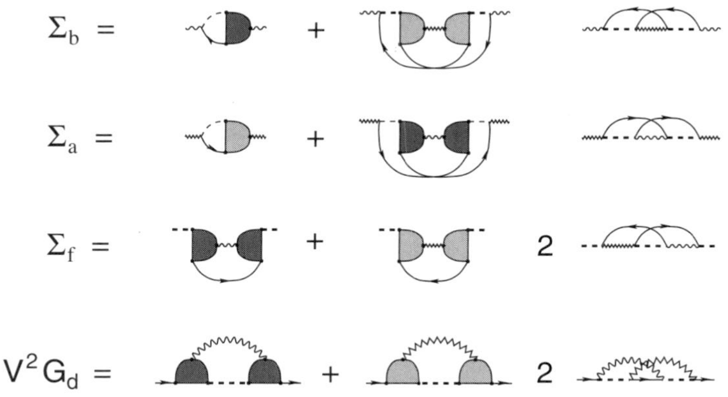

FIGURE 7.13. Diagrammatic representation of the auxiliary particle self-energies of SUNCA in terms of the renormalized hybridization vertices, defined in Fig. 7.12. Note that in each line the third diagram is subtracted in order to avoid double counting of terms within the first two diagrams.

rule argument based on the assumed Fermi liquid ground state Section 3.2), are given by

$$\alpha_f = n_d - \frac{n_d^2}{2} \tag{7.5}$$

$$\alpha_b = 1 - \frac{n_d^2}{2} \tag{7.6}$$

$$\alpha_a = -1 + 2n_d - \frac{n_d^2}{2}. \tag{7.7}$$

The threshold exponents obtained from the SUNCA solution are shown in Fig. 7.14 as functions of the average impurity occupation number n_d. They agree with the exact results in the Kondo limit ($n_d \to 1$) where $\alpha_f = \alpha_b = \alpha_a = \frac{1}{2}$, but deviate for $n_d < 1$. It is expected that the SUCTMA will recover the correct exponents, as does the CTMA in the case of infinite U. The d-electron spectral function for the symmetric Anderson model as obtained in SUNCA is shown in Fig. 7.15 (left panel). In this case the Kondo resonance is located exactly at $\omega = 0$. Its width is a measure of the Kondo temperature T_K. The three curves shown correspond to the elementary NCA (Fig. 7.10a), the UNCA (Fig. 7.10b), and the SUNCA. The inset shows that for the parameters chosen the T_K obtained in NCA is too small by a factor 10^{-2}, whereas the UNCA is only a factor of $\frac{1}{3}$ off. Contrary to [49], we do not find that the UNCA is sufficient to recover T_K. In Fig. 7.15 (right panel) the results of the SUNCA for T_K as determined from the width of the Kondo resonance are compared to the exact result,

$$T_K = \min\left\{\frac{1}{2\pi}U\sqrt{I}, \sqrt{D\Gamma}\right\}e^{(-\pi/I)}, \tag{7.8}$$

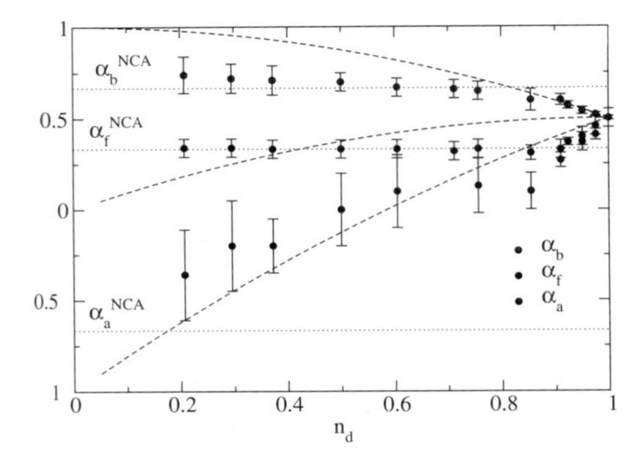

FIGURE 7.14. Infrared threshold exponents of the auxiliary particle spectral functions for the case of finite U as a function of the the impurity occupation n_d. Dashed curved lines: exact results (Eqs. 7.5–7.7); horizontal lines: NCA results; data points with error bars: SUNCA results. In the Kondo limit ($n_d \to 1$) the exact exponents are recovered, while in the mixed valence and empty impurity regime the SUNCA results for α_f and α_b cross over to the NCA values.

where $I = (2\Gamma U)/(|E_d|(U + E_d))$. The agreement is seen to be very good for a large range of parameters I and E_d.

8 Conclusion

We have reviewed a technique to describe correlated quantum impurity systems with strong on-site repulsion, which is based on a conserving formulation of the auxiliary boson method. The conserving scheme allows us to implement the conservation of the local charge Q without taking into account time-dependent fluctuations of the gauge field λ, while exactly projecting the quantum dynamics onto the physical subspace of no double occupancy of sites (in the limit of infinitely strong on-site repulsion). Any spurious condensation of the auxiliary boson field is avoided in this way.

By including the physically dominant contributions, spin flip processes in the Kondo regime, and spin as well as charge fluctuation processes in the mixed valence and empty impurity regimes, the method recovers the Fermi liquid ground state of the single-channel Anderson impurity model as well as the non-Fermi-liquid low-temperature behavior of the two-channel Anderson model: The correct infrared threshold exponents of the auxiliary particle propagators, which are identified as indicators for Fermi or non-Fermi liquid behavior, are obtained in both cases. Physical quantities, like the magnetic susceptibility, are correctly described both in the Fermi and in the non-Fermi liquid cases of the model over the complete temperature range, including the crossover to the correlated many-body state at

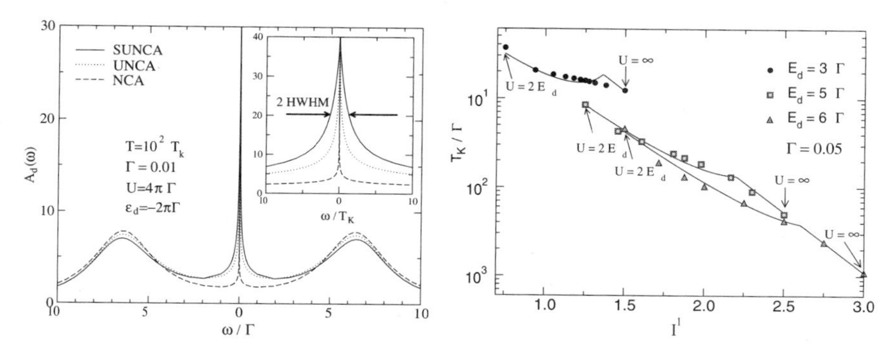

FIGURE 7.15. Left panel: Local electron spectral function calculated using NCA, UNCA, and SUNCA. The Kondo temperature is determined by the peak width. It is seen that in NCA it comes out orders of magnitude too low.—Right panel: Kondo Temperature for various parameters E_d, U and fixed Γ. Solid lines represent the results from the Bethe ansatz solutions (Eq. 7.8). Data points are the SUNCA results determined from the width of the Kondo peak in the d-electron spectral function.

the lowest temperatures, which has previously been notoriously difficult to obtain within a diagrammatic many-body theory.

We have also presented a generalization of the auxiliary particle method for finite on-site repulsion U and solved the corresponding equations on the level of a generalized noncrossing approximation (SUNCA), where virtual excitations into the empty or into the doubly occupied impurity state are treated in a fully symmetric way. In this way the correct low-temperature scale of the finite-U Anderson model was obtained. The controlled treatment of a finite on-site repulsion is a precondition for applying the auxiliary particle method to models of the Hubbard type or—within the scheme of the dynamical mean field theory (limit of infinite dimensions)—of the Anderson impurity type, where a Mott–Hubbard metal-insulator transition occurs as a function of band filling for finite U. As a standard diagram technique the conserving auxiliary particle method has the potential to be applicable to problems of correlated systems on a lattice as well as to mesoscopic systems out of equilibrium via the Keldysh technique. These developments are currently under investigation.

Collaboration and discussions with S. Böcker, T. A. Costi, K. Haule, S. Kirchner, and Th. Schauerte are gratefully acknowledged. This work is supported in part by DFG through SFB 195 and by a grant of computer time of the High-Performance Computing Center Stuttgart.

9 References

[1] P.A. Lee, T.M. Rice, J.W. Serene, L.J. Sham, and J.W. Wilkins, Comments on Condensed Matter Physics **12** (1986), 99.

[2] P.W. Anderson, Science **235** (1987), 1196; and *The Theory of Superconductivity in the High-T_c Cuprates* (Princeton University Press, Princeton, NJ, 1997).

[3] For a review see S. Das Sarma and A. Pinczuk, eds., *Perspectives in Quantum Hall Effects* (Wiley, New York, 1996).

[4] A.A. Abrikosov, Physics **2** (1965), 21.

[5] S.E. Barnes, J. Phys. F **6** (1976), 1375; **7** (1977), 2637.

[6] P. Coleman, Phys. Rev. B **29** (1984), 2.

[7] D.M. Newns, N. Read, J. Phys. C **16** (1983), 3273; Adv. Phys. **36** (1988), 799.

[8] A.C. Hewson, *The Kondo Problem to Heavy Fermions* (Cambridge University Press, Cambridge, 1993).

[9] D.L. Cox and A. Zawadowski, Adv. Phys. **47** (1998), 599.

[10] See, e.g., related articles in *Mesoscopic Electron Transport*, eds. L.L. Sohn, L.P. Kouwenhoven and G.Schön, Vol. 345 of *NATO ASI Series E: Applied Sciences* (Kluwer Science Publishers, Dordrecht, 1997).

[11] P.W. Anderson, Phys. Rev. **124** (1961), 41.

[12] J.R. Schrieffer and P.A. Wolff, Phys. Rev. **149** (1966), 491.

[13] K.G. Wilson, Rev. Mod. Phys. **47** (1975), 773.

[14] N. Andrei, K. Furuya, and J.H. Löwenstein, Rev. Mod. Phys. **55** (1983), 331.

[15] P.B. Wiegmann and A.M. Tsvelik, J. Phys. C **12** (1983), 2281 and 2321.

[16] P. Nozières, J. Low Temp. Phys. **17** (1974), 31.

[17] N. Andrei and C. Destri, Phys. Rev. Lett. **52** (1984), 364.

[18] A.M. Tsvelik, J. Phys.C **18** (1985), 159.

[19] I. Affleck and A.W.W. Ludwig, Nucl. Phys. **352** (1991), 849; **B360** (1991), 641; Phys. Rev. B **48** (1993), 7297.

[20] K. Yamada, Prog. Theor. Phys. **53** (1975), 970; **54** (1975), 316; **55** (1976), 1345; K. Yosida and K. Yamada, Prog. Theor. Phys. **53** (1975), 1286.

[21] D.L. Cox, N.E. Bickers, and J.W. Wilkins, Phys. Rev. B **36** (1987), 2036.

[22] P.W. Anderson, Phys. Rev. Lett. **18** (1967), 1049.

[23] P. Nozières and C.T. De Dominicis, Phys. Rev. **178** (1969), 1073; 1084; 1097.

[24] G.D. Mahan, *Many-Particle Physics*, 2nd ed. (Plenum Press, New York, 1990), gives an overview, p. 732.

[25] K.D. Schotte and U. Schotte, Phys. Rev. **185** (1969), 509.

[26] B. Menge and E. Müller-Hartmann, Z. Phys. B **73** (1988), 225.

[27] T.A. Costi, P. Schmitteckert, J. Kroha, and P. Wölfle, Phys. Rev. Lett. **73** (1994), 1275.

[28] T.A. Costi, P. Schmitteckert, J. Kroha, and P. Wölfle, Physica (Amsterdam) **235–240C** (1994), 2287.

[29] S. Fujimoto, N. Kawakami, and S.K. Yang, J. Phys. Korea **29** (1996), S136.

[30] D.M. Newns and N. Read, J. Phys. **C16** (1983), 3273; Adv. Phys. **36** (1988), 799.

[31] S. Elitzur, Phys. Rev. D **12** (1975), 3978.

[32] A. Jevicki, Phys. Lett. B **71** (1977), 327.

[33] F. David, Comm. Math. Phys. **81** (1981), 149.

[34] I.D. Lawrie, J. Phys. A **18** (1985), 1141.

[35] B. Jin and Y. Kuroda, J. Phys. Soc. Japan **57** (1988), 1687.

[36] T. Matsuura et al., J. Phys. Soc. Japan **66** (1997), 1245.

[37] G. Sellier, Diploma Thesis, Universität Karlsruhe, 1998, unpublished.

[38] G. Baym and L.P. Kadanoff, Phys. Rev. **124** (1961), 287; G. Baym, Phys. Rev. **127** (1962) 1391.

[39] H. Keiter and J.C. Kimball, J. Appl. Phys. **42** (1971), 1460; N. Grewe and H. Keiter, Phys. Rev B **24** (1981), 4420.

[40] Y. Kuramoto, Z. Phys. B **53** (1983), 37; H. Kojima, Y. Kuramoto and M Tachiki, **54** (1984), 293; Y. Kuramoto and H. Kojima, **57** (1984), 95; Y. Kuramoto, **65** (1986), 29.

[41] E. Müller-Hartmann, Z. Phys. B **57** (1984), 281.

[42] T.A. Costi, J. Kroha, and P. Wölfle, PRB **53** (1996), 1850.

[43] N.E. Bickers, Rev. Mod. Phys. **59** (1987), 845; N.E. Bickers, D.L. Cox, and J.W. Wilkins, Phys. Rev. B **36** (1987), 2036.

[44] D.L. Cox and A.E. Ruckenstein, Phys. Rev. Lett. **71** (1993), 1613.

[45] F. Anders and N. Grewe, Europhys. Lett. **26** (1994), 551; F. Anders, J. Phys. Cond. Mat. **7** (1995), 2801.

[46] J. Kroha, P. Wölfle and T.A. Costi, Phys. Rev. Lett. **79** (1997), 261.

[47] Th. Schauerte, J. Kroha, and P. Wölfle, Phys. Rev. B **62** (2000), 4394.

[48] S. Kirchner, J. Kroha, and P. Wölfle, in preparation.

[49] Th. Pruschke and N. Grewe, Z. Phys. B **74** (1989), 439.

[50] I. Holm and K. Schönhammer, Solid State Comm. **69** (1989), 10, 969.

[51] K. Haule, S. Kirchner, J. Kroha, and P. Wölfle, Proceedings of the NATO Advanced Research Workshop on "Size Dependent Magnetic Scattering," Pecs, Hungary, 2000 (Kluwer Academic Publishers, in press).

[52] K. Haule, S. Kirchner, J. Kroha, and P. Wölfle, `cond-mat/0105490`, 2001.

Appendix A Infrared Cancellation of Non-CTMA Diagrams

The CTMA is not only justified on physical grounds by the inclusion of the maximum number of spin flip and charge fluctuation processes at any given order of perturbation theory, but also by an infrared cancellation of all diagrams not included in the CTMA. In the following we will prove this cancellation theorem.

A.1 Power Counting

Each auxiliary particle loop carries a factor of the fugacity $\exp(-\beta\lambda)$, which vanishes upon projection onto the $Q = 1$ subspace, $\lambda \to \infty$. Therefore, an arbitrary f or b self-energy diagram consists of one single line of alternating fermion and boson propagators, with the hybridization vertices connected by conduction electron lines in any possible way, as shown in Fig. 7.16 (see also [39]). Such a fermion self-energy skeleton diagram of loop order L is calculated as

$$\Sigma_f^{(L)}(\omega) = (-1)^{3L-1+L_{\rm sp}} N^{L_{\rm sp}} M^{L_{\rm ch}} \Gamma^L \\ \times \int \frac{d\varepsilon_1}{\pi} \cdots \frac{d\varepsilon_L}{\pi} f(\varepsilon_1) \ldots f(\varepsilon_L) A_c^0(s_1\varepsilon_1) \ldots A_c^0(s_L\varepsilon_L) \\ \times G_b(\omega+\omega_1) G_f(\omega+\omega_1+\omega_1') \cdots G_b\left(\omega + \sum_{i=1}^{k} \omega_i + \sum_{i=1}^{k-1} \omega_i'\right) \\ \times G_f\left(\omega + \sum_{i=1}^{k} (\omega_i + \omega_i')\right) \cdots G_b(\omega+\omega_L), \quad \text{(A.1)}$$

where $G_{f,b}$ are the *renormalized*, i.e., power law divergent auxiliary particle propagators, and $L_{\rm sp}$ and $L_{\rm ch}$ denote the number of spin (or fermion, c–f) loops and the number of channel (or c–b) loops contained in the diagram, respectively. Spin and channel indices are not shown for simplicity. Each of the auxiliary particle frequencies ω_i, ω_i' coincides with one of the integration variables ε_j, $j = 1, \ldots, L$, in such a way that energy is conserved at each hybridization vertex. This implies that the sign of the frequency carried by a c-electron line is $s_i = +$, if the c-electron line runs from right to left, and $s_i = -$, if it runs from left to right in Fig. 7.16. An analogous expression holds for the slave boson self-energy diagrams.

By substituting $x_j = \varepsilon_j/\omega$, $j = 1, \ldots, L$ and factoring out $\omega^{-\alpha_{f,(b)}}$ from each fermion (boson) propagator, the infrared behavior of the term (A.1) is deduced as

$$\mathrm{Im}\, \Sigma_{f,b}^{(L)}(\omega) = C\omega^{\alpha_{f,b}+L(1-\alpha_f-\alpha_b)}, \quad \text{(A.2)}$$

where C is a finite constant. Clearly, when the NCA solutions are inserted for the propagators $G_{f,b}$, i.e., $\alpha_f + \alpha_b = 1$, their power law behavior is just reproduced by any term of the form (A.1). However, this is no longer the case for the exact propagators in the Fermi liquid regime ($M < N$), where in general $\alpha_f + \alpha_b > 1$. Thus, the infinite resummation of terms to arbitrary loop order is unavoidable in this case.

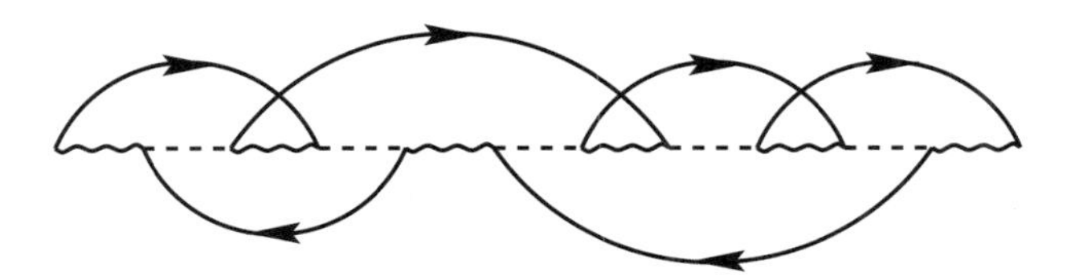

FIGURE 7.16. Typical pseudo-fermion self-energy skeleton diagram of loop order $L = 6$, containing $L_{\rm sp} = 1$ spin (or fermion) loop and $L_{\rm ch} = 1$ channel loop.

A.2 Infrared Cancellation

As discussed in Section 6.2, the CTMA is equivalent to the self-consistent summation of all skeleton free energy diagrams, where a conduction electron line spans at most two hybridization vertices (Fig. 7.5). Thus, any skeleton self-energy diagram *not* included in CTMA contains at least one conduction electron "arch" which spans four (or more) vertices, with four conduction lines reaching from inside to outside of the arch as shown in Fig. 7.17a. For each such diagram there exists another skeleton, which differs from Fig. 7.17a only in that the end points of two conduction lines inside the arch are interchanged (Fig. 7.17b). The corresponding permutation of fermionic operators implies a relative sign between the terms Fig. 7.17a and Fig. 7.17b. Without loss of generality we now assume $\omega > 0$ for the external frequency of the self-energy. The leading infrared singular behavior of the term (A.1) arises from those parts of the integrations, where the arguments of the G_f, G_b are such that the divergences of all propagators lie within the integration range. This implies at least $-\omega \leq \varepsilon_j \leq 0$, $j = 1, \ldots, L$. Therefore, the terms corresponding to Fig. 7.17a,b differ only in the frequency arguments of the Green's functions inside the arch, and

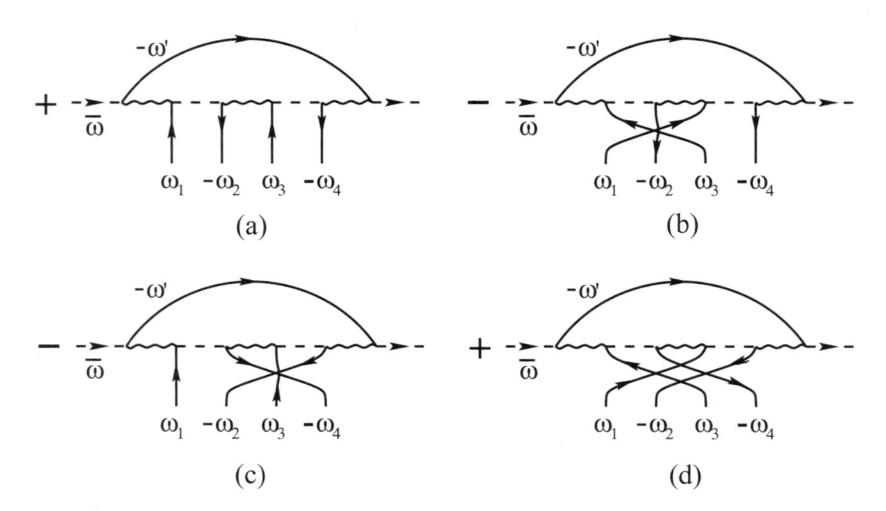

FIGURE 7.17. Set of contributions to skeleton diagrams *not* contained in CTMA which cancel in the infrared limit to leading and subleading order in the external frequency, $\omega \to 0$.

at temperature $T = 0$ the leading infrared behavior of their sum reads

$$\Sigma_f^{(L,a)}(\omega) + \Sigma_f^{(L,b)}(\omega) \overset{\omega \to 0}{=} (-1)^{3L-1+L_{\rm sp}} N^{L_{\rm sp}} M^{L_{\rm ch}} \Gamma^L$$
$$\times \int_{-\omega}^{0} \frac{d\varepsilon_1}{\pi} \cdots \frac{d\varepsilon_L}{\pi} F(\omega, \{\varepsilon_j\}) [G_f(\bar{\omega}+\omega'+\omega_1) G_b(\bar{\omega}+\omega'+\omega_1+\omega_2)$$
$$- G_f(\bar{\omega}+\omega'+\omega_3) G_b(\bar{\omega}+\omega'+\omega_3+\omega_2)]. \quad (A.3)$$

Here $\bar{\omega}$ denotes the sum of all frequencies ω, ε_j entering the diagrammatic part (Fig. 7.17), from the left, and $F(\omega, \{\varepsilon_j\})$ consists of all terms which are not altered by interchanging the c-electron lines. In the infrared limit, $\omega_1 - \omega_3 \to 0$, the term in square brackets may be written as

$$\frac{d}{d\bar{\omega}} [G_f(\bar{\omega} + \omega') G_b(\bar{\omega} + \omega' + \omega_2)] (\omega_1 - \omega_3) \quad (A.4)$$

and upon performing the integrations over ω_1, ω_3 the difference $(\omega_1 - \omega_3)$ leads to an additional factor of ω. A similar cancellation of the leading infrared singularity occurs between the terms shown in Fig. 7.17c,d. In an analogous way it may be shown that combining the terms Fig. 7.17(a)–(d) leads to a factor of ω^2 compared to the power counting result for one single term. Thus, the infrared singularity of all non-CTMA terms of loop order L is weaker than the Lth order CTMA terms by at least $O(\omega^2)$,

$$\Sigma_{f,b}^{(L,a)}(\omega) + \cdots + \Sigma_f^{(L,d)}(\omega) \overset{\omega \to 0}{\propto} \omega^{\alpha_{f,b} + L(1-\alpha_f-\alpha_b)+2}. \quad (A.5)$$

It should be emphasized that in the above derivation, L appears only as a parameter and, thus, the cancellation theorem holds for arbitrarily high loop order L. This proves that the CTMA captures the leading and subleading infrared singularities ($\omega \to 0$) at any given order L.

8

Conserving Approximations vs. Two-Particle Self-Consistent Approach

S. Allen, A.-M.S. Tremblay, and Y.M. Vilk

ABSTRACT The conserving approximation scheme to many-body problems was developed by Kadanoff and Baym using the functional-derivative approach. Another approach for the Hubbard model also satisfies conservation laws, but in addition it satisfies the Pauli principle and a number of sum rules. A concise formal derivation of that approach, using functional derivatives, is given in this conference paper to highlight formal analogies and differences with conserving approximations.

1 Introduction

Although quantum many-body theory is in principle exact, any practical application of the formalism necessarily leads to the violation of some exact results. For example, it was noticed long ago that whenever infinite subsets of diagrams are resummed, conservation laws will be violated if certain rules are not followed. As another example of a possible inconsistency, one can obtain a relation between number of particles, chemical potential, and temperature from either a Green function or from a derivative of the free energy. How can we make sure that the diagrammatic series for the Green function and that for the free energy give the same result? Both of these problems, namely, obtaining response functions that satisfy conservation laws and making sure that the free energy is "thermodynamically consistent" with the Green function, can be solved using an approach that was developed by Kadanoff and Baym [1, 2] many years ago. This approach, described in the paper by Bickers (Chapter 6) in these proceedings, allows one to obtain so-called "conserving approximations" [3] that have been extensively used in the last decade. However, as in any approximate calculations, some exact identities are not satisfied by conserving approximations [4, 6]. For example, the Pauli principle is violated, integration over the coupling constant of the potential energy obtained from susceptibilities does not lead back to the starting free energy, and the irreducible verticies generated by a functional derivative of the self-energy are not the same as the vertex corrections entering the self-energy itself.[1] In addition, since there is an infinite number of choice of diagrams for the Luttinger–Ward functional, there is also an infinite number of conserving approximations and it is difficult to decide on the validity of an approximation based on the criterion of conservation only. Further criticism has appeared in [4] and [5].

In this chapter, we wish to present a new derivation of a *nondiagrammatic* approach to the *repulsive* Hubbard model that was developed a few years ago [4]. An analogous derivation in the case of the *attractive* Hubbard model was presented at this workshop by Bumsoo Kyung and has already appeared [7, 8]. Here, we go back to our earlier results on the repulsive Hubbard model, but we use a formal approach that highlights analogies and differences with conserving approximations. This chapter will strive to be concise with the explicit purpose of being clear without getting lost in details. The published derivation for the attractive Hubbard model contains more of these details [7].

In Section 2 we first present a derivation of conserving approximations that can be extended, in Section 3, to the derivation of our nondiagrammatic approach. This derivation focuses first on response functions and on the self-consistent determination of irreducible spin and charge vertices. That determination is done in such a way that the Pauli principle in its simplest form is satisfied. In other words, crossing symmetry (Chapter 6) is satisfied only for field operators on a single site, all at the same time. This is the so-called two-particle self-consistent (TPSC) part of the approach . We then improve our approximation for the self-energy by using the TPSC results in the exact non Hartree–Fock part of the self-energy. Vertices and Green function entering the final expression are at the same level of approximation. No Migdal theorem is assumed. The effects of transverse and longitudinal fluctuations are considered in separate subsections before the final expression can be written down. The internal-accuracy check [4] is also presented and used in the derivation. We end up in Section 4 with a discussion of extensions of the approach that are now being developed. The appendix addresses a technical detail.

2 Functional Derivative Formalism and Conserving Approximations

In this section, we first discuss single-particle properties, then response functions and, finally, we present the Hartree–Fock approximation as an example of a conserving approximation. The latter will be helpful as a

[1] In FLEX for example, the structure of the self-energy is the same as if Migdal's theorem was applicable.

starting point for the derivation of the two-particle self-consistent (TPSC) approach in the next section.

2.1 Single-Particle Properties

Following functional methods of the Schwinger school [1, 2, 9], we begin with the generating function with source fields ϕ_σ and field destruction operators ψ in the grand canonical ensemble

$$\ln Z[\phi] = \ln \mathrm{Tr}\big[e^{-\beta(\widehat{H}-\mu\widehat{N})}\mathrm{T}_\tau\big(e^{-\psi^\dagger_{\bar\sigma}(\bar 1)\phi_{\bar\sigma}(\bar 1,\bar 2)\psi_{\bar\sigma}(\bar 2)}\big)\big]. \tag{2.1}$$

We adopt the convention that 1 stands for the position and imaginary time indices $(\mathbf{r}_1, \tau_1)$. The overbar means summation over every lattice site and integration over imaginary-time from 0 to β. T_τ is the time-ordering operator.

The propagator in the presence of the source field is obtained from functional differentiation

$$G_\sigma(1,2;\{\phi\}) = -\langle\psi_\sigma(1)\psi^\dagger_\sigma(2)\rangle_\phi = -\frac{\delta \ln Z[\phi]}{\delta\phi_\sigma(2,1)}. \tag{2.2}$$

From now on, *the time-ordering operator in averages,* $\langle\ \rangle$, *is implicit.* Physically relevant correlation functions are obtained for $\{\phi\} = 0$ but it is extremely convenient to keep finite $\{\phi\}$ in intermediate steps of the calculation.

Using the equation of motion for the field ψ and the definition of the self-energy, one obtains the Dyson equation in the presence of the source field [2, p. 43]

$$(G_0^{-1} - \phi)G = 1 + \Sigma G; \quad G^{-1} = G_0^{-1} - \phi - \Sigma \tag{2.3}$$

where, from the commutator of the interacting part of the Hubbard Hamiltonian H, one obtains

$$\Sigma_\sigma(1,\bar 1;\{\phi\})G_\sigma(\bar 1,2;\{\phi\}) = -U\langle\psi^\dagger_{-\sigma}(1^+)\psi_{-\sigma}(1)\psi_\sigma(1)\psi^\dagger_\sigma(2)\rangle_\phi. \tag{2.4}$$

The imaginary time in 1^+ is infinitesimally larger than in 1.

2.2 Response Functions

Response (four-point) functions for spin and charge excitations can be obtained from functional derivatives $(\delta G/\delta\phi)$ of the source-dependent propagator. Following the standard approach and using matrix notation to abbreviate the summations and integrations we have

$$\begin{gathered} GG^{-1} = 1 \\ \frac{\delta G}{\delta\phi}G^{-1} + G\frac{\delta G^{-1}}{\delta\phi} = 0. \end{gathered} \tag{2.5}$$

Using the Dyson equation (2.3) $G^{-1} = G_0^{-1} - \phi - \Sigma$,S we may rewrite this as

$$\frac{\delta G}{\delta \phi} = -G \frac{\delta G^{-1}}{\delta \phi} G = G \,\hat{}\, G + G \frac{\delta \Sigma}{\delta \phi} G, \tag{2.6}$$

where the symbol $\hat{}$ reminds us that the neighboring labels of the propagators have to be the same as those of the ϕ in the functional derivative. If perturbation theory converges, we may write the self-energy as a functional of the propagator. From the chain rule, one then obtains an integral equation for the response function in the particle-hole channel that is the analog of the Bethe–Salpeter equation in the particle-particle channel

$$\frac{\delta G}{\delta \phi} = G \,\hat{}\, G + G \left[\frac{\delta \Sigma}{\delta G} \frac{\delta G}{\delta \phi} \right] G. \tag{2.7}$$

The labels of the propagators in the last term are attached to the self-energy, as in (2.6).[2] Vertices appropriate for spin and charge responses are given, respectively, by

$$U_{\mathrm{sp}} = \frac{\delta \Sigma_\uparrow}{\delta G_\downarrow} - \frac{\delta \Sigma_\uparrow}{\delta G_\uparrow}; \quad U_{\mathrm{ch}} = \frac{\delta \Sigma_\uparrow}{\delta G_\downarrow} + \frac{\delta \Sigma_\uparrow}{\delta G_\uparrow}. \tag{2.8}$$

2.3 Hartree–Fock and RPA as an Example

As an example of calculation of response functions, consider the Hartree–Fock approximation which corresponds to factoring the four-point function in the definition of the self-energy (2.4) as if there were no correlations,

$$\Sigma_\sigma^{\mathrm{H}}(1, \bar{1}; \{\phi\}) G_\sigma^{\mathrm{H}}(\bar{1}, 2; \{\phi\}) = U G_{-\sigma}^{\mathrm{H}}(1, 1^+; \{\phi\}) G_\sigma^{\mathrm{H}}(1, 2; \{\phi\}). \tag{2.9}$$

Multiplying the above equation by $(G_\sigma^{\mathrm{H}})^{-1}$, we are left with

$$\begin{aligned} \Sigma_\sigma^{\mathrm{H}}(1, 2; \{\phi\}) &= U G_{-\sigma}^{\mathrm{H}}(1, 1^+; \{\phi\}) \delta(1-2), \\ \left. \frac{\delta \Sigma_\uparrow^{\mathrm{H}}(1, 2; \{\phi\})}{\delta G_\downarrow^{\mathrm{H}}(3, 4; \{\phi\})} \right|_{\{\phi\}=0} &= U \delta(1-2) \delta(3-1) \delta(4-2), \end{aligned} \tag{2.10}$$

which, when substituted in the integral equation (2.7) for the response function, tells us that we have generated the random phase approximation (RPA) with, from (2.8), $U_{\mathrm{sp}} = U_{\mathrm{ch}} = U$.

3 Another Approach

The approach developed in [4], consists of two steps, corresponding to the following two subsections. First, one computes response functions from the

[2] To remind ourselves of this, we may also adopt an additional "vertical matrix notation" convention and write (2.6) as $\delta G/\delta \phi = G \,\hat{}\, G + G \left[\frac{\delta \Sigma / \delta G}{\delta G / \delta \phi} \right] G$.

TPSC approach, inspired by earlier work of Singwi [10]. Second, one generates an improved approximation for the self-energy starting from an exact expression for Σ that explicitly separates the infinite-frequency limit from the lower-frequency contribution. The method also contains an internal-accuracy check that is discussed in the last subsection.

3.1 First Step: Two-Particle Self-Consistency for $G^{(1)}$, $\Sigma^{(1)}$, $\Gamma_{\rm sp}^{(1)} = U_{\rm sp}$, *and* $\Gamma_{\rm ch}^{(1)} = U_{\rm ch}$

In conserving approximations, the self-energy is obtained from a functional derivative $\Sigma[G] = \delta\Phi[G]/\delta G$ of Φ the Luttinger–Ward functional, which is itself computed from a set of diagrams. To liberate ourselves from diagrams, we start instead from the exact expression for the self-energy (Eq. 2.4) and notice that when label 2 equals 1^+, the right-hand side of this equation is equal to double-occupancy $\langle n_\uparrow n_\downarrow \rangle$. Factoring as in Hartree–Fock amounts to assuming no correlations. Instead, we should insist that $\langle n_\uparrow n_\downarrow \rangle$ should be obtained self-consistently. After all, in the Hubbard model, there are only two local four point functions: $\langle n_\uparrow n_\downarrow \rangle$ and $\langle n_\uparrow^2 \rangle = \langle n_\downarrow^2 \rangle$. The latter is given exactly, through the Pauli principle, by $\langle n_\uparrow^2 \rangle = \langle n_\downarrow^2 \rangle = \langle n_\uparrow \rangle = \langle n_\downarrow \rangle = n/2$, when the filling n is known. In a way, $\langle n_\uparrow n_\downarrow \rangle$ in the self-energy equation (2.4), can be considered as an initial condition for the four point function when one of the points, 2, separates from all the others which are at 1. When that label 2 does not coincide with 1, it becomes more reasonable to factor *à la* Hartree–Fock. These physical ideas are implemented by postulating

$$\Sigma_\sigma^{(1)}(1,\bar{1};\{\phi\})G_\sigma^{(1)}(\bar{1},2;\{\phi\}) = A_{\{\phi\}}G_{-\sigma}^{(1)}(1,1^+;\{\phi\})G_\sigma^{(1)}(1,2;\{\phi\}), \quad (3.1)$$

where $A_{\{\phi\}}$ depends on external field and is chosen such that the exact result[3]

$$\Sigma_\sigma(1,\bar{1};\{\phi\})G_\sigma(\bar{1},1^+;\{\phi\}) = U\langle n_\uparrow(1)n_\downarrow(1)\rangle_\phi \quad (3.2)$$

is satisfied. It is easy to see that the solution is

$$A_{\{\phi\}} = U\frac{\langle n_\uparrow(1)n_\downarrow(1)\rangle_\phi}{\langle n_\uparrow(1)\rangle_\phi\langle n_\downarrow(1)\rangle_\phi}. \quad (3.3)$$

Substituting $A_{\{\phi\}}$ back into our *ansatz* (3.1) we obtain the self-energy by right-multiplying by $\left(G_\sigma^{(1)}\right)^{-1}$:

$$\Sigma_\sigma^{(1)}(1,2;\{\phi\}) = A_{\{\phi\}}G_{-\sigma}^{(1)}(1,1^+;\{\phi\})\delta(1-2). \quad (3.4)$$

[3] If we had chosen $2 \to 1^-$ instead of $2 \to 1^+$ we would have found $\langle n_\downarrow(1)(n_\uparrow(1)-1)\rangle$ on the right-hand side of (3.2), as discussed in the Appendix. Since we are interested in using the local limit of ΣG as an initial condition for spin and charge correlations, it is more natural to focus on $2 \to 1^+$ since in that case it is $\langle n_\downarrow(1)n_\uparrow(1)\rangle$ that appears on the right-hand side, a quantity that follows directly from the spin and charge correlations and where spin up and spin down are treated on equal footing. See also the appendix.

We are now ready to obtain irreducible vertices using the prescription of the previous section (Eq. 2.8), namely, through functional derivatives of Σ with respect to G. In the calculation of U_{sp}, the functional derivative of $\langle n_\uparrow n_\downarrow\rangle/(\langle n_\uparrow\rangle\langle n_\downarrow\rangle)$ drops out, so we are left with[4]

$$U_{\mathrm{sp}} = \left.\frac{\delta\Sigma_\uparrow^{(1)}}{\delta G_\downarrow^{(1)}}\right|_{\{\phi\}=0} - \left.\frac{\delta\Sigma_\uparrow^{(1)}}{\delta G_\uparrow^{(1)}}\right|_{\{\phi\}=0} = A_{\{\phi\}=0} = U\frac{\langle n_\uparrow n_\downarrow\rangle}{\langle n_\uparrow\rangle\langle n_\downarrow\rangle}. \qquad (3.5)$$

The renormalization of this irreducible vertex may be physically understood as coming from Kanamori–Brueckner screening [4]. To close the system of equations, we need to know double-occupancy. It may be found self-consistently using the fluctuation-dissipation theorem and the Pauli principle. First notice that the Pauli principle, $\langle n_\sigma^2\rangle = \langle n_\sigma\rangle$, implies that

$$\langle(n_\uparrow - n_\downarrow)^2\rangle = \langle n_\uparrow\rangle + \langle n_\downarrow\rangle - 2\langle n_\uparrow n_\downarrow\rangle \qquad (3.6)$$

while the fluctuation-dissipation theorem tells us that $\langle(n_\uparrow - n_\downarrow)^2\rangle$ is given by the equal-time equal-position imaginary-time susceptibility χ_{sp}. Since $\chi_{\mathrm{sp}}^{-1}(q) = \chi_0^{-1}(q) - \frac{1}{2}U_{\mathrm{sp}}$ with $q \equiv (\mathbf{q}, 2\pi nT)$, this is equivalent to the equation

$$\frac{T}{N}\sum_q \frac{\chi_0(q)}{1 - \frac{1}{2}U_{\mathrm{sp}}\chi_0(q)} = n - 2\langle n_\uparrow n_\downarrow\rangle \qquad (3.7)$$

that, with (3.5) for U_{sp}, gives us double-occupancy.

The functional-derivative procedure generates an expression for the charge vertex U_{ch} which involves the functional derivative of $\langle n_\uparrow n_\downarrow\rangle/(\langle n_\uparrow\rangle\langle n_\downarrow\rangle)$ which contains six point functions that one does not really know how to evaluate. But, if we again assume that the vertex U_{ch} is a constant, it is simply determined by the requirement that charge fluctuations also satisfy the fluctuation-dissipation theorem and the Pauli principle,

$$\frac{T}{N}\sum_q \frac{\chi_0(q)}{1 + \frac{1}{2}U_{\mathrm{ch}}\chi_0(q)} = n + 2\langle n_\uparrow n_\downarrow\rangle - n^2. \qquad (3.8)$$

Note that, in principle, $\Sigma^{(1)}$ also depends on double-occupancy, but since $\Sigma^{(1)}$ is a constant, it is absorbed in the definition of the chemical potential, and we do not need to worry about it in this case. That is why the noninteracting irreducible susceptibility $\chi_0(q)$ appears in the expressions for the susceptibility, even though it should be evaluated with $G^{(1)}$ that contains $\Sigma^{(1)}$. One can check that spin and charge conservation are satisfied by our susceptibilities.

Detailed comparisons with quantum Monte Carlo simulations (QMC) [4,5,11,12] have shown that (3.5), (3.7), and (3.8) give predictions that are

[4]For $n > 1$, all particle occupation numbers must be replaced by hole occupation numbers.

quantitative at the few percent level, in regions of parameter space where size effects are negligible. This remains true even at couplings of the order of the bandwidth and when second-neighbor hopping t' is present [13]. QMC size effects become important at half-filling below a crossover temperature T_X where the renormalized-classical regime appears. Even though (3.5) for $U_{\rm sp}$ fails at $n=1$, $t'=0$ in this regime, by assuming that $\langle n_\uparrow n_\downarrow \rangle$ is temperature independent below T_X one obtains a qualitatively correct description of the renormalized-classical regime. The universality class of our theory, however, is $O(N=\infty)$ instead of $O(N=3)$ [14]. A rough estimate of the renormalized chemical potential (or equivalently of $\Sigma^{(1)}$), is given in the appendix.

3.2 Second Step: An Improved Self-Energy $\Sigma^{(2)}$

Collective modes are less influenced by details of the single-particle properties than the other way around. We thus wish now to obtain an improved approximation for the self-energy that takes advantage of the fact that we have found accurate approximations for the low-frequency spin and charge fluctuations. We begin from the general definition of the self-energy (Eq. 2.4) obtained from Dyson's equation. The right-hand side of that equation can be obtained either from a functional derivative with respect to an external field that is diagonal in spin, as in our generating function (Eq. 2.1), or by a functional derivative of $\langle \psi_{-\sigma}(1)\psi_\sigma^\dagger(2)\rangle_{\phi_t}$ with respect to a transverse external field ϕ_t. These two approaches will be considered in turn below. They give a self-energy formula that takes into account, respectively, longitudinal and transverse fluctuations. Crossing symmetry, rotational symmetry and sum rules will dictate the final formula for the improved self-energy $\Sigma^{(2)}$ that will be presented at the end of the subsection on the consistency check.

3.2.1 Longitudinal Channel

The right-hand side of the general definition of the self-energy (Eq. 2.4) may be written as

$$\begin{aligned}\Sigma_\sigma(1,\bar{1})G_\sigma(\bar{1},2)\\ = -U\left[\left.\frac{\delta G_\sigma(1,2;\{\phi\})}{\delta\phi_{-\sigma}(1^+,1)}\right|_{\{\phi\}=0} - G_{-\sigma}(1,1^+)G_\sigma(1,2)\right].\end{aligned} \tag{3.9}$$

The last term is the Hartree–Fock contribution. It gives the exact result for the self-energy in the limit $\omega\to\infty$ [4]. The $\delta G_\sigma/\delta\phi_{-\sigma}$ term is thus a contribution to lower frequencies and it comes from the spin and charge fluctuations. Right-multiplying the last equation by G^{-1} and replacing the lower energy part $\delta G_\sigma/\delta\phi_{-\sigma}$ by its general expression in terms of irreducible vertices (Eq. 2.7) we find

$$\Sigma_\sigma^{(2)}(1,2) = UG_{-\sigma}^{(1)}(1,1^+)\delta(1-2) - UG_\sigma^{(1)}(1,\bar{3})\left[\left.\frac{\delta\Sigma_\sigma^{(1)}(\bar{3},2;\{\phi\})}{\delta G_{\overline{\sigma}}^{(1)}(\bar{4},\bar{5};\{\phi\})}\right|_{\{\phi\}=0}\left.\frac{\delta G_{\overline{\sigma}}^{(1)}(\bar{4},\bar{5};\{\phi\})}{\delta\phi_{-\sigma}(1^+,1)}\right|_{\{\phi\}=0}\right]. \quad (3.10)$$

Every quantity appearing on the right-hand side of that equation has been taken from the TPSC results. This means in particular that the irreducible vertices $\delta\Sigma_\sigma^{(1)}/\delta G_{\sigma'}^{(1)}$ are at the same level of approximation as the Green functions $G_\sigma^{(1)}$ and self-energies $\Sigma_\sigma^{(1)}$. In approaches that assume that Migdal's theorem applies to spin and charge fluctuations, one often sees renormalized Green functions $G^{(2)}$ appearing on the right-hand side along with unrenormalized vertices, $\delta\Sigma_\sigma/\delta G_{\sigma'} \to U$.

In terms of U_{sp} and U_{ch} in Fourier space, the above formula [5] reads

$$\Sigma_\sigma^{(2)}(k)_{\mathrm{long}} = Un_{-\sigma} + \frac{U}{4}\frac{T}{N}\sum_q \left[U_{\mathrm{sp}}\chi_{\mathrm{sp}}^{(1)}(q) + U_{\mathrm{ch}}\chi_{\mathrm{ch}}^{(1)}(q)\right]G_\sigma^{(1)}(k+q). \quad (3.11)$$

3.2.2 Transverse Channel

In the transverse channel, the calculation basically has to be redone from scratch. It is closely analogous to the attractive Hubbard model case, for which detailed calculations have been published [7]. We will thus be concise. The generating function in a transverse field is

$$\ln Z[\phi_t] = \ln \mathrm{Tr}\left[e^{-\beta(\widehat{H}-\mu\widehat{N})}\mathrm{T}_\tau\left(e^{-\psi_\uparrow^\dagger(\bar{1})\phi_-(\bar{1},\bar{2})\psi_\downarrow(\bar{2})-\psi_\downarrow^\dagger(\bar{1})\phi_+(\bar{1},\bar{2})\psi_\uparrow(\bar{2})}\right)\right]. \quad (3.12)$$

The corresponding spin-space matrix Green function

$$\mathbf{G}(1,2;\{\phi_t\}) = -\begin{pmatrix} \langle\psi_\uparrow(1)\psi_\uparrow^\dagger(2)\rangle_{\phi_t} & \langle\psi_\uparrow(1)\psi_\downarrow^\dagger(2)\rangle_{\phi_t} \\ \langle\psi_\downarrow(1)\psi_\uparrow^\dagger(2)\rangle_{\phi_t} & \langle\psi_\downarrow(1)\psi_\downarrow^\dagger(2)\rangle_{\phi_t} \end{pmatrix} \quad (3.13)$$

obeys the matrix Dyson equation

$$\mathbf{G}^{-1}(1,2;\{\phi_t\}) = \mathbf{G}_0^{-1}(1-2) - \mathbf{\Sigma}(1,2;\{\phi_t\}) - \Phi_t \quad (3.14)$$

with

$$\Phi_t = \begin{pmatrix} 0 & \phi_- \\ \phi_+ & 0 \end{pmatrix} \quad (3.15)$$

and

$$\mathbf{\Sigma}\left(1,\bar{1};\{\phi_t\}\right)\mathbf{G}\left(\bar{1},2;\{\phi_t\}\right)$$
$$\equiv -U\begin{pmatrix} \langle\psi_\downarrow^\dagger(1^+)\psi_\downarrow(1)\psi_\uparrow(1)\psi_\uparrow^\dagger(2)\rangle_{\phi_t} & \langle\psi_\downarrow^\dagger(1^+)\psi_\downarrow(1)\psi_\uparrow(1)\psi_\downarrow^\dagger(2)\rangle_{\phi_t} \\ \langle\psi_\uparrow^\dagger(1^+)\psi_\uparrow(1)\psi_\downarrow(1)\psi_\uparrow^\dagger(2)\rangle_{\phi_t} & \langle\psi_\uparrow^\dagger(1^+)\psi_\uparrow(1)\psi_\downarrow(1)\psi_\downarrow^\dagger(2)\rangle_{\phi_t} \end{pmatrix}. \tag{3.16}$$

We need to obtain, in turn, the renormalized vertex U_{sp} (Eq. 3.5), and finally find an improved formula for the self-energy $\Sigma^{(2)}$. Starting from the matrix equation $\mathbf{G}\mathbf{G}^{-1} = \mathbf{1}$ and following the procedure below (Eq. 2.5), one of the two response functions that do not vanish in zero external field obeys

$$\left.\frac{\delta G_{21}(1,2;\{\phi_t\})}{\delta\phi_+(3,4)}\right|_{\{\phi_t\}=0} = G_{22}(1,3)G_{11}(4,2)$$
$$+\, G_{22}(1,\bar{2})\left.\frac{\delta\Sigma_{21}(\bar{2},\bar{3};\{\phi_t\})}{\delta G_{21}(\bar{6},\bar{7};\{\phi_t\})}\right|_{\{\phi_t\}=0}\left.\frac{\delta G_{21}(\bar{6},\bar{7};\{\phi_t\})}{\delta\phi_+(3,4)}\right|_{\{\phi_t\}=0} G_{11}(\bar{3},2), \tag{3.17}$$

where the subscripts denote matrix elements in spin-space. There is an analogous equation for $\delta G_{12}(1,2)/\delta\phi_-(3,4)$. Note that

$$\delta G_{21}(1,1)/\delta\phi_+(2,2) = -\langle S_+(1)S_-(2)\rangle$$
$$= -\langle\psi_\uparrow^\dagger(1)\psi_\downarrow(1)\psi_\downarrow^\dagger(2)\psi_\uparrow(2)\rangle = -\chi_{+-}(1,2). \tag{3.18}$$

The usual Hartree–Fock factorization for the self-energy would transform (3.17) into the RPA equation for transverse spin fluctuations $\langle S_+(1)S_-(2)\rangle$.

The value of $\Sigma^{(1)}$ and of the corresponding vertices at the TPSC level are obtained using steps analogous to those in the longitudinal channel. We factor the self-energy in the Hartree–Fock manner,

$$\mathbf{\Sigma}^{(1)}(1,\bar{1};\{\phi_t\})\mathbf{G}^{(1)}(\bar{1},2;\{\phi_t\})$$
$$= -A_{\{\phi_t\}}\begin{pmatrix} G_{22}^{(1)}(1,1^+;\{\phi_t\}) & -G_{12}^{(1)}(1,1^+;\{\phi_t\}) \\ -G_{21}^{(1)}(1,1^+;\{\phi_t\}) & G_{11}^{(1)}(1,1^+;\{\phi_t\}) \end{pmatrix}$$
$$\begin{pmatrix} G_{11}^{(1)}(1,2;\{\phi_t\}) & G_{12}^{(1)}(1,2;\{\phi_t\}) \\ G_{21}^{(1)}(1,2;\{\phi_t\}) & G_{22}^{(1)}(1,2;\{\phi_t\}) \end{pmatrix} \tag{3.19}$$

but we correct this factorization by the factor $A_{\{\phi_t\}}$ which is determined in such a way that when $2 \to 1^+$ the exact result for the four point function on the right-hand side of (3.16) is recovered. Using

$$\langle\psi_\downarrow^\dagger(1^+)\psi_\downarrow(1)\psi_\uparrow(1)\psi_\downarrow^\dagger(1^+)\rangle_{\phi_t} = \langle\psi_\downarrow^\dagger(1^+)\psi_\downarrow^\dagger(1^+)\psi_\downarrow(1)\psi_\uparrow(1)\rangle_{\phi_t} = 0$$

and the analogous result for the other off-diagonal element, we obtain the exact result for $2 \to 1^+$:

$$\mathbf{\Sigma}^{(1)}(1,\bar{1};\{\phi_t\})\mathbf{G}^{(1)}(\bar{1},1^+;\{\phi_t\}) \\ = U\begin{pmatrix} \langle n_\downarrow(1)n_\uparrow(1)\rangle_{\phi_t} & 0 \\ 0 & \langle n_\uparrow(1)n_\downarrow(1)\rangle_{\phi_t} \end{pmatrix}. \quad (3.20)$$

Equating with the right-hand side of the approximate result (Eq. 3.19) when $2 \to 1^+$ determines the value of $A_{\{\phi_t\}}$ by a simple 2×2 matrix inversion. From this, one extracts the off-diagonal component of the self-energy and the corresponding irreducible vertex that is needed for response functions in (3.17):

$$\left.\frac{\delta\Sigma_{21}^{(1)}(1,2;\{\phi_t\})}{\delta G_{21}^{(1)}(3,4;\{\phi_t\})}\right|_{\{\phi_t\}=0} = -U\frac{\langle n_\uparrow n_\downarrow\rangle}{\langle n_\uparrow\rangle\langle n_\downarrow\rangle}\delta(1-3)\delta(1-2)\delta(1-4). \quad (3.21)$$

Substituting in the equation for the response function (Eq. 3.17), one precisely recovers for the transverse spin fluctuations the same TPSC result as for the longitudinal fluctuations: $\chi_{\mathrm{sp}}^{-1}(q) = \chi_0^{-1}(q) - \frac{1}{2}U_{\mathrm{sp}}$ with U_{sp} given by (3.5). The determination of U_{sp} through the fluctuation-dissipation theorem leads again to (3.7). We thus recover the results expected from rotational invariance.

To move to the second level of approximation for the self-energy, we return to the exact definition for the self-energy (3.16) but this time in zero applied external field. The first diagonal component can be written as

$$\Sigma_{11}(1,2) = U\left.\frac{\delta G_{21}(1,\bar{2};\{\phi_t\})}{\delta\phi_+(1^+,1)}\right|_{\{\phi_t\}=0} G_{11}^{-1}(\bar{2},2). \quad (3.22)$$

If we use the exact result for the transverse response (Eq. 3.17), this takes the form

$$\Sigma_{11}(1,2) = UG_{22}(1,1^+)\delta(1-2) \\ + UG_{22}(1,\bar{3})\left.\frac{\delta\Sigma_{21}(\bar{3},2;\{\phi_t\})}{\delta G_{21}(\bar{6},\bar{7};\{\phi_t\})}\right|_{\{\phi_t\}=0}\left.\frac{\delta G_{21}(\bar{6},\bar{7};\{\phi_t\})}{\delta\phi_+(1^+,1)}\right|_{\{\phi_t\}=0}, \quad (3.23)$$

where, as in the longitudinal case, the high-frequency Hartree–Fock result is now explicit. Substituting on the right-hand side the TPSC (i.e., level 1) results, we obtain an improved approximation for the self-energy due to transverse spin fluctuations. In momentum space, it reads

$$\Sigma_\sigma^{(2)}(k)_{\mathrm{trans}} = Un_{-\sigma} + \frac{U}{2}\frac{T}{N}\sum_q [U_{\mathrm{sp}}\chi_{\mathrm{sp}}^{(1)}(q)t]G_{-\sigma}^{(1)}(k+q). \quad (3.24)$$

3.3 Internal Accuracy Check

The equation (3.2) that relates $\mathrm{Tr}[\Sigma G]$ to double occupancy (potential energy) relates purely single-particle quantities, Σ and G, to a quantity that may be computed from two-particle quantities, namely, spin and charge correlation functions. In our approach, it can be checked [4] that $\frac{1}{2}\mathrm{Tr}[\Sigma^{(2)}G^{(1)}]$ is exactly equal to $U\langle n_\uparrow n_\downarrow\rangle$, with $\langle n_\uparrow n_\downarrow\rangle$ computed at the TPSC level. More specifically, we find,

$$\frac{1}{2}\mathrm{Tr}[\Sigma^{(2)}G^{(1)}] = \lim_{\tau\to 0^-}\frac{T}{N}\sum_k \Sigma^{(2)}_\sigma(k)G^{(1)}_\sigma(k)e^{-ik_n\tau} = U\langle n_\uparrow n_\downarrow\rangle. \tag{3.25}$$

One can use the difference between $\frac{1}{2}\mathrm{Tr}[\Sigma^{(2)}G^{(1)}]$ and $\frac{1}{2}\mathrm{Tr}[\Sigma^{(2)}G^{(2)}]$ as an internal accuracy check of the theory. In the pseudogap regime at $n = 1$ for example, the breakdown of the theory is clearly indicated by the growing difference between the two quantities. Reference [8] gives a detailed table that illustrates these facts in the case of the attractive Hubbard model.

The $\mathrm{Tr}[\Sigma^{(2)}G^{(1)}]$ formula can also be used to help in the interpretation of the two different results obtained above for the self-energy (Eqs. 3.10 and 3.23). Without the approximation that the TPSC results should be used on the right-hand side, the results that follow from the corresponding exact expressions (Eqs. 3.10 and 3.23) would be identical. To resolve this problem, we follow [17]. Figure 1 of this paper shows the self-energy in terms of the fully reducible vertex $\Gamma(q, k-k', k+k'-q)$. In both formulas for the self-energy (Eqs. 3.11 and 3.24), the dependence of Γ on the particle-particle channel center of mass momentum $k+k'-q$ is neglected since this channel is not singular. The longitudinal version of the self-energy (Eq. 3.11) takes good care of the singularity of Γ when its first argument q is near (π,π). The transverse version does the same for the singular dependence near (π,π) of the second argument $k-k'$, which corresponds to the other particle-hole channel. One then expects that averaging the two possibilities gives a better approximation for Γ since it preserves crossing symmetry in the two particle-hole channels. Furthermore, one can verify that the longitudinal spin fluctuations in (3.11) contribute an amount $U\langle n_\uparrow n_\downarrow\rangle/2$ to the consistency condition $\frac{1}{2}\mathrm{Tr}(\Sigma^{(2)}_{\mathrm{long}}G^{(1)}) = U\langle n_\uparrow n_\downarrow\rangle$ and that each of the two transverse spin components also contribute $U\langle n_\uparrow n_\downarrow\rangle/2$ to $\frac{1}{2}\mathrm{Tr}(\Sigma^{(2)}_{\mathrm{trans}}G^{(1)}) = U\langle n_\uparrow n_\downarrow\rangle$. Hence, averaging (3.11) and the expression in the transverse channel (3.24) also preserves rotational invariance. In addition, one verifies numerically that the exact sum rule [4] $-\int d\omega' \,\mathrm{Im}[\Sigma_\sigma(\mathbf{k},\omega')]/\pi = U^2 n_{-\sigma}(1-n_{-\sigma})$ determining the high-frequency behavior is satisfied to a higher degree of accuracy. We thus obtain a self-energy formula that we called [17] "symmetric":

$$\Sigma_\sigma^{(2)}(k)_{\rm sym} = Un_{-\sigma} + \frac{U}{8}\frac{T}{N}\sum_q [3U_{\rm sp}\chi_{\rm sp}^{(1)}(q) + U_{\rm ch}\chi_{\rm ch}^{(1)}(q)]G_\sigma^{(1)}(k+q). \quad (3.26)$$

$\Sigma_\sigma^{(2)}(k)_{\rm sym}$ is different from so-called Berk–Schrieffer type expressions [15] that do not satisfy [4] the consistency condition between one- and two-particle properties, $\frac{1}{2}\,\mathrm{Tr}(\Sigma G) = U\langle n_\uparrow n_\downarrow\rangle$.

4 Discussion and Extensions

The approach described above is valid in the weak to intermediate coupling regime. It involves two steps. First, a self-consistent determination of double-occupancy and renormalized vertices that enter the dynamical susceptibilities of the most important four-point correlation functions (density-density and spin-spin correlation functions). This is summarized by (3.5), (3.7), and (3.8). Conservation laws, such as charge and spin conservation, are satisfied. These results are then used to obtain an improved approximation for single-particle properties through the self-energy (Eq. 3.26). An internal accuracy check allows one to decide on the validity of the results in cases where QMC or other exact results are not available as references. The approach satisfies the Pauli principle, the Mermin–Wagner theorem, contains Kanamori–Brueckner screening, and does not assume a Migdal theorem in the calculation of the self-energy.

In addition to provide an accurate calculational tool for the Hubbard model, this methodology has allowed us to develop insight into the physics of this model. The physically most important result obtained to date is probably the detailed description of the physics of the pseudogap that appears *in two dimensions* in the renormalized-classical regime ($\hbar\omega \ll k_B T$) when the antiferromagnetic correlation length (superconducting correlation length in the case of the attractive model) becomes larger than the single particle thermal de Broglie wavelength [4, 5]. This physics explains a possible route to the destruction of the Fermi liquid in *two dimensions.* These results have been confirmed by extensive QMC calculations [12, 16, 17]. The pseudogap (depletion near $\omega = 0$) in $\mathrm{Im}\, G_\sigma^R(\mathbf{k}_F, \omega)$ appears along with precursors of the Bogoliubov quasiparticles (finite ω peaks) of the ordered state. To extrapolate more deeply in the pseudogap regime, where strictly speaking the above method fails, one assumes that double-occupancy becomes temperature independent below the crossover temperature T_X where one enters the renormalized-classical regime.

The above methodology has been applied successfully to the attractive Hubbard model [7, 8, 18]. It trivially applies to the Hubbard model with an arbitrary hopping matrix. $\Sigma^{(2)}$ can also be used to obtain consistent thermodynamic predictions [19]. Extensions to multiband problems are non-trivial but are being developed [20]. Recently an extension of this approach has been used to obtain quantitative results for the spin and charge susceptibilities in the attractive Hubbard model [21]. Proceeding along the same lines for the repulsive Hubbard model, pairing correlations have been calculated [22]. They yield a dome shape dependence on doping of the d-wave superconductivity transition temperature T_c. The decrease of T_c near half-filling comes from the detrimental effect of opening a pseudogap. Phenomenological extensions to more complicated models with d-wave superconductivity and antiferromagnetism have also been proposed [23]. Future directions include generalizations to the pseudogap regime, to states with broken symmetry, [24] to longer-range interactions and to impurity models. It would also be extremely valuable to obtain the irreducible vertices $\delta\Sigma_{\sigma}^{(2)}/\delta G_{\sigma'}^{(2)}$ that are consistent with the best estimate of the self-energy.

Acknowledgments: This paper is based in part on a course given by A.-M.S.T. at Université de Provence in 1999 and on seminars at the Newton Institute for Mathematical Sciences and at the Institute for Theoretical Physics in Santa Barbara in 2000. Work there was partially supported by the National Science Foundation under grant No. PHY94-07194. A.-M.S.T. and S. Allen are grateful to B. Kyung, F. Lemay and A.-M. Daré for numerous discussions. A.-M.S.T. thanks Gilbert Albinet and his group for hospitality. This work was supported by a grant from the Natural Sciences and Engineering Research Council (NSERC) of Canada and the Fonds pour la formation de Chercheurs et l'Aide à la Recherche (FCAR) of the Québec government. We thank the Centre de recherches mathématiques for its hospitality. A.-M.S.T. holds a Tier I Canada Research Chair in Condensed Matter Physics.

5 References

[1] G. Baym, Phys. Rev. **127** (1962), 1391.

[2] L.P. Kadanoff and G. Baym, *Quantum Statistical Mechanics*, Benjamin, Menlo Park, 1962.

[3] N.E. Bickers and D.J. Scalapino, Ann. Physics **193** (1989), 206.

[4] Y.M. Vilk and A.-M.S. Tremblay, J. Physique I France **7** (1997), 1309.

[5] Y.M. Vilk and A.-M.S. Tremblay, Europhys. Lett. **33** (1996), 159; J. Phys. Chem. Solids **56** (1995), 1769.

[6] P.C.E. Stamp, J. Phys. F: Met. Phys. **15** (1985), 1829.

[7] S. Allen and A.-M.S. Tremblay, Phys. Rev. B **64** (2001), 075115/1-14.

[8] B. Kyung, S. Allen, and A.-M.S. Tremblay, Phys. Rev. B **64** (2001), 075116.

[9] P.C. Martin and J. Schwinger, Phys. Rev. **115** (1959). This paper also contains numerous references to previous work.

[10] For a review, see K.S. Singwi and M.P. Tosi, in eds. H. Ehrenreich, F. Seitz, and D. Turnbull, *Solid State Physics*, Vol. 36 (Academic, New York, 1981), p. 177; S. Ichimaru, Rev. Modern Phys. **54** (1982), 1017.

[11] Y.M. Vilk, L. Chen, and A.-M.S. Tremblay, Phys. Rev. B Rapid Comm. **49** (1994), 13267.

[12] B. Kyung, J.-S. Landry, D. Poulin, and A.-M.S. Tremblay, Phys. Rev. Lett. **90** (2003), 099702.

[13] A.F. Veilleux, A.-M. Daré, L. Chen, Y.M .Vilk, and A.-M.S. Tremblay, Phys. Rev. B **52** (1995), 16255.

[14] A.-M. Daré, Y.M. Vilk, and A.-M.S. Tremblay, Phys. Rev. B **53** (1996), 14236.

[15] N. Bulut, D.J. Scalapino, and S.R. White, Phys. Rev. B **47** (1993), 2742; N.F. Berk and J.R. Schrieffer, Phys. Rev. Lett. **17** (1966), 433.

[16] S. Allen, H. Touchette, S. Moukouri, Y.M. Vilk, and A.-M.S. Tremblay, Phys. Rev. Lett. **83** (1999), 4128; Y.M. Vilk, S. Allen, H. Touchette, S. Moukouri, L. Chen, and A.-M.S. Tremblay, J. Phys. Chem. Solids **59** (1998), 1873.

[17] S. Moukouri, S. Allen, F. Lemay, B. Kyung, D. Poulin, Y.M. Vilk, and A.-M. S. Tremblay, Phys. Rev. B **61** (2000), 7887.

[18] S. Allen, Ph.D. Thesis, Université de Sherbrooke, 2000 (unpublished).

[19] S. Roy, M.Sc. thesis, Université de Sherbrooke, 2002 (unpublished).

[20] A.-M. Daré (unpublished).

[21] S. Allen, B. Kyung, and A.-M.S. Tremblay (unpublished).

[22] B. Kyung, J.-S. Landry, and A.-M.S. Tremblay, cond-mat/0205165.

[23] B. Kyung, Phys. Rev. B **63** (2000), 014502; Phys. Rev. B **64** (2001), 104512.

[24] S. Allen (unpublished).

Appendix A An Approximate Formula for $\Sigma^{(1)}$

The equation (3.2) that was used to obtain finally

$$U_{\rm sp} = U\langle n_\uparrow n_\downarrow\rangle / (\langle n_\downarrow\rangle\langle n_\uparrow\rangle)$$

would be different if we had taken the limit $2 \to 1^-$ instead of $2 \to 1^+$ in the general expression for ΣG (Eq. 2.4). More specifically, at $\{\phi\} = 0$,

$$\begin{aligned}\Sigma_\sigma(1,\bar{1})G_\sigma(\bar{1},1^+) &= U\langle n_\downarrow(1)n_\uparrow(1)\rangle;\\ \Sigma_\sigma(1,\bar{1})G_\sigma(\bar{1},1^-) &= U\langle n_\downarrow(1)(n_\uparrow(1)-1)\rangle.\end{aligned} \tag{A.1}$$

Any approximation for the self-energy that has Hartree–Fock as its infinite-frequency limit will be such that the difference between the above two results, $\Sigma_\sigma(1,\bar{1})G_\sigma(\bar{1},1^+) - \Sigma_\sigma(1,\bar{1})G_\sigma(\bar{1},1^-) = U\langle n_\downarrow(1)\rangle$ is satisfied. The proof [7, (44)] is as follows:

$$\begin{aligned}&\frac{T}{N}\sum_{\mathbf{k}}\sum_{ik_n}\left(\frac{\Sigma(\mathbf{k},ik_n)}{ik_n-(\varepsilon_{\mathbf{k}}-\mu)-\Sigma(\mathbf{k},ik_n)} - \frac{U\langle n_\downarrow\rangle}{ik_n}\right)\left(e^{-ik_n0^-}-e^{-ik_n0^+}\right)\\ &\qquad + \frac{T}{N}\sum_{\mathbf{k}}\sum_{ik_n}\left[\frac{U\langle n_\downarrow\rangle}{ik_n}\left(e^{-ik_n0^-}-e^{-ik_n0^+}\right)\right] = U\langle n_\downarrow\rangle.\end{aligned} \tag{A.2}$$

In this expression, the first sum vanishes because we have added and subtracted a term that makes it convergent at infinity without the need for convergence factors $e^{-ik_n0^\pm}$. Hence, only the last sum survives.

Since $\Sigma_\sigma^{(1)}(1,2)$ is a constant times $\delta(1-2)$, one can obtain two estimates of the constant depending on which of the two equations in (A.1) one starts from. By analogy with (A.2) one expects that the difference between these two estimates is related to the high-frequency behavior of the true result while the average of the two estimates is related to the low-frequency behavior. This average is

$$\frac{n}{2}\left(U + U_{\rm sp}(1-n)\right)\frac{1}{2-n}. \tag{A.3}$$

The exact result, $U/2$, is recovered at $n = 1$. For other fillings, the above formula gives a very rough estimate of the chemical potential shift induced by interactions. For example, for U up to $4t$ on 8×8 lattices and temperatures of order $t/4$ in energy units, one finds results that deviate by up to 20% from the QMC results. The corresponding procedure in the *attractive* Hubbard model case [7] seems to work better.

Index